AF411449

Forecasting and Risk Management Techniques for Electricity Markets

Forecasting and Risk Management Techniques for Electricity Markets

Editor

Yuji Yamada

MDPI • Basel • Beijing • Wuhan • Barcelona • Belgrade • Manchester • Tokyo • Cluj • Tianjin

Editor
Yuji Yamada
University of Tsukuba
Japan

Editorial Office
MDPI
St. Alban-Anlage 66
4052 Basel, Switzerland

This is a reprint of articles from the Special Issue published online in the open access journal *Energies* (ISSN 1996-1073) (available at: https://www.mdpi.com/journal/energies/special_issues/ solar_power_forecasting_risk_management_techniques_electricity_markets).

For citation purposes, cite each article independently as indicated on the article page online and as indicated below:

LastName, A.A.; LastName, B.B.; LastName, C.C. Article Title. *Journal Name* **Year**, *Volume Number*, Page Range.

ISBN 978-3-0365-5183-8 (Hbk)
ISBN 978-3-0365-5184-5 (PDF)

Contents

About the Editor

Yuji Yamada

Yuji Yamada was born in Matsumoto-City, Nagano, Japan, in 1969. He received a B.S. and a M.S. degrees from University of Chiba and Tokyo Institute of Technology, Japan, respectively, in 1993 and 1995. In 1998, he defended his Ph.D. thesis on global optimization for robust controls under supervision by Dr. Shinji Hara and received a Ph.D. in Engineering from Tokyo Institute of Technology. He became a postdoctoral scholar at the Control and Dynamical Systems, Caltech, USA, in 1998, and worked with his host adviser, Dr. John Doyle, from 1998 to 2001. He also served as a course lecturer in 2001 and taught a Ph.D. course entitled "Stochastic Simulation in Finance" with Dr. Peter Bossaerts at Caltech. In 2002, he joined the University of Tsukuba, Japan, as an associate professor and has been a full professor since 2013 at the Faculty of Business Sciences. He was the Dean of the Faculty of Business Sciences, and an executive officer of the University of Tsukuba from 2018 to 2021. His current research interests are at the intersection between finance, optimization, and control theory with applications for electricity and financial markets. He has been a PI of several projects for renewable energy and electricity trading including Grant-in-Aid for Scientific Research (A) 16H01833 and 20H00285 from Japan Society for the Promotion of Science (JSPS).

Preface to "Forecasting and Risk Management Techniques for Electricity Markets"

The construction of sustainable energy systems is currently one of the most important issues to achieve a resilient society. In particular, while the expansion of the use of renewable energy toward carbon neutrality has actively been promoted worldwide, it is essential to design risk management techniques and transaction schemes with a focus on renewable energy trading for stable and creative social environments. However, electricity market participants are traditionally exposed to many risk sources. For example, the rapid introduction of solar power and other renewable electricity generation brings a growing impact of weather and climate changes on electricity markets for both price and volume executions. As a result, the system operator (or an aggregator in the region) needs to prepare a sufficient capacity of backup thermal generators to match real-time power production with electricity consumption, which varies with, e.g., solar radiation, temperature, humidity, and other conditions. This leads to an additional cost or a loss for both/either consumers and/or power producers in the network. The use of thermal power provides another source of uncertainty in electricity markets as well because the generation cost largely depends on fuel prices and type of energy. Under these circumstances, forecasting and risk management techniques have become more and more important not only for traditional centralized electricity markets, but also for decentralized energy resources.

This book focuses on the recent development of forecasting and risk management techniques for electricity markets. In addition, we discuss research on new trading platforms and environment using blockchain-based peer-to-peer (P2P) market and computer agents.

The book consists of two parts. The first part is entitled "Forecasting and Risk Management Techniques" and contains the following five chapters:

Customized yet Standardized Temperature Derivatives: A Non-Parametric Approach with Suitable Basis Selection for Ensuring Robustness, by Takuji Matsumoto and Yuji Yamada

Comprehensive and Comparative Analysis of GAM-Based PV Power Forecasting Models Using Multidimensional Tensor Product Splines against Machine Learning Techniques, by Takuji Matsumoto and Yuji Yamada

Going for Derivatives or Forwards? Minimizing Cashflow Fluctuations of Electricity Transactions on Power Markets, by Yuji Yamada and Takuji Matsumoto

Comprehensive Review on Electricity Market Price and Load Forecasting Based on Wind Energy, by Hakan Acaroğlu and Fausto Pedro García Márquez

Short-Term Electricity Prices Forecasting Using Functional Time Series Analysis, by Faheem Jan, Ismail Shah and Sajid Ali

The second part is entitled "Peer-to-Peer (P2P) Electricity Trading System and Strategy" and contains the following five chapters:

Designing a User-Centric P2P Energy Trading Platform: A Case Study—Higashi-Fuji Demonstration, by Yasuhiro Takeda, Yoichi Nakai, Tadatoshi Senoo and Kenji Tanaka,

Feasibility Conditions for Demonstrative Peer-to-Peer Energy Market, by Reo Kontani, Kenji Tanaka and Yuji Yamada

Demonstration of Blockchain Based Peer to Peer Energy Trading System with Real-Life Used PHEV and HEMS Charge Control, by Yuki Matsuda, Yuto Yamazaki, Hiromu Oki, Yasuhiro Takeda, Daishi Sagawa and Kenji Tanaka

Bidding Agents for PV and Electric Vehicle-Owning Users in the Electricity P2P Trading Market, by Daishi Sagawa, Kenji Tanaka, Fumiaki Ishida, Hideya Saito, Naoya Takenaga, Seigo Nakamura, Nobuaki Aoki, Misuzu Nameki and Kosuke Saegusa

Effectiveness and Feasibility of Market Makers for P2P Electricity Trading, by Shinji Kuno, Kenji Tanaka and Yuji Yamada

Special thanks to all the authors and contributors involved, and much appreciation to Ms. Reka Kovacs, Deputy Office Manager, MDPI Romania, who helped organizing this Special Issue. Last but not least, I would like to thank my wife, Yumiko, my son, Yukei, and my daughter, Sakura, for their tremendous support in my life.

Yuji Yamada
Editor

Article

Customized yet Standardized Temperature Derivatives: A Non-Parametric Approach with Suitable Basis Selection for Ensuring Robustness

Takuji Matsumoto [1] and Yuji Yamada [2],*

[1] Socio-Economic Research Center, Central Research Institute of Electric Power Industry, Tokyo 100-8126, Japan; matsumoto3832@criepi.denken.or.jp

[2] Faculty of Business Sciences, University of Tsukuba, Tokyo 112-0012, Japan

* Correspondence: yuji@gssm.otsuka.tsukuba.ac.jp

Abstract: Previous studies have demonstrated that non-parametric hedging models using temperature derivatives are highly effective in hedging profit/loss fluctuation risks for electric utilities. Aiming for the practical applications of these methods, this study performs extensive empirical analyses and makes methodological customizations. First, we consider three types of electric utilities being exposed to risks of "demand", "price", and their "product (multiplication)", and examine the design of an appropriate derivative for each utility. Our empirical results show that non-parametrically priced derivatives can maximize the hedge effect when a hedger bears a "price risk" with high nonlinearity to temperature. In contrast, standard derivatives are more useful for utilities with only "demand risk" in having a comparable hedge effect and in being liquidly traded. In addition, the squared prediction error derivative on temperature has a significant hedge effect on both price and product risks as well as a certain effect on demand risk, which illustrates its potential as a new standard derivative. Furthermore, spline basis selection, which may be overlooked by modeling practitioners, improves hedge effects significantly, especially when the model has strong nonlinearities. Surprisingly, the hedge effect of temperature derivatives in previous studies is improved by 13–53% by using an appropriate new basis.

Keywords: electricity markets; non-parametric regression; minimum variance hedge; spline basis functions; cyclic cubic spline; weather derivatives

Citation: Matsumoto, T.; Yamada, Y. Customized yet Standardized Temperature Derivatives: A Non-Parametric Approach with Suitable Basis Selection for Ensuring Robustness. *Energies* **2021**, *14*, 3351. https://doi.org/10.3390/en14113351

Academic Editor: François Vallée

Received: 26 April 2021
Accepted: 3 June 2021
Published: 7 June 2021

Publisher's Note: MDPI stays neutral with regard to jurisdictional claims in published maps and institutional affiliations.

1. Introduction

Electric utilities are generally exposed to the risk of daily fluctuations in price and demand, and constructing an efficient hedging methodology is an extremely important management issue. To this end, "electricity derivatives" may be introduced to prevent price fluctuations in electricity businesses. However, there is a potential problem that electricity derivatives may not be effective for "volume" (demand) risks. Moreover, for "price" risks as well, electricity derivatives may be unavailable in some markets, or their efficient use may be impossible because of low liquidity (especially for electricity derivatives with short-time granularity). In response to this awareness of issues, some studies have demonstrated the effectiveness of using "weather derivatives" instead of electric power derivatives. As examples of previous studies that verified the hedge effect of weather derivatives, Lee and Oren [1,2] discussed the effect of introducing standard weather derivatives into the hedging portfolio using equilibrium pricing models; however, their studies focused on the theory of suitable pricing instead of the empirical evaluation. Bhattacharya et al. [3] proposed the optimal trading strategy for standard derivatives based on heating degree days (HDD) and cooling degree days (CDD) using a data-driven approach. Their study empirically examined hedge effects, but as it optimized relatively simple two-dimensional vectors for the hedge weight of two different derivatives, the inherent nonlinear relationship between price/demand and temperature was not necessarily incorporated. (Note that for existing standard weather derivatives discussed by these studies, many methods have been proposed in the context of "pricing" [4–7]).

To fill the research area of weather derivatives being uncovered in these past studies, non-parametric optimal hedging techniques for finding arbitrarily derivative payoff functions have been proposed in some studies (e.g., pricing method for derivatives on monthly average temperature) [8]; hedging for loss of power prediction errors for either wind [9,10] or solar power [11]; and simultaneous hedging method for electricity price and volume risks [12–14]). Of these, the most recent study [14] demonstrated that portfolios of weather derivatives may be constructed by applying generalized additive models (GAMs [15]), which provide a significantly high hedge effect for the fluctuation risks in electricity sales profit/loss defined by the "product of price and demand". The profit/loss to be hedged, which was defined as such, corresponds to the fluctuation risk of the procurement costs of "aggregators" who are procuring variable demand from the wholesale market at variable prices (if they re-sell the procured electricity to consumers at a fixed price, the hedged target corresponds to the fluctuation risk of the excess profits). However, some electric utilities are exposed only to price risk, while some are exposed only to volume risk; hence, it is necessary to pay particular attention to the fact that practically, each electricity business has different exposed risks. For instance, the electricity sales revenue of IPPs (independent power producers) that sell fixed power outputs generated by "base-load power plants" (i.e., coal-fired or nuclear power generation operated at rated output) at the wholesale market price bears only price risk. Similarly, retailers who supply power at a fixed price and volume to consumers (e.g., large factories) under a "base-load contract" and procure that volume from the wholesale market at a variable price also bear only price risks. On the contrary, electric utilities whose price risks are hedged by forward contracts want to hedge only volume risks. Especially in immature markets such as Japan, fixed-price bilateral contracts are widely concluded, wherein the daily supply volume can be flexibly changed, among retailers who have newly entered the market [16]; retailers who have such contracts are completely exposed only to volume risk. Hence, different types of power utilities have special needs to hedge individual risks, namely, for volume only or price only, and the development of a more customized hedging method proposed by [14] is still an open question. Therefore, the main purpose of this study is to conduct extensive empirical analysis for different types of business risks described above and to demonstrate practical applicability of the derivative models by using the empirical data from the PJM market, which is the world's largest regional transmission organization (RTO).

In order to deal with these different types of business risks, we explore how appropriate weather derivative product can be designed for each business risk. Our previous study [14] has provided the following two approaches using non-parametric hedging models to minimize fluctuations in daily revenues (cash flow) in terms of product design: (i) apply standard derivatives of weather values in which the number of contracts is optimized, and (ii) synthesize optimal derivative contracts using arbitrary payoff functions of weather values given the profit/loss structure of a hedger. There is a trade-off between the above two approaches, wherein case (i) has the advantage that the "customized yet standardized" derivatives can be traded liquidly among multiple players; while case (ii) uses a "made-to-order" derivative for each hedger, so it cannot be liquidly traded, but the hedge effect might be enhanced compared with case (i). In fact, the study [14] has demonstrated that case (ii) generally has a higher hedge effect than case (i) for the fluctuation risk defined by the "product" of price and demand. However, if the hedged target contains either only demand or price risk, the nonlinear effect of temperature or other weather indices is supposedly weakened; therefore, rather than designing a completely made-to-order derivative, as in previous studies [14], using "customized yet standardized" derivatives are expected to have the advantages of having sufficient hedging effects (or maybe comparable to the "made-to-order" type), as well as high versatility in that they can be traded among multiple players. Note that the detailed transaction flow of "customized yet standardized" derivatives is found in Appendix A.

Furthermore, to ensure the out-of-sample hedging effect while using such standardized derivatives, methodological ingenuities to ensure robustness are considered very useful. Since non-parametric regressions with GAMs can express nonlinearities with more flexibility than parametric methods, we need to be aware of possible "overtraining"; therefore, when constructing various nonlinear models, it is necessary to estimate a function that expresses the nonlinearities "appropriately." For estimating nonlinear trends (i.e., payoffs or contract volumes) in hedging models, the R package "mgcv" [17] is very useful for practitioners in terms of its implementability and interpretability, as detailed in [18]. In "mgcv," various spline basis functions and smoothing methods are implemented, wherein the popular "cubic spline" (see, e.g., [19]) and "thin plate spline" [20] are set as the default basis (detailed in Section 3.3.1). However, the available basis functions include other advanced types, which may be overlooked by practitioners, such as "P-spline" [21], which has the advantage of "avoiding overtraining" [22], and "cyclic cubic spline" [23], which can robustly model periodic trends. Hence, if these bases are used instead, it is expected that robustness will be ensured, and the extrapolated hedging effects will be enhanced. In fact, our empirical result demonstrates that this hypothesis is correct and reveals that the hedge effect of temperature derivatives in previous studies is surprisingly improved by 13% to 53% by using an appropriate new basis function.

As described above, this study explores the important issues for decision-makers in the derivative contract practice, such as (i) what is an appropriate hedge product design for different business risks? and (ii) what is the appropriate spline basis function to enhance hedge effects? Then, by clarifying the interesting empirical results along with the theoretical interpretations, useful suggestions for practical application are provided.

This paper is organized as follows: Section 2 provides an overview of the background data of the PJM market, especially focusing on the nonlinearity of the data; Section 3 outlines the techniques used in this study; Section 4 formulates a specific hedging model treated in this paper. Section 5 examines the hedge effects of derivatives using empirical data. We also add considerations in the context of comparing business risk models or choice of bases in this section. Finally, Section 6 concludes the paper.

2. Overview of Background Data

This section provides an overview of the price and demand data in the PJM market, focusing on the "nonlinearities" that exist between data. First, Figure 1 illustrates a plot of demand in the PJM area with respect to the minimum and maximum temperatures. Both have a downwardly convex U-shaped curve, but they are relatively sensitive, especially when the temperature rises (i.e., in the summer). This figure suggests the existence of a strong nonlinear relationship between temperature and demand, but it is the story of looking at all the observation samples for one year. Considering that the temperature on each date fluctuates around the "climatological normal value," it is assumed that the fluctuation range of a certain date may be about 20 °F (11.1 °C) at the largest. This plot also indicates that the nonlinearity of the demand to temperature seems to be not very large, as long as temperature fluctuates within such range on the specific date.

Next, regarding the nonlinearity of price to demand, we overview the typical PJM "generation stack," shown in Figure 2. The generation stack is a curve wherein the marginal costs of supply capacity are arranged in ascending order, and it means the "supply curve." In other words, the generation stack corresponds to the plot of market price to demand (especially when assuming that the price elasticity of demand is 0, i.e., the demand curve is a vertical line). In practice, the generation stack can be frequently changed in the long- or short-term because of the termination/suspension of thermal power generation and the dependence of renewable power output on weather conditions; however, the point is that it is generally curved like a convex hockey stick. This strong nonlinearity of marginal cost curves is consistent with the fact that extreme price spikes occur when demand exceeds a certain level in winter or summer. Note that although the generation stack implies that price is nonlinear to demand, because there is a strong correlation between temperature and demand, as seen in Figure 1, price is also inferred to be nonlinear to temperature.

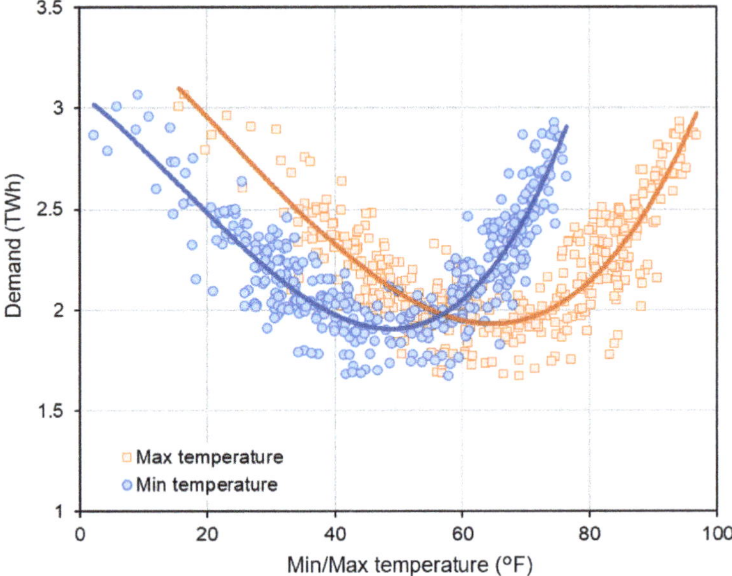

Figure 1. Relationship between temperature and electricity demand in PJM (2018).

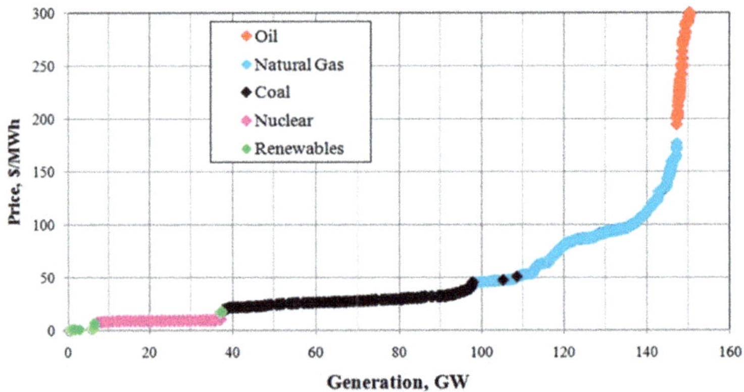

Figure 2. Typical PJM generation stack. Source: The Pennsylvania State University (https://www.e-education.psu.edu/ebf200/node/151) (accessed on 11 April 2021).

It should also be noted that the PJM price has a strong correlation with the Henry Hub (HH) natural gas price (see "Figure 5" in [14]). This corresponds to the fact that the natural gas part (blue line in Figure 2) of the generation stack shifts vertically because of fluctuations in the HH price. Because the demand curve and this supply curve intersect at the natural gas part in many time zones, electricity price may well be linked to the HH price as well. Additionally, considering that natural gas-fired power generation has increased significantly in recent years in the PJM area to replace coal-fired power generation, we will construct a hedge model that incorporates both the annual changes and the seasonal changes in the sensitivity of the HH price to the PJM price. Note that to target the PJM market, this paper treats HH price as the representative fuel price that explains the electricity spot price (i.e., we use HH futures as hedge products for the fuel-linked price risk). However, when constructing a hedging model for another market, it is necessary to properly select fuel futures that have a strong correlation with the electricity price in that country/area. For instance, WTI crude oil futures have been reported to be effective in hedging Japanese electricity market prices [14].

3. Minimum Variance Hedging Problem

In this study, we consider the problem of minimizing the cash flow variance of a portfolio consisting of the sales revenue (or procurement cost) of an electric utility and the payoffs of derivatives on temperature and fuel price. In the first half of this section, the previous method [14] applied to the empirical analysis in this study will be briefly explained while supplementing the methodological and theoretical background. In the second half, we will elaborate on the spline basis functions used for non-parametric hedge models.

We first assume that electric utilities (hedgers) can use two types of temperature derivatives. One is the temperature "futures," whose payoff is defined as the observed temperature T_t minus its (predicted) seasonal trend $h_T(t)$ at date t (i.e., the prediction error $\varepsilon_{T,t} := T_t - h_T(t)$). As the temperature futures designed in this manner can be regarded as having an expected payoff value of 0 at the time of the prior derivative contract, they are practically easy to handle because, for example, they do not necessarily require premium payments between risk-neutral players.

The other is the temperature "derivatives" on the prediction errors, whose payoff is expressed as a form of an arbitrary function of $\varepsilon_{T,n}$ (if expressed as a univariate function, it will be a payoff function of the form $\psi(\varepsilon_{T,n})$). As with temperature "futures," temperature "derivatives" can also be designed so that the expected payoffs are 0; and this issue will be dealt with in Section 3.2. The main problem to be considered here is that electric utilities seek the optimal contract volumes of futures or payoff functions of derivatives, aiming to suppress their fluctuation risks of sales revenues.

3.1. Optimal Futures Contract Volume Calculation Problem

Of the two types of derivative products mentioned above, the hedging problem for temperature futures is considered in this subsection. For example, when an electric utility wants to hedge the daily fluctuation of sales revenue π_t with HH and temperature futures (whose payoffs are HH_t and $\varepsilon_{T,t}$, respectively), the minimum variance hedging problem to be solved is as follows:

$$\min_{f(\cdot)\in\mathcal{S}_{\lambda_f},\,\Delta(\cdot)\in\mathcal{S}_{\lambda_\Delta},\,\gamma(\cdot)\in\mathcal{S}_{\lambda_\gamma}} \text{Var}[\pi_t - f(t) - \Delta(t)HH_t - \gamma(t)\varepsilon_{T,t}] \tag{1}$$

where $\text{Var}[\cdot]$ denotes the sample variance; $f(t)$, $\Delta(t)$, and $\gamma(t)$ are the contract volumes of the discount bond, HH futures, and temperature futures, respectively, at date t; and $\mathcal{S}_\lambda$ is a set of smoothing spline functions, with the smoothing parameters λ that control the tradeoff between model fit and smoothness [15] (detailed in Section 3.3.1). Thus, (1) depicts a problem that minimizes the cash flow variance of a portfolio; it consists of the sales revenue, discount bonds, and futures on HH and temperature under the smoothing parameters. Importantly, this optimization problem corresponds to constructing the following prediction formula for π_t and applies a GAM to estimate the smoothing spline functions f, Δ, and γ:

$$\pi_t = f(t) + \Delta(t)HH_t + \gamma(t)\varepsilon_{T,t} + \eta_t \tag{2}$$

where η_t is the residual term with an average of 0. As proven in [10], estimating the GAM (2) corresponds to minimizing the variance of η_t under the smoothing conditions of f, Δ, and γ; hence, it is synonymous with solving (1) (here, η_t can be interpreted as the "hedging error"). Note that the theoretical explanation for estimating the smoothing spline function using the GAM will be detailed in Section 3.3.

3.2. Optimal Derivative Payoff Calculation Problem

Considering that price has strong nonlinearity with respect to temperature, as explained in Section 2, this section introduces temperature "derivatives" with nonlinear payoff functions. Assuming the optimal payoff is given as a smooth function of temperature (when viewed on a specific date), changing smoothly according to the season, we formulate the following optimal payoff function calculation problem:

$$\underset{\widetilde{f}(\cdot)\in\mathcal{S}_{\lambda_{\widetilde{f}}},\ \Delta(\cdot)\in\mathcal{S}_{\lambda_{\Delta}},\ \widetilde{\psi}(\cdot)\in\mathcal{S}_{\lambda_{\psi 1},\ \lambda_{\psi 2}}}{\text{Min}}\quad \text{Var}\left[\pi_t - \widetilde{f}(t) - \Delta(t)HH_t - \widetilde{\psi}(t,\ \varepsilon_{T,t})\right] \tag{3}$$

where $\widetilde{f}(t)$ is the contract volume of discount bonds and $\widetilde{\psi}(t,\ \varepsilon_{T,t})$ is the payoff function of weather derivatives, which does not include the date-dependent trend. Here, the payoff function $\widetilde{\psi}(t,\ \varepsilon_{T,t})$ is estimated as a two-dimensional tensor product spline function [23] (detailed in Section 3.3.2) with ANOVA decomposition [24] (see Appendix B) by applying the following GAM, as was done in (2):

$$\pi_t = \widetilde{f}(t) + \Delta(t)HH_t + \widetilde{\psi}(t,\ \varepsilon_{T,t}) + \eta_t \tag{4}$$

We can obtain the temperature derivative payoff, which does not include the deterministic date-dependent trend, by ANOVA decomposition, and the expected value can be regarded as 0 at each date t. In other words, the temperature derivatives here can be treated as those that may not require a premium payment at the contract time, as with the temperature futures introduced in the previous section.

3.3. Spline Function Estimation Procedure

In this section, to understand how the spline functions estimated by the GAM are defined and calculated, the bases of spline functions and their estimation algorithms are briefly explained.

3.3.1. Univariate Smoothing Spline Function

First, the univariate smoothing spline function is estimated as the function s that minimizes the penalized residual sum of squares (PRSS) given by:

$$\text{PRSS} = \sum_{n=1}^{N}\{y_n - s(x_n)\}^2 + J(s,\ \lambda),\ \text{where } J(s,\ \lambda) = \lambda \int \{s''(x)\}^2 dx. \tag{5}$$

In (5), the first term measures the approximation of the data, and the second term ("penalty term") $J(s,\ \lambda)$ adds penalties according to the curvature of the function. In this study, we estimate GAMs using the R 3.6.1 package "mgcv" to obtain the series of smoothing spline functions, wherein the smoothing parameter is calculated by general cross-validation criterion.

In particular, this study pays attention to the fact that different basis functions can be applied when estimating the function $s(x)$, wherein the basis represents the function $b_i(x)$ in the following formula:

$$s(x) = \sum_{i=1}^{k} b_i(x)\beta_i \tag{6}$$

where β_i is the coefficient of the basis function.

The basis functions (smoothing methods) have some variations, such as "cubic spline" (see, e.g., [19]), "cyclic cubic spline" [23], "P-spline" [21], and "thin plate spline" [20]. Of these, all bases other than "thin plate spline" are expressed as "piecewise polynomials" joined at the points called "knots." Each of them has the following characteristics:

- "Cubic spline": It is one of the most popular basis functions and is defined based on the third order "truncated power basis functions." In other words, it is expressed by cubic polynomials, and each piecewise polynomial smoothly connects at each knot (i.e., the value, the first derivative, and the second derivative are continuous; see Appendix C.1 for the concrete formulas).
- "Cyclic cubic spline": It has a basis function that is defined to be smoothly connected not only at each knot, but also at the start and end points of the domain (see Appendix C.2 for the concrete formulas). For this reason, it is suitable for robustly estimating the trend of periodic data.
- "P-spline": While using B-spline basis [25] (which is based on "a special parametrization of a cubic spline" [26]), P-spline uses the unique penalty term called "discrete penalty" [21]. Unlike the other basis functions having "continuous penalties" with "integral squared curvature," as shown in (5), P-spline penalizes the changes in discrete coefficients of adjacent bases of B-spline (see [21] for the formula of penalty term). Initially, B-spline bases are arranged so that adjacent bell-shape curves overlap each other (e.g., see "Figure 1" of [21]); therefore, even though the discrete penalty is imposed for coefficients of the bases, smoothness is ensured, as in the case of continuous penalties.
- "Thin plate spline": Also called "radial basis functions," the basis of the thin plate spline depends only on the distance (norm) from each control point rather than the coordinates for each dimension. Therefore, unlike the previous three bases, the "thin plate spline" does not have "knots" as connection points (see [20,23] for more detail).

In the R package "mgcv," "cubic spline" or "thin plate spline" is given as the default basis, and the implementer has the option of reselecting other bases. This study particularly examines the above-mentioned "cyclic cubic spline" and "P-spline," comparing them with these default bases; the result will be detailed in Section 5.2.

Note that although the "mgcv" package also implements dozens of smoothing methods, such as "adaptive smoothers," an extension of P-spline (see "smooth.terms" of [17]), the comparison in this study focuses on the above-mentioned (basic) four bases. Regarding those four bases, Wood [17], who implemented "mgcv", introduces them as representatives of smoothers and provides detailed explanations (see the first part of "Section 5.3" in [17]), and there are many applied researches compared to other bases; therefore, we decided to choose those basis functions. That is, this study aims to explore the degree of improvement caused by reselection of the bases, and identifying the best basis functions is a future task.

Note also that there are multiple applied research cases in fields such as meteorology for "cyclic cubic spline," which is suitable for modeling periodic trends, but as far as we know, previous research applied in the field of energy does not exist, excluding the literature [27] that modeled electricity demand. Hence, this study is the first attempt to apply the "cyclic cubic spline" to model periodic electricity prices (for which trigonometric-function-based Fourier series expansion has been adopted in many studies).

3.3.2. Multivariate Smoothing Spline Function

Next, we describe the multivariate smoothing spline function. When spline functions are extended to multiple dimensions, their smoothing approaches are roughly divided into "tensor product smoothing" (tensor product spline) and "isotropic smoothing" [23].

3.3.2.1. Tensor Product Smoothing (Tensor Product Spline)

The tensor product spline function has different basis functions for each dimension, and the basis is given by the tensor product. For example, in the case of the bivariate function $s(x,z)$, the tensor product spline is written as the sum of the products of the basis functions, such as $a_i(x)$ and $b_j(z)$, as follows (note that for $a_i(x)$ and $b_j(z)$, it is possible to specify the different types of basis separately [17,23]):

$$s(x,z) = \sum_{i=1}^{k_1} \sum_{j=1}^{k_2} \beta_{i,j} a_i(x) b_j(z). \tag{7}$$

For estimating the function $s(x, z)$, the following penalty term $J_{te}(s, \lambda_x, \lambda_z)$ is included in the PRSS (i.e., (5) for the univariate case) that should be minimized [23]:

$$J_{te}(s, \lambda_x, \lambda_z) = \int_{x,z} \lambda_x \left(\frac{\partial^2 s}{\partial x^2} \right)^2 + \lambda_z \left(\frac{\partial^2 s}{\partial z^2} \right)^2 dxdz. \tag{8}$$

3.3.2.2. Isotropic Smoothing

Among the multivariate spline functions, the concept opposite to the tensor product spline is "isotropic smoothing" [23], and its representative one is the thin plate spline (note that there are other isotropic smoothing approaches, such as "Duchon splines" [28], which are a generalization of thin plate splines, and "soap film smoothing" [29], which is based on the idea of constructing a 2-D as smooth as a film of soap). As mentioned in the previous section, the basis of the thin plate spline is given as a function based on the distance (norm) from specific points, so when it is extended in multiple dimensions, the penalty term is different from that of the tensor product spline. For example, the penalty term $J_{tp}(s, \lambda)$ for the bivariate thin plate spline function is given as the following Equation [23]:

$$J_{tp}(s, \lambda) = \lambda \int_{x,z} \left(\frac{\partial^2 s}{\partial x^2} \right)^2 + 2 \left(\frac{\partial^2 s}{\partial x \partial z} \right)^2 + \left(\frac{\partial^2 s}{\partial z^2} \right)^2 dxdz. \tag{9}$$

As is clear from the comparison between (8) and (9), the term related to the mixed partial derivative $\partial^2 s / \partial x \partial z$ is added to the penalties for the thin plate spline. The thin plate spline is characterized by the isotropic addition of smoothing penalties at each point, as referenced by its name, which comes from its resemblance to the bent shape of a thin elastic plate.

While the thin plate spline is suitable if different axis units are the same, the tensor product spline, which can independently incorporate the smoothing conditions for each axis, is more suitable if the axis units are different. Therefore, this study adopts the tensor product spline for the pricing of the derivative $\widetilde{\psi}(t, \varepsilon_{T,t})$, with smooth trends in directions with different units such as date and temperature.

4. Construction of Hedging Models

In this section, we construct the concrete hedge models using the methods introduced in Section 3. Because the models have the same forms as the ones used in [14], only the outline is provided in this section.

4.1. Base Model Consisting of Fuel Price and Calendar Trend

First, considering that PJM electricity prices are strongly linked to HH prices and day type, and that they have annual change trends, as explained in Section 2, the following GAM is constructed, referred to as the "base model":

$$\pi_t = f(t) + \Delta(t)HH_t + \eta_t$$
$$\text{where} \begin{cases} f(t) := f_O(t) + f_H(t)I_{H,t} + f_P(t)Period_t \\ \Delta(t) := \Delta_O(t) + \Delta_P(t)Period_t \end{cases} \tag{10}$$

where f and Δ are yearly cyclical trends estimated as spline functions by the GAM (as $f(Seasonal_t)$ with yearly cyclical dummy variables $Seasonal_t (= 1, \ldots, 365(or\ 366))$ [30], denoted as $f(t)$ and $\Delta(t)$ for concise notation), $I_{H,t}$ is a dummy variable for holiday, and $Period_t$ is the elapsed day of date t (annualized) from the beginning of the starting year of the data. Of these, the term $f_P(t)Period_t$ ($\Delta_P(t)Period_t$) is introduced because the calendar trend (sensitivity of HH to π_t) is assumed to have yearly cyclical trends, even when viewed at the rate of annual change. When estimating the model, three sets of the same data sample are used side by side so that the start and end points of the estimated cyclical trends f and Δ are approximately connected; by doing so, the desired yearly cyclical trends can be obtained as the estimated function in the middle domain (see "Appendix" in [14]).

4.2. Temperature Futures

Next, for the case wherein temperature futures can be used, the following hedging model is considered:

$$\pi_t = f(t) + \Delta(t)HH_t + \gamma_1(t)\varepsilon_{Tmin,t} + \gamma_2(t)\varepsilon_{Tmax,t} + \eta_t \tag{11}$$

where $\varepsilon_{Tmin,t}$ and $\varepsilon_{Tmax,t}$ are the payoffs of minimum and maximum temperature futures on day t, respectively, and $\gamma.(t)$ is the estimated yearly cyclical trend, corresponding to the contract volume of those futures.

4.3. Temperature Derivatives Estimated by the Tensor Product Spline

Similarly, a hedging model for temperature derivatives is constructed as follows:

$$\pi_t = f(t) + \Delta(t)HH_t + \widetilde{\psi_1}(t,\varepsilon_{Tmin,t}) + \widetilde{\psi_2}(t,\varepsilon_{Tmax,t}) + \eta_t \tag{12}$$

where $\widetilde{\psi}.(t,\varepsilon_{.,t})$ are the smooth payoff functions of temperature derivatives, which change smoothly depending on the date t, estimated as tensor-product spline functions from which yearly cyclical trends have been removed via ANOVA decomposition, as explained in Section 3.2. Notably, the two payoff functions on minimum and maximum temperature derivatives can be uniquely estimated here because the date-dependent trend is unified into the identical term $f(t)$ through ANOVA decomposition. (More specifically, if ANOVA decomposition is not applied, f, ψ_1, ψ_2 all contain trends with respect to t (i.e., they have overlapping degrees of freedom with respect to t), and each function shape cannot be determined. Conversely, in (12) with ANOVA decomposition applied, the trend for t is removed from the derivative payoff functions $\widetilde{\psi_1}$, $\widetilde{\psi_2}$ and is explained only by f, which solves the problem of overlapping degrees of freedom. See also Appendix B for details on ANOVA decomposition.)

4.4. Temperature Derivatives for the Squared Prediction Error

The temperature derivatives' payoffs in (12) are estimated differently by each hedger, but here, we consider a "standard" derivative on temperature, which can be commonly traded by multiple hedgers. From the idea of approximating $\widetilde{\psi}(t,\varepsilon_{.,t})$ with a quadratic function for $\varepsilon_{.,t}$, we introduce temperature derivatives on squared prediction errors and construct the following model:

$$\pi_t = f(t) + \Delta(t)HH_t + \gamma(t)\varepsilon_{T,t} + \tau_1(t)\left(\varepsilon_{Tmin,t}{}^2 - \overline{\varepsilon_{Tmin,t}{}^2}\right) + \tau_2(t)\left(\varepsilon_{Tmax,t}{}^2 - \overline{\varepsilon_{Tmax,t}{}^2}\right) + \eta_t. \tag{13}$$

where $\varepsilon_{.,t}{}^2 - \overline{\varepsilon_{.,t}{}^2}$ are the payoffs of the squared prediction error derivatives on temperatures ($\overline{\varepsilon_{.,t}{}^2}$ is the predicted value (sample mean) of $\varepsilon_{.,t}{}^2$), and $\tau.(t)$ are the spline functions representing the contract volumes of the derivatives estimated by the GAM. These temperature squared error derivatives also have zero expected payoffs for each t, and similar to other derivatives, they do not require premium payments.

5. Empirical Analysis

This section empirically validates the hedging models introduced in Section 4 by using PJM market data. First, Section 5.1 applies those models to each of the three different "business risk models" and compares hedge effects as well as the shapes of the estimated derivative payoffs. Then, Section 5.2 compares the hedge effect by using the different bases described in Section 3.3.1. The empirical data used are as follows:

(a)   Demand $D_{t,h}$ (TWh): the hourly load of the entire PJM-RTO [31].

(b)   Electricity price $S_{t,h}$ (USD/MWh): day-ahead hourly spot price of PJM-RTO [31].

(c)   Maximum and minimum temperatures $Tmax_t$, $Tmin_t$: population-weighted average of four main cities (Philadelphia, Pittsburgh, Baltimore, and Newark) published by the National Oceanic and Atmospheric Administration (NOAA) [32].

(d)   Henry Hub natural gas price HH_t (USD/million BTU): historical daily HH spot price FOB [33].

The model parameters and functions are estimated from the in-sample period data (1 January 2011–31 December 2017), and the hedge effects are calculated from the out-of-sample data (1 January 2018–31 December 2018). In this study, we choose only the above four major cities in the PJM area (as with [34] for example) for the temperature index, but it may be possible to construct a more fitted hedge model (i.e., obtain higher hedge effect) by increasing the number of temperature observation points. However, it should be also noted that when assuming that the temperature derivatives are traded in practice, the smaller the number of points, the easier it is for traders to understand and handle.

5.1. Empirical Analysis by Business Risk Models

This section verifies estimated trend functions and hedge effects in the context of comparing business risk models. As introduced in Section 1, we deal with the three business risk models as exposed to (i) both price and demand risks, (ii) price risk only, and (iii) demand risk only. That is, for each case, the hedged target (i.e., the hedger's fluctuating revenue/cost) is expressed, respectively, as (i) the sum of the product of hourly spot price $S_{t,h}$ and demand $D_{t,h}$ ($\pi_t = \sum_h S_{t,h} \times D_{t,h}$, referred to as the "product model"); (ii) price ($\pi_t = (1/24) \sum_h S_{t,h}$; "price model"); and (iii) demand ($\pi_t = \sum_h D_{t,h}$; "demand model").

Note that because the superiority of the cyclic cubic spline over different bases is revealed by the empirical analysis in Section 5.2, this section uses that basis to compare business risk models.

5.1.1. Trend Estimation of Hedge Models

5.1.1.1. Optimal Payoff Function of the Temperature Derivatives

First, Figure 3 displays the min/max temperature derivatives' payoff functions $\widetilde{\psi}_1(t, \varepsilon_{Tmin,t})$ and $\widetilde{\psi}_2(t, \varepsilon_{Tmax,t})$ of the "product model," which were simultaneously estimated as a tensor product spline function using ANOVA decomposition in hedge model (12). In both cases, it can be confirmed that the trends in the seasonal direction are removed (e.g., having shapes with zero mean at each date t) via ANOVA decomposition. In addition, the payoff of the derivative of minimum temperature (corresponding to the sensitivity of the minimum temperature prediction error to the sales revenue) is specifically increased as temperatures drop in winter, whereas that of maximum temperature is increased as temperatures rise significantly in summer, reflecting that both distinctive effects complement each other.

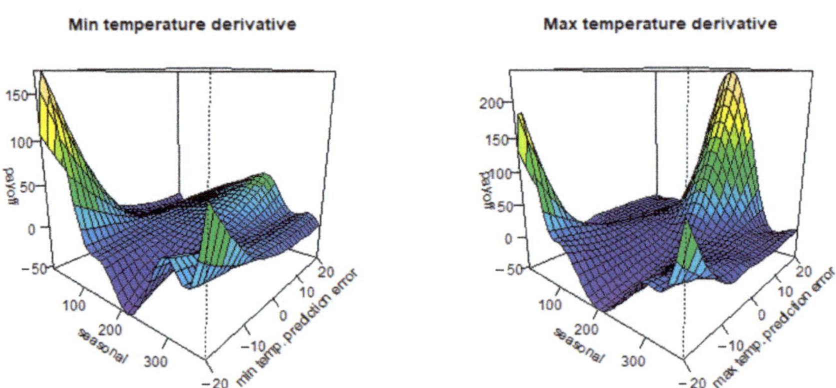

Figure 3. Estimated temperature derivatives' payoffs in the product model (12).

Next, Figure 4 shows the payoff functions for the temperature derivatives of the "price model." The shapes of the payoff functions are not significantly different from those of the product model shown in Figure 3, so it can be inferred that the nonlinearities of the "product" model's derivatives mostly result from those of the "price" to the temperature (to put it in detail, the daily change in the slope of the maximum temperature derivative during summer in Figure 3 is slightly more rapid than that in Figure 4, which may indicate that the product model has slightly stronger nonlinearity than the price model).

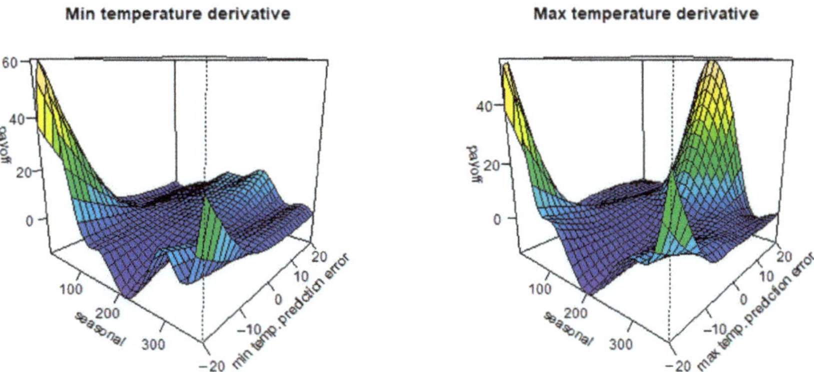

Figure 4. Estimated temperature derivatives' payoffs in the price model (12).

The payoff functions for the temperature derivatives of the demand model are shown in Figure 5. Similar to that of the product model, the slope of the payoff function is positive in summer and negative in winter, but notably, the payoff function of the demand model is smoother than that of the product model (i.e., the nonlinearity is relatively small). This probably indicates that the temperature sensitivity to the demand with the normal temperature on a specific date is relatively small, as seen in Figure 1.

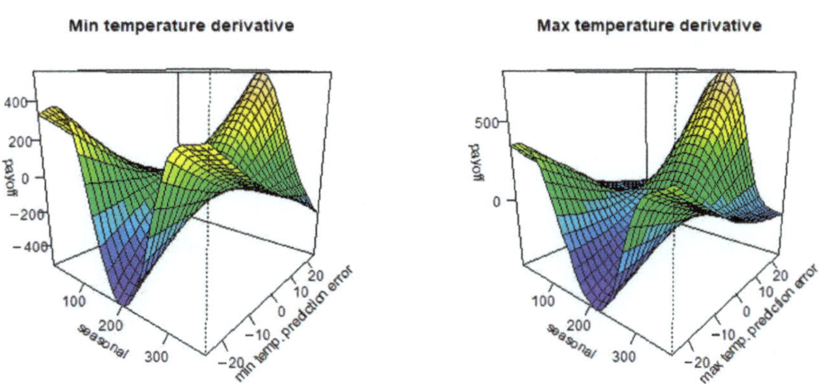

Figure 5. Estimated temperature derivatives' payoffs in the demand model (12).

5.1.1.2. Optimal Contract Volume of the Squared Temperature Prediction Error Derivatives

Figure 6 displays the estimated contract volume trends (the dotted line indicates 95% confidence interval) of the squared prediction error derivatives of the min/max temperatures in the "product model" (13). They can be rephrased as trends that reflect the magnitudes of the (downward) convexities of temperature sensitivities to the sales revenue. Both trends rise in summer and winter, indicating that the maximum temperature's nonlinearity to sales revenue is particularly strong in summer, while that of the minimum temperature is particularly strong in winter (consistent with the non-parametrically-priced derivatives shown in Figure 3). Similarly, Figure 7 shows the same trends for the price model. The shapes are not significantly different from the product model shown in Figure 6, probably because of the same reason explained in Section 5.1.1.1.

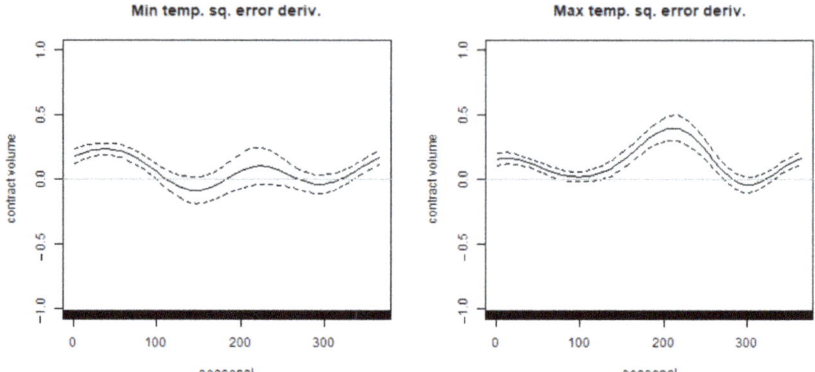

Figure 6. Estimated contract volume of the temperature squared error derivatives in the product model (13).

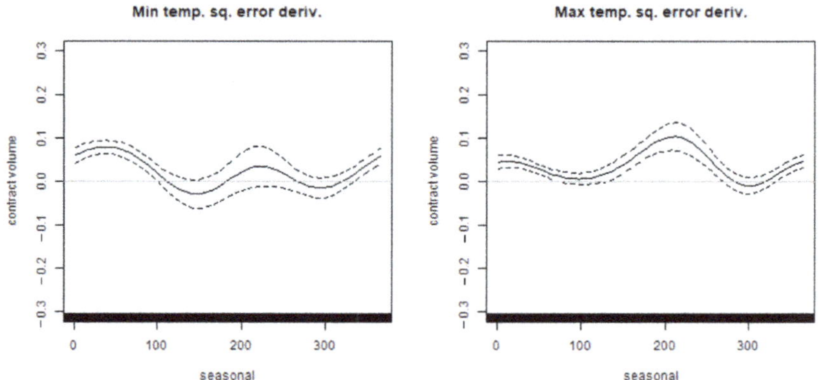

Figure 7. Estimated contract volume of the temperature squared error derivatives in the price model (13).

Figure 8 shows the estimated contract volume of the temperature squared error derivatives in the demand model. The derivative contract volumes of the minimum and maximum temperatures approaching 0 is common in winter, but in summer, that of the maximum temperature is significantly higher while that of the minimum temperature approaches 0. This may be due to the following reasons: In the PJM area, the absolute value of the temperature sensitivity (slope) to power demand tends to be higher in summer than in winter, as seen in Figure 1, and the "change rates" in temperature sensitivity have a similar seasonal tendency (i.e., convexity to temperature). Probably reflecting these tendencies, the maximum temperature strongly explains such convexity around the summer season, while the minimum temperature complementarity explains the (rest of the) seasonal changes in other convexities. In addition, this estimated result is consistent with the fact that the downward convexity of the maximum temperature derivative payoff becomes relatively larger during the summer, as confirmed in Figure 5.

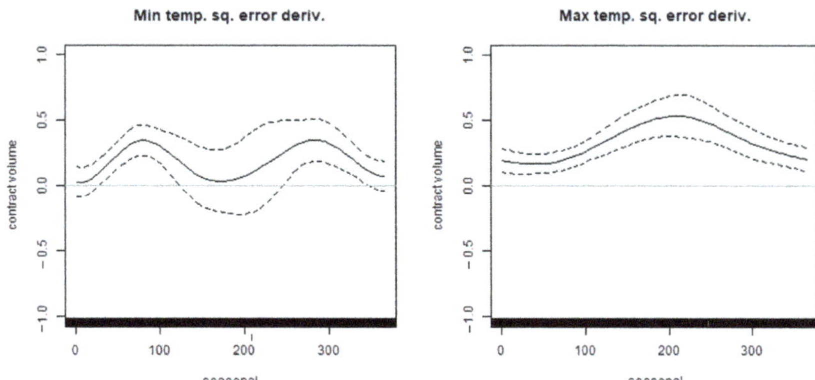

Figure 8. Estimated contract volume of the temperature squared error derivatives in the demand model (13).

5.1.2. Measurement of Hedge Effects

In the following, we measure the hedge effect of each derivative for each business risk model. This study uses the variance reduction rate (VRR), defined as follows, and it refers to 1-VRR as the hedge effect:

$$\text{VRR} := \frac{\text{Var}[\text{hedge error of the target model}]}{\text{Var}[\text{hedge error of the base model}]}. \tag{14}$$

5.1.2.1. Cumulative/Individual Hedge Effects by Derivatives

Here, we analyze changes in hedge effects when each derivative (hedge model term) is cumulatively combined; the results are summarized in Figures 9–11 for "product model," "price model," and "demand model," respectively. Each figure illustrates the "single contribution ratio" when each derivative is used alone (bar graph), the "cumulative contribution ratio" (corresponding to the R-squared; see, e.g., ref. [35] for "out-of-sample R-squared statistic") when the derivatives are combined in order from the top (blue line graph), and the "cumulative hedge effect" of the temperature derivatives compared with the "base model" (3) (red line graph). All are measured for the three models (product, price, and demand models).

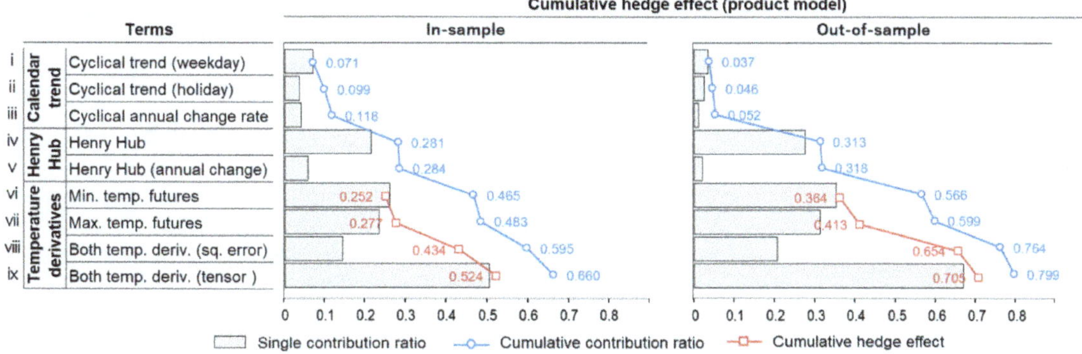

Figure 9. Contribution ratio and hedge effect of product model.

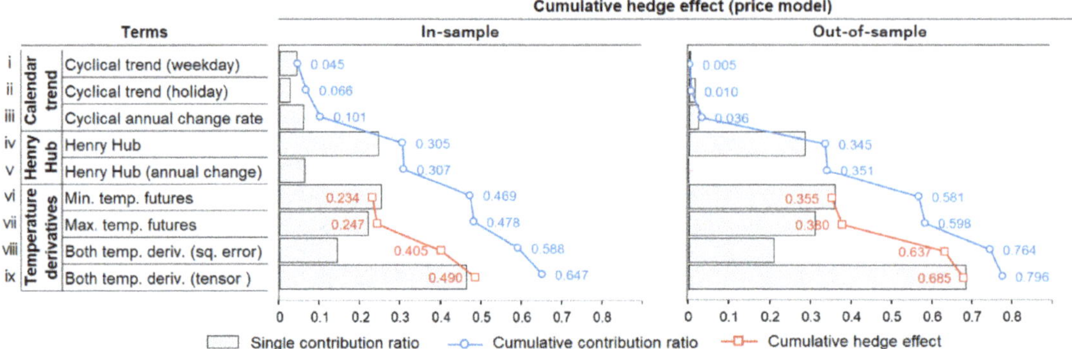

Figure 10. Contribution ratio and hedge effect of price model.

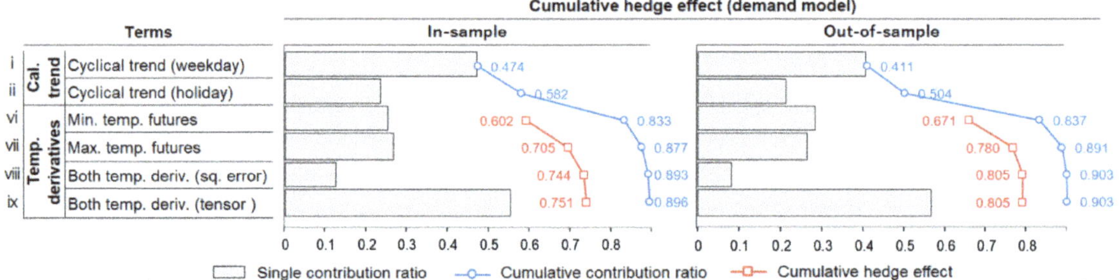

Figure 11. Contribution ratio and hedge effect of demand model.

Note that the Roman numerals of the terms in Figures 9–11 correspond to the following: i. $f_O(t)$; ii. $f_H(t)I_{H,t}$; iii. $f_P(t)Period_t$; iv. $\Delta_O(t)HH_t$; v. $\Delta_P(t)Period_tHH_t$; vi. $\gamma_1(t)\varepsilon_{Tmin,t}$; vii. $\gamma_2(t)\varepsilon_{Tmax,t}$; viii. $\tau_1(t)(\varepsilon_{Tmin,t}^2 - \overline{\varepsilon_{Tmin,t}^2}) + \tau_2(t)(\varepsilon_{Tmax,t}^2 - \overline{\varepsilon_{Tmax,t}^2})$; ix. $\widetilde{\psi_1}(t,\varepsilon_{Tmin,t}) + \widetilde{\psi_2}(t,\varepsilon_{Tmax,t})$ (in the cumulative usage case including up to ix, terms vi–viii are excluded). Note also that in the demand model, terms regarding HH are excluded since no correlation is assumed.

First, for all three models, the temperature derivatives had the highest single contribution ratios (around 60–70%) among all derivatives (terms) in the out-of-sample period. Next, the cumulative contribution ratios increased monotonously with the inclusion of each term, even in the out-of-sample case, and they reached close to 80% for both the product and price models and over 90% for the demand model. Similarly, the cumulative hedge effects increased monotonically and reached approximately 70% for both the product and price models, and over 80% for the demand model. When the maximum temperature futures were combined with the minimum temperature futures, the hedge effect was improved by approximately 3–5 percentage points for the product and price models and approximately 11 percentage points for the demand model, respectively. Hence, it is suggested that the two different temperature products have complementary effects for all models.

Regarding the product and price models, when compared with using only temperature futures, the combined use of the squared error temperature derivatives further improved the hedge effect by about 24–26 percentage points, and further improvement of approximately 5 percentage points occurred when using the derivatives of the tensor product splines. This result reflects the strong nonlinear correlation between temperature and PJM price. (This may be easier to understand when the price (product) is simply regressed by temperature regardless of date t. In such case, the quadratic function (corresponding to the payoff of squared error temperature derivatives) fits better than the linear function (corresponding to the payoff of temperature futures), and arbitrary spline function (corresponding to the payoff of tensor product spline derivatives) fits further better than both, thereby reducing the variance of the residuals (i.e., variance of hedging error; see the explanation of (2)). At this time, the numerator of VRR (4) becomes smaller, so the hedging effect (1-VRR) becomes larger.

On the other hand, regarding the demand model, the high hedge effect of 67% was confirmed only by using the minimum temperature futures with a linear payoff function. Presumably, it indicates that the temperature sensitivity to demand has a small nonlinearity, as was confirmed with the derivative payoff shape in Figure 5. Regarding the hedge effect of nonlinear derivatives, a 2 percentage-point improvement was confirmed for the squared error derivatives, but no improvement was seen for the tensor-product derivatives. This result implies that "customized yet standardized" square error derivatives (combined with temperature futures) may be superior to "made-to-order" tensor-product derivatives, in that the standardized derivatives allow for liquid trading.

In this way, the squared error derivative significantly improves the hedging model of both the product and price models, and improves even the hedge effect of the demand model, which has relatively weak nonlinearity. It is suggested that for many risk types, electric businesses may be able to trade it in common for efficiently hedging "nonlinearity-derived" fluctuation risks. Note that although this study verified the hedging effects of different types of "electric utilities," since the payoff function of "customized yet standardized derivative" is defined only by the (public) measured temperature, it could also be used for businesses in other sectors affected by weather, such as the agriculture and leisure industries.

5.1.2.2. Monthly Hedge Effect

Figure 12 demonstrates the monthly hedge effects (1-VRR) of the temperature derivatives (estimated by the tensor-product spline functions) for each of the three models for both in-sample and out-of-sample periods. For each period, the hedge effects have generally similar seasonal tendencies. The hedge effect tends to increase in summer (June–July) and winter (December–January) for each model, which corresponds to the payoff functions of each model's derivatives having an extremely steep slope during these seasons, as seen in Figures 3–5. The demand model has a higher hedge effect throughout the period (for all months) than the price or product model. It is suggested that the electricity demand tends to fluctuate relatively greatly because of temperature. The hedge effect for the price and product models is not so large during spring and autumn (around April–May or September–November) because the effect of temperature on price fluctuations during these periods is smaller than other factors, such as changes in the market environment and power supply operation (e.g., in the price model of September 2018, the hedge effect of the out-of-sample period has a negative value of -0.55). Hence, derivative trading strategies limited to summer and winter may be effective for price risk and product risk.

See Appendix C as well, wherein we verify the accuracy of the model by month from the perspective of "hedge error."

5.2. Comparison between Basis Functions

This section examines the extent to which the hedge effects measured in Section 5.1.2 may change when using different basis functions, as introduced in Section 3.3.1. The examined basis functions include: (a) thin plate spline / cubic spine ("tp/cr"), which is the default case of the R package "mgcv" (wherein the thin plate spline is used for univariate splines, and the cubic spline is used for tensor product splines); (b) P-spline ("ps"); and (c) cyclic cubic spline ("cc"). We target all hedge cases involving the cumulative hedge effect of the weather derivative seen in Figures 9–11, and we compare the basis functions by changing only those for seasonal trends in the date direction in common for all cases (i.e., for the temperature direction in the tensor product spline, the default basis "cr" is used as is in all cases). For each of these bases, we compare the cumulative hedge effect and contribution ratio for all business risk models. The result is shown in Table 1 (the red gradation is colored by comparing the three values of each basis within the same model).

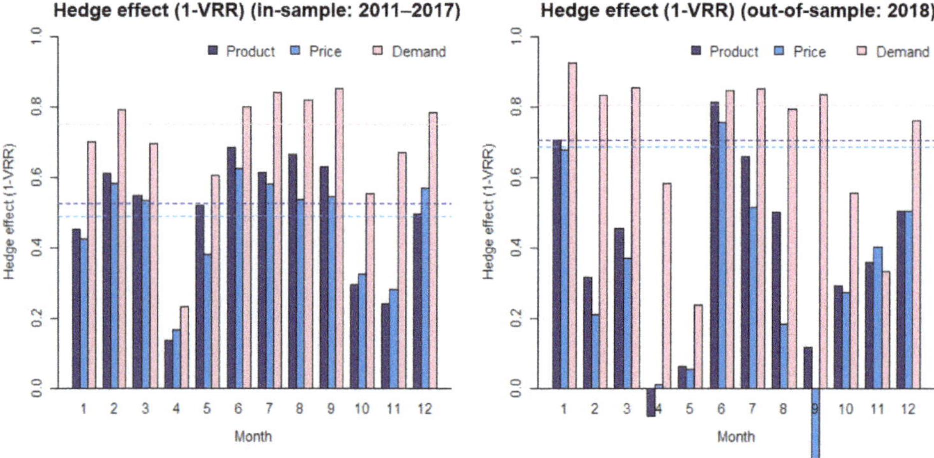

Figure 12. Monthly hedge effect for each business risk model.

Table 1. Comparison of cumulative hedge effect/cumulative contribution ratio by basis.

Business Risk Model		In Sample						Out-of-Sample							
		Cumulative Contribution Ratio			Cumulative Hedge Effect			Cumulative Contribution Ratio			Cumulative Hedge Effect			Δ (to tp/cr)	
		tp/cr	ps	cc	tp/cr	ps	cc	tp/cr	ps	cc	tp/cr	ps	cc	ps	cc
Product	vi	0.480	0.477	0.465	0.267	0.262	0.252	0.521	0.546	0.566	0.238	0.289	0.364	21%	53%
	vii	0.495	0.492	0.483	0.289	0.283	0.277	0.548	0.571	0.599	0.281	0.329	0.413	17%	47%
	viii	0.601	0.598	0.595	0.438	0.433	0.434	0.720	0.722	0.764	0.555	0.565	0.654	2%	18%
	ix	0.660	0.641	0.660	0.520	0.493	0.524	0.762	0.788	0.799	0.621	0.668	0.705	8%	13%
Price	vi	0.484	0.483	0.469	0.246	0.244	0.234	0.545	0.568	0.581	0.248	0.290	0.355	17%	43%
	vii	0.494	0.491	0.478	0.259	0.257	0.247	0.564	0.585	0.598	0.278	0.318	0.380	14%	37%
	viii	0.596	0.592	0.588	0.409	0.404	0.405	0.731	0.733	0.764	0.555	0.561	0.637	1%	15%
	ix	0.647	0.633	0.647	0.484	0.463	0.490	0.781	0.782	0.796	0.637	0.642	0.685	1%	8%
Demand	vi	0.837	0.826	0.833	0.607	0.589	0.602	0.835	0.837	0.837	0.668	0.667	0.671	0%	0%
	vii	0.881	0.870	0.877	0.712	0.694	0.705	0.887	0.894	0.891	0.772	0.783	0.780	1%	1%
	viii	0.896	0.888	0.893	0.748	0.735	0.744	0.902	0.907	0.903	0.802	0.810	0.805	1%	0%
	ix	0.898	0.891	0.896	0.753	0.744	0.751	0.903	0.905	0.903	0.804	0.804	0.805	0%	0%

As can be seen from this table, the value of "tp/cr" in all cases is the highest of the three basis functions for the in-sample period, but it is the lowest for the out-of-sample period overall. On the contrary, "cc" tends to have a small value for the in-sample, but a high value in the out-of-sample period. Moreover, "ps" is almost in the middle of the two. In other words, such a phenomenon of value reversal between the in-sample and the out-of-sample suggests that "cc" and "ps" are superior to "tp/cr" in terms of robustness. Looking at the improvement of the hedge effect of "cc" and "ps" in particular, when compared with the default case "tp/cr" in the out-of-sample (shown in the right two columns), relatively large improvements are observed for the "product model" (and the "price model") with strong nonlinearities (note that in Section 5.1.2.1, when measuring the improvement of the cumulative hedge effect by additionally incorporating hedge products, it was meaningful to use "percentage point," which measures the differentials of the hedge effects; however, in this section, the improvement "ratio" from the default case is measured to compare the methodologies in the same hedge model). In particular, "cc" has larger hedge effects than the default basis in all cases, and in the product model, it improves by no less than 13–53% (Note 1: The results of the product model in Table 1 can be easily verified by using the R source code published in [14] and by only changing each of the default basis functions to "ps" or "cc"; Note 2: The values of the hedge effect shown in [14] exactly match the values in the crossing sells of the "tp/cr" columns and the "product" rows in Table 1; Note 3: Although not shown in this table, when the "normal" cubic spline "cr" was used instead of the thin plate spline "tp," the change was very slight in all cases, and in fact, more cases worsened).

The technical consideration of the above results is as follows. First, the high robustness of the P-spline is likely because the "discrete penalty" of the P-spline contributes to the avoidance of overfitting. In fact, the P-spline has an advantage in that the "loss of control" problem that tends to occur when using "continuous penalty" can be avoided [36], and such a mechanism may have worked. Regarding the cyclic cubic spline, its high robustness probably comes from the constraint that the start and end points are smoothly connected. The hedge model method proposed in [14] tried to impose this constraint through a data-driven manner using three sets of the same data sample side by side, so that the start and end points of the yearly cyclical trend were smoothly connected (see, e.g., "Appendix" in [14]). However, we found in this study that such ingenuity is not always sufficient, and that by incorporating similar constraints as well into the basis functions used in the model, robustness is further ensured, and the out-of-sample hedge effect can be significantly improved. More interestingly, in the comparison among business risk models, the improvement resulting from appropriate basis selection is remarkably large, especially for the product model (and price model) with strong nonlinearity, and small for the demand model with weak nonlinearity. This result suggests that the stronger the nonlinearity inherent in the model, the more important the robustness (constraint strength) to be incorporated into the functional expression (Appendix D).

Although the basis function selection may be overlooked by modeling practitioners, it may be a highly critical issue in robustly estimating a model in which strong nonlinearities are intertwined in a complex manner, such as the hedging models treated in this study.

6. Conclusions

In this study, paying attention to the fact that different types of electric utilities are exposed to risks of demand, price, and both, we verified the hedge effects for each of the three business risk models ("demand model," "price model," and "product model") using a previously proposed temperature derivative portfolio estimated using non-parametric hedging models. In addition, we found that choosing the appropriate basis for spline function can ensure the robustness of the model and significantly improve the out-of-sample hedge effects.

First, regarding the comparison between the three business risk models, the following empirical results and suggestions were obtained:

- The nonlinearity of the temperature derivative payoffs by the business risk model is strong in the product model and the price model, and relatively weak in the demand model.
- Reflecting this, the non-parametrically-priced derivative payoff function, which can flexibly express strong nonlinearity, has the highest hedge effect on the product model and the price model.
- On the contrary, for the demand model, the hedge effect of the non-parametric derivatives does not exceed that of the standard derivative on the squared temperature prediction error; hence, the squared error derivatives may be superior in that it allows for liquid trading.

It was confirmed that the squared error derivative has high hedge effects on both product and price models, which are comparable to the non-parametric derivatives. This result also suggests that this "customized yet standardized" squared error derivative is promising as a new standard derivative that can be traded between players exposed to many different business risks. On the other hand, it is also true that the "made-to-order" non-parametric derivatives have the highest hedge effect for both product and price models. Therefore, in practical decision-making scenes, derivative contracts may need to be made after considering the trade-off between liquid tradability and maximization of the hedge effect.

Next, regarding the basis selection for the spline functions, we obtained the following implications:

- When estimating the series of trends existing in the hedge models, using the P-spline or the cyclic cubic spline instead of the thin plate spline or cubic spline set as the default in the R "mgcv" package can secure the robustness of the models; as a result, the out-of-sample hedge effect may be significantly improved.
- The improvement of the hedge effect by appropriate basis selection is larger for the product and price models than for the demand model. This means that the stronger the nonlinearity in the model, the more critical the basis selection that can robustly express the inherent trends of the data.

When the cyclic cubic spline was applied to the seasonal trend of the "product model" with the strongest nonlinearity, surprisingly, the hedge effect improved by 13–53%, compared with the previous empirical results using the default case that was demonstrated in [14]. Although the selection of basis functions seems to be overlooked in practice, we conclude that it is extremely important to keep in mind for the robust estimation of models with strong nonlinearities, as treated in this study.

The non-parametric hedging models we have proposed have been evolving in demonstrating applicability to different empirical data and devising methodologies for ensuring robustness. In the electricity market of the future, wherein transactions of decentralized players are assumed to increase significantly, it is expected that there will be increasing needs for financial instruments that can flexibly hedge fluctuation risks in finer time granularity (daily and hourly), such as the weather derivatives used in this study. Our future task is to expand the empirical analysis further and refine the model aiming for the practical application of this unique non-parametric hedging model and the high-resolution weather derivatives.

Author Contributions: Conceptualization, Y.Y.; methodology, T.M.; software, T.M.; validation, T.M.; data curation, T.M.; writing—original draft preparation, T.M.; writing—review and editing, Y.Y.; visualization, T.M.; supervision, Y.Y.; project administration, Y.Y.; funding acquisition, T.M. and Y.Y. Both authors have read and agreed to the published version of the manuscript.

Funding: This work was funded by a Grant-in-Aid for Scientific Research (A) 16H01833, Grant-in-Aid for Scientific Research (A) 20H00285, Grant-in-Aid for Challenging Research (Exploratory) 19K22024, and Grant-in-Aid for Young Scientists 21K14374 from the Japan Society for the Promotion of Science (JSPS).

Conflicts of Interest: The authors declare no conflict of interest.

Nomenclature

t	Date
h	Hour
π_t	Sales revenue (or procurement cost) of an electric utility
$D_{t,h}$	Demand
$S_{t,h}$	Spot price
HH_t	Henry Hub natural gas price
T_t	Temperature
$Tmax_t, Tmin_t$	Maximum/Minimum temperature

$\varepsilon_{T,t}$	Payoff of the temperature futures (temperature prediction error)
$\varepsilon_{T,t}{}^2 - \overline{\varepsilon_{T,t}{}^2}$	Payoff of the squared prediction error derivative on temperature
$\psi(t,\,\varepsilon_{T,t}),\ \widetilde{\psi}(t,\,\varepsilon_{T,t})$	Payoff of the temperature derivative estimated by tensor product spline function
$f(t),\ \widetilde{f}(t)$	Contract volume of discount bonds
$\Delta(t)$	Contract volume of HH futures
$\gamma(t)$	Contract volume of the temperature futures
$\tau(t)$	Contract volume of the squared prediction error derivatives on temperature
η_t	Residual term (hedging error of derivative portfolio]
λ	Smoothing parameter
$\mathcal{S}_\lambda$	A set of smoothing spline functions with smoothing parameter λ
$s.(x)$	Spline function
J	Penalty term of penalized residual sum of squares
$b_i(x)$	Basis functions of spline function
β_i	Coefficients of the basis functions

Appendix A. Transaction Flow of "Customized Yet Standardized" Derivatives

Figure A1 shows the transaction flowchart of the "customized yet standardized" derivatives. Although the figure is created for an electric utility (retailer) exposed to "product risk," if it is not exposed to price or demand risk, the flow excluding (fixing) that risk may be considered. The transaction procedure is as follows:

1. The electric utility optimizes the contract volume (such as $\gamma(t)$ and $\tau(t)$) of the standard derivatives for each future delivery date based on the past profit/loss function (for the retailer in the figure, it is procurement cost $\pi_{t,\,h} := S_{t,\,h} \times D_{t,\,h}$).
2. The utility makes a contract (transaction) of the temperature derivatives of the volume calculated in 1. in the derivative market. At this time, no premium payment is made.
3. The utility purchases electricity at the spot market and pays the corresponding procurement cost (or records the cost in its account books) on the day before the delivery date of electricity.
4. The utility receives (or pays) a payoff of temperature derivatives calculated based on the measured temperature on the delivery day of electricity. (Since this payoff is greatly linked to the procurement cost of 3., the net cash flow, which is the sum of 3. and 4., will be less volatile than the original cash flow of 3.)

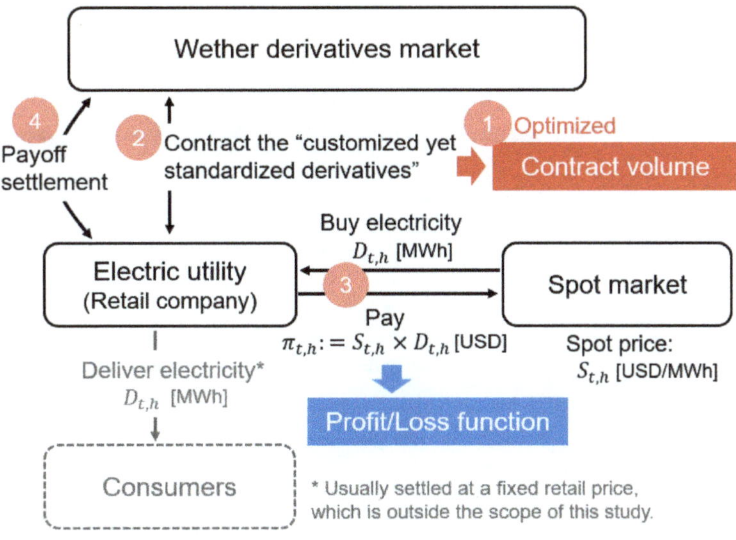

Figure A1. Transaction flowchart of "customized yet standardized derivatives".

Note that when trading "made-to-order derivatives" that cannot be fluidly traded, what the utility should optimize is the "payoff function" itself of temperature derivatives, as illustrated in "Figure 6" of [14]. In this case, the utility would contract with a specific insurer (who agreed on the optimized payoff function) rather than trading derivatives in the derivatives market. In contrast, in the case of "customized yet standardized derivatives" (and if the derivatives are traded in a fluid manner), the utility does not need to negotiate a bilateral contract with such a particular insurer; thus, the advantage of reducing trading costs may be also expected.

Appendix B. Separation of Deterministic Trends by ANOVA Decomposition

To understand ANOVA decomposition, first consider the following minimum variance hedge problem without ANOVA decomposition:

$$\min_{\Delta(\cdot)\in\mathcal{S}_{\lambda_\Delta},\ \psi(\cdot)\in\mathcal{S}_{\lambda_{\psi 1},\ \lambda_{\psi 2}}} \mathrm{Var}[\pi_t - \Delta(t)HH_t - \psi(t,\ \varepsilon_{T,t})]. \tag{A1}$$

where $\psi(t,\ \varepsilon_{T,t})$ is the payoff function of the temperature derivative estimated as the tensor product spline function. However, since this function $\psi(t,\ \varepsilon_{T,t})$ contains a trend related to the date t, the problem that it is difficult to grasp the structure as a hedge model arises. Here, if ANOVA decomposition is applied to the bivariate tensor spline $\psi(t,\ \varepsilon_{T,t})$, the following equation can be obtained (note that in the R package "mgcv," ANOVA decomposition for the tensor product spline function can be easily calculated using the "ti" term used in the function gam() [17]):

$$\psi(t,\ \varepsilon_{T,t}) = c + \psi_t(t) + \psi_\epsilon(\varepsilon_{T,t}) + \psi_{te}(t,\ \varepsilon_{T,t}) \tag{A2}$$

where c is a constant term, and $\psi_t(t)$, $\psi_\epsilon(\varepsilon_{T,t})$ and $\psi_{te}(t,\ \varepsilon_{T,t})$ are obtained as zero mean functions. The univariate spline functions $\psi_t(t)$ and $\psi_\epsilon(\varepsilon_{T,t})$ are called the "main effects," which correspond to the trends that the date and temperature contribute to independently, among the original tensor product spline functions. On the other hand, the bivariate spline function $\psi_{te}(t,\ \varepsilon_{T,t})$ is called "interaction," which corresponds to the interaction trend of the date and temperature wherein the main effect was removed from the original function. Here, the function $\tilde{\psi}(t,\ \varepsilon_{T,t}) := \psi_\epsilon(\varepsilon_{T,t}) + \psi_{te}(t,\ \varepsilon_{T,t})$ is the desired derivative payoff function in which just the deterministic date trend was removed from the original tensor product spline function. Additionally, by setting the contract volumes of discount bonds that depend only on t to $\tilde{f}(t) := c + \psi_t(t)$, the minimum variance hedge problem (A1) can be modified to (3).

Appendix C. Basis Functions of the Cubic Spline and the Cyclic Cubic Spline

In this section, in order to understand the basis functions of the "cyclic cubic spline" dealt with in this study, the definition formula is briefly described along with the normal "cubic spline" (here, the contents described in [25] are summarized with supplementary explanation).

Appendix C.1. Cubic Spline Function

Cubic (regression) spline function $s_{cr}(x)$ with k knots, $x_1 \ldots x_k$ can be defined as follows by using cubic truncated power basis functions:

$$s_{cr}(x) = a_j^-(x)\beta_j + a_j^+(x)\beta_{j+1} + c_j^-(x)\delta_j + c_j^+(x)\delta_{j+1} \text{ if } x_j \leq x \leq x_{j+1} \tag{A3}$$

where $\beta_j = s_{cr}(x_j)$, $\delta_j = s_{cr}''(x_j)$, and the basis functions a_j^-, a_j^+, c_j^-, and c_j^+ are defined as follows:

$$\begin{aligned} a_j^-(x) &= \frac{x_{j+1}-x}{h_j}, \quad c_j^- t(x) = \frac{1}{6}\left[\frac{(x_{j+1}-x)^3}{h_j} - h_j(x_{j+1}-x)\right], \\ a_j^+(x) &= \frac{x-x_j}{h_j}, \quad c_j^+(x) = \frac{1}{6}\left[\frac{(x-x_j)^3}{h_j} - h_j(x-x_j)\right] \end{aligned} \tag{A4}$$

where $h_j = x_{j+1} - x_j$.

Here, from (A3) and (A4), this function already satisfies the condition that the value and the second derivative are equal at each knot (this fact can be proved inductively because it satisfies $s_{cr}(x_j) = \beta_j$, $s_{cr}(x_{j+1}) = \beta_{j+1}$; $s''_{cr}(x_j) = \delta_j$, $s''_{cr}(x_{j+1}) = \delta_{j+1}$). However, in order for this function to connect smoothly, the first derivative is also required to be equal at each knot. This condition can be expressed as the following matrix equation, which is derived by expanding (A3) and (A4):

$$\mathbf{B}\delta^- = \mathbf{D}\beta \tag{A5}$$

where $\delta^- = (\delta_2, \ldots, \delta_{k-1})^\top$, $\delta_1 = \delta_k = 0$; $\mathbf{B}$ and $\mathbf{D}$ are defined as follows:

$$D_{i,i} = \tfrac{1}{h_i}, \; D_{i,i+1} = -\tfrac{1}{h_i} - \tfrac{1}{h_{i+1}}, \; D_{i,i+2} = \tfrac{1}{h_{i+1}}, \; B_{i,i} = \tfrac{h_i + h_{i+1}}{3} \; (i = 1 \ldots k - 2);$$
$$B_{i,i+1} = \tfrac{h_{i+1}}{6}, \; B_{i+1,i} = \tfrac{h_{i+1}}{6} (i = 1 \ldots k - 3) \tag{A6}$$

Here, each element of δ can be obtained by the matrix transformation of (A5). By substituting them into (A3) and rearranging the equation by β_i, the cubic spline function $s_{cr}(x)$ can be re-written as follows:

$$s_{cr}(x) = \sum_{i=1}^{k} b_i(x)\beta_i \tag{A7}$$

Appendix C.2. Cyclic Cubic Spline Function

Regarding the cyclic cubic spline function $s_{cc}(x)$, the condition that the value, first derivative and second derivative, be equal is imposed even at the start and end points of the domain (that is, knots x_1 and x_k). Even in this case, the spline can still be written in the form of (A3) and (A4), and the additional required conditions are $\beta_1 = \beta_k$, $\delta_1 = \delta_k$, and the following equations:

$$\tilde{\mathbf{B}}\delta = \tilde{\mathbf{D}}\beta \tag{A8}$$

where $\beta^\top = (\beta_1, \ldots, \beta_{k-1})$, $\delta^\top = (\delta_1, \ldots, \delta_{k-1})$; $\tilde{\mathbf{B}}$ and $\tilde{\mathbf{D}}$ are defined as follows:

$$\tilde{B}_{i-1,i} = \tilde{B}_{i,i-1} = \tfrac{h_{i-1}}{6}, \; \tilde{B}_{i,i} = \tfrac{h_{i-1} + h_i}{3},$$
$$\tilde{D}_{i-1,i} = \tilde{D}_{i,i-1} = \tfrac{1}{h_{i-1}}, \; \tilde{D}_{i,i} = -\tfrac{1}{h_{i-1}} - \tfrac{1}{h_i} \; (i = 2 \ldots k - 1);$$
$$\tilde{B}_{1,1} = \tfrac{h_{k-1} + h_1}{3}, \; \tilde{B}_{1,k-1} = \tfrac{h_{k-1}}{6}, \; \tilde{B}_{k-1,1} = \tfrac{h_{k-1}}{6},$$
$$\tilde{D}_{1,1} = -\tfrac{1}{h_1} - \tfrac{1}{h_{k-1}}, \; \tilde{D}_{1,k-1} = \tfrac{1}{h_{k-1}}, \; \tilde{D}_{k-1,1} = \tfrac{1}{h_{k-1}} \tag{A9}$$

Then, like the cubic spline, the cyclic cubic spline function $s_{cc}(x)$ can also be re-written as follows:

$$s_{cc}(x) = \sum_{i=1}^{k-1} \tilde{b}_i(x)\beta_i \tag{A10}$$

Appendix D. Monthly RMSE and Daily Fitting Curves

Appendix D.1. Monthly RMSE

Here, the hedge errors are measured using RMSE (root mean square error). Contrary to the hedge effects, the RMSE is small for both the in-sample and out-of-sample periods in the order of demand, price, and product model, as shown in Figure A2; the annual averages of the out-of-sample period were 4.6%, 26.2%, and 33.2%, respectively. Regarding seasonality, similar shapes were confirmed for all three models; however, unlike the hedge effect, which was high in both summer and winter, the RMSE was large in winter but smaller in summer. This may be because the PJM price is prone to significant spikes in winter (while the price fluctuations in summer are relatively mild). Hence, the hedge error is large in winter, but the correlation with temperature is strong in both summer and winter. Note that the reason the RMSE in January–February of in-sample period is significantly higher than the out-of-sample period is that in-sample data include an extreme price spike in January 2014, during which time "PJM experienced tight operational conditions and a significantly higher number of forced generator outages due to the extreme weather" [37], which is also shown in the daily price fluctuation graph in Appendix A3.

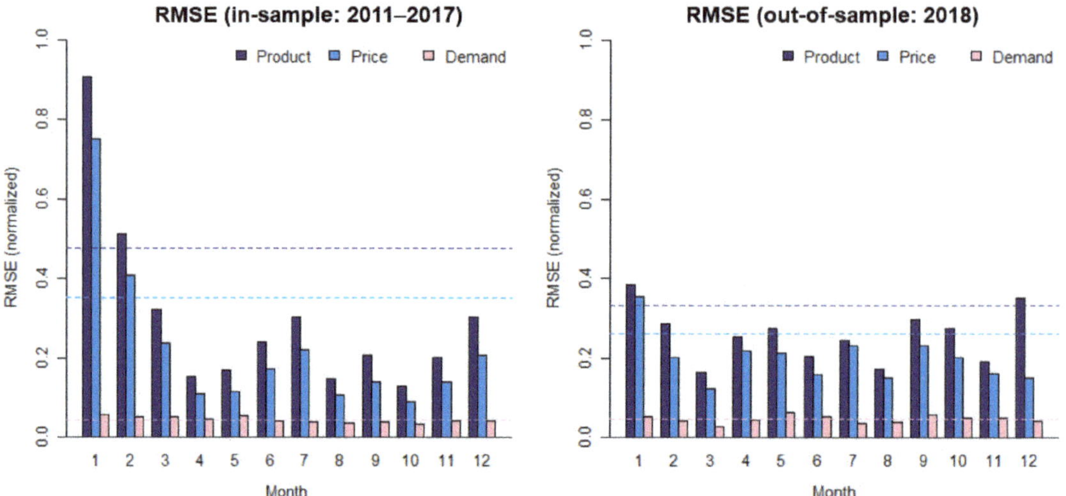

Figure A2. Monthly RMSE for each hedging target.

Appendix D.2. Daily Fitting Curves

Here, the observed value (black line) and the estimated payoff of the derivative portfolio (red line) when using all derivatives are compared for the product, price, and demand models. It can be confirmed that the derivatives' payoffs follow daily fluctuations in general, even during periods in which significant fluctuations occur, such as summer and winter.

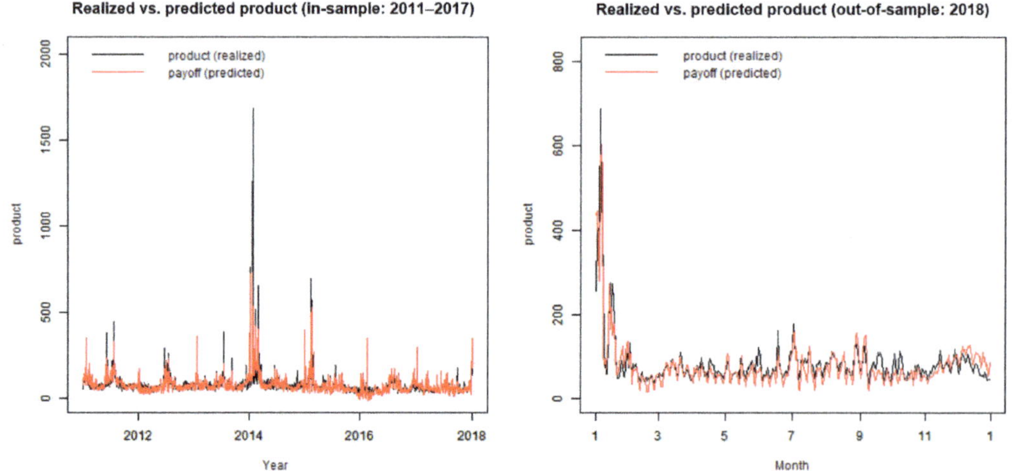

Figure A3. Comparison of realized and predicted values for the product model.

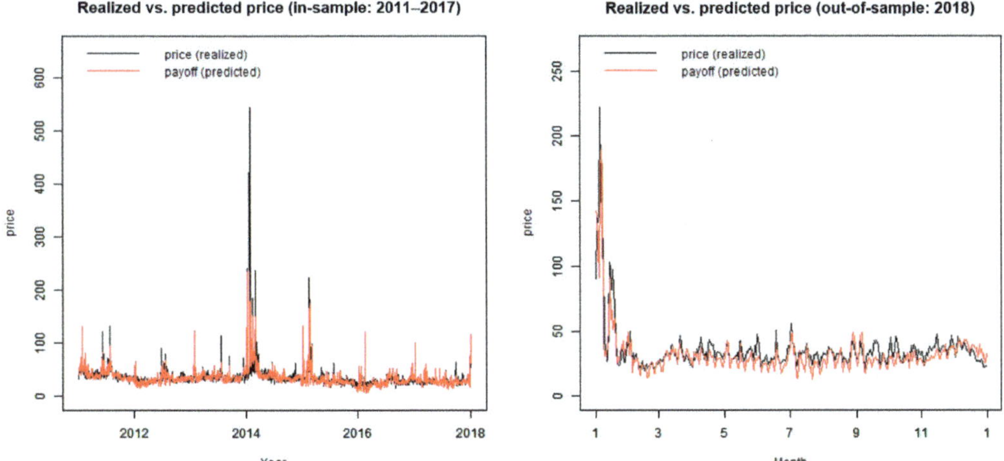

Figure A4. Comparison of realized and predicted values for the price model.

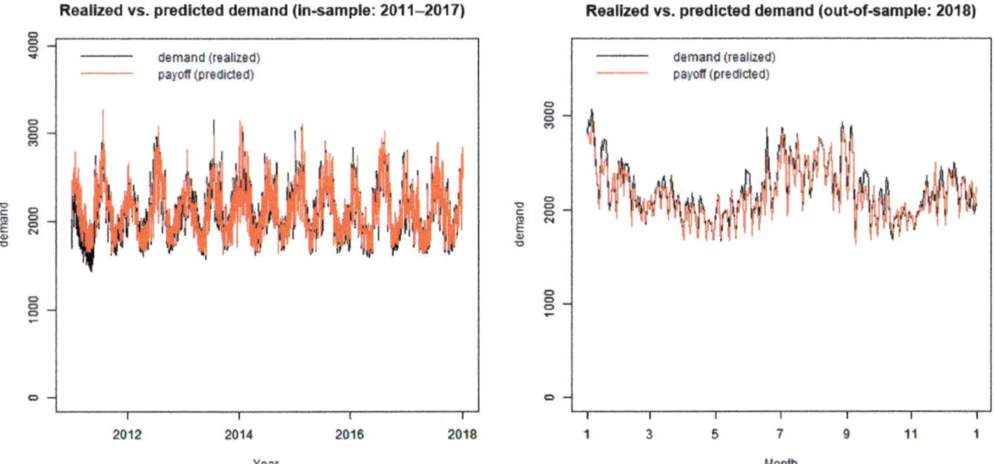

Figure A5. Comparison of realized and predicted values for the demand model.

References

1. Lee, Y.; Oren, S.S. An equilibrium pricing model for weather derivatives in a multi-commodity setting. *Energy Econ.* **2009**, *31*, 702–713. [CrossRef]
2. Lee, Y.; Oren, S.S. A multi-period equilibrium pricing model of weather derivatives. *Energy Syst.* **2010**, *1*, 3–30. [CrossRef]
3. Bhattacharya, S.; Gupta, A.; Kar, K.; Owusu, A. Risk management of renewable power producers from co-dependencies in cash flows. *Eur. J. Oper. Res.* **2020**, *283*, 1081–1093. [CrossRef]
4. Davis, M. Pricing weather derivatives by marginal value. *Quant. Financ.* **2001**, *1*, 305–308. [CrossRef]
5. Platen, E.; West, J. A Fair Pricing Approach to Weather Derivatives. *Asia Pac. Financ. Mark.* **2004**, *11*, 23–53. [CrossRef]
6. Brockett, P.L.; Wang, M.; Yang, C.; Zou, H. Portfolio Effects and Valuation of Weather Derivatives. *Financ. Rev.* **2006**, *41*, 55–76. [CrossRef]
7. Kanamura, T.; Ohashi, K. Pricing summer day options by good-deal bounds. *Energy Econ.* **2009**, *31*, 289–297. [CrossRef]
8. Yamada, Y. Valuation and hedging of weather derivatives on monthly average temperature. *J. Risk* **2007**, *10*, 101–125. [CrossRef]
9. Yamada, Y. Simultaneous optimization for wind derivatives based on prediction errors. In Proceedings of the 2008 American Control Conference, Seattle, WA, USA, 11–13 June 2008; pp. 350–355.
10. Yamada, Y. Optimal Hedging of Prediction Errors Using Prediction Errors. *Asia Pac. Financ. Mark.* **2008**, *15*, 67–95. [CrossRef]

11. Matsumoto, T.; Yamada, Y. Cross Hedging Using Prediction Error Weather Derivatives for Loss of Solar Output Prediction Errors in Electricity Market. *Asia Pac. Financ. Mark.* **2018**, *26*, 211–227. [CrossRef]
12. Yamada, Y. Simultaneous hedging of demand and price by temperature and power derivative portfolio: Consideration of periodic correlation using GAM with cross variable. In Proceedings of the 50th JAFEE Winter Conference, Tokyo, Japan, 22–23 February 2019.
13. Matsumoto, T.; Yamada, Y. Hedging strategies for solar power businesses in electricity market using weather derivatives. In Proceedings of the 2019 IEEE 2nd International Conference on Renewable Energy and Power Engineering (REPE), Toronto, ON, Canada, 2–4 November 2019; pp. 236–240.
14. Matsumoto, T.; Yamada, Y. Simultaneous hedging strategy for price and volume risks in electricity businesses using energy and weather derivatives. *Energy Econ.* **2021**, *95*, 105101. [CrossRef]
15. Hastie, T.; Tibshirani, R. *Generalized Additive Models*; Chapman & Hall: Boca Raton, FL, USA, 1990.
16. EMSC, Electricity and Gas Market Surveillance Commission. Current Status of Activation of Wholesale Power Trading, 21st Institutional Design Special Meeting Secretariat Submission Materials. 2017. Available online: https://www.emsc.meti.go.jp/activity/emsc_system/pdf/021_04_00.pdf (accessed on 11 April 2021).
17. Wood, S.N. Package mgcv v. 1.8-34. Available online: https://cran.r-project.org/web/packages/mgcv/mgcv.pdf (accessed on 11 April 2021).
18. Matsumoto, T. Forecast Based Risk Management for Electricity Trading Market. Ph.D. Thesis, University of Tsukuba, Tokyo, Japan, 2020.
19. Hastie, T.; Tibshirani, R.; Friedman, J. *The Elements of Statistical Learning: Data Mining, Inference, and Prediction*; Springer: Berlin/Heidelberg, Germany, 2009.
20. Wood, S.N. Thin plate regression splines. *J. R. Stat. Soc. Ser. B Stat. Methodol.* **2003**, *65*, 95–114. [CrossRef]
21. Eilers, P.H.C.; Marx, B.D. Flexible smoothing with B-splines and penalties. *Stat. Sci.* **1996**, *11*, 89–121. [CrossRef]
22. Eilers, P.H.; Marx, B.D. *Practical Smoothing: The Joys of P-Splines*; Cambridge University Press: Cambridge, MA, USA, 2021.
23. Wood, S.N. *Generalized Additive Models: An Introduction with R*, 2nd ed.; Chapman & Hall: Boca Raton, FL, USA, 2017.
24. Efron, B.; Stein, C. The Jackknife Estimate of Variance. *Ann. Stat.* **1981**, *9*, 586–596. [CrossRef]
25. De Boor, C. *A Practical Guide to Splines*; Springer: New York, NY, USA, 1978; Volume 27, p. 325.
26. Perperoglou, A.; Sauerbrei, W.; Abrahamowicz, M.; Schmid, M. A review of spline function procedures in R. *BMC Med. Res. Methodol.* **2019**, *19*, 1–16. [CrossRef]
27. McCulloch, J.; Ignatieva, K. Intra-day Electricity Demand and Temperature. *Energy J.* **2020**, *41*. [CrossRef]
28. Duchon, J. Splines minimizing rotation-invariant semi-norms in Sobolev spaces. In *Constructive Theory of Functions of Several Variables*; Springer: Berlin/Heidelberg, Germany, 1977; pp. 85–100.
29. Wood, S.N.; Bravington, M.V.; Hedley, S.L. Soap film smoothing. *J. R. Stat. Soc. Ser. B Stat. Methodol.* **2008**, *70*, 931–955. [CrossRef]
30. Yamada, Y.; Makimoto, N.; Takashima, R. JEPX price predictions using GAMs and estimations of volume-price functions. *JAFEE J.* **2015**, *14*, 8–39.
31. PJM Data Miner 2. Available online: http://dataminer2.pjm.com/ (accessed on 25 June 2019).
32. NOAA Climate Data Online Search. Available online: https://www.ncdc.noaa.gov/cdo-web/search (accessed on 25 June 2019).
33. EIA Henry Hub Natural Gas Spot Price. Available online: https://www.eia.gov/dnav/ng/hist/rngwhhdd.htm (accessed on 25 June 2019).
34. Thompson, T.; Webber, M.; Allen, D.T. Air quality impacts of using overnight electricity generation to charge plug-in hybrid electric vehicles for daytime use. *Environ. Res. Lett.* **2009**, *4*, 014002. [CrossRef]
35. Campbell, J.Y.; Thompson, S.B. Predicting Excess Stock Returns Out of Sample: Can Anything Beat the Historical Average? *Rev. Financ. Stud.* **2007**, *21*, 1509–1531. [CrossRef]
36. Aguilera, A.; Aguilera-Morillo, M. Comparative study of different B-spline approaches for functional data. *Math. Comput. Model.* **2013**, *58*, 1568–1579. [CrossRef]
37. PJM. Analysis of Operational Events and Market Impacts during the January 2014 Cold Weather Events. Available online: https://www.hydro.org/wp-content/uploads/2017/08/PJM-January-2014-report.pdf (accessed on 11 April 2021).

Article

Comprehensive and Comparative Analysis of GAM-Based PV Power Forecasting Models Using Multidimensional Tensor Product Splines against Machine Learning Techniques

Takuji Matsumoto [1,*] **and Yuji Yamada** [2,*]

1 Socio-Economic Research Center, Central Research Institute of Electric Power Industry, Chiyoda-ku, Tokyo 100-8126, Japan
2 Faculty of Business Sciences, University of Tsukuba, Bunkyo-ku, Tokyo 112-0012, Japan
* Correspondence: mtakuji1122@gmail.com (T.M.); yuji@gssm.otsuka.tsukuba.ac.jp (Y.Y.)

Abstract: In recent years, as photovoltaic (PV) power generation has rapidly increased on a global scale, there is a growing need for a highly accurate power generation forecasting model that is easy to implement for a wide range of electric utilities. Against this background, this study proposes a PV power forecasting model based on the generalized additive model (GAM) and compares its forecasting accuracy with four popular machine learning methods: k-nearest neighbor, artificial neural networks, support vector regression, and random forest. The empirical analysis provides an intuitive interpretation of the multidimensional smooth trends estimated by the GAM as tensor product splines and confirms the validity of the proposed modeling structure. The effectiveness of GAM is particularly evident in trend completion for missing data, where it is able to flexibly express the tangled trend structure inherent in time series data, and thus has an advantage not only in interpretability but also in improving forecast accuracy.

Keywords: forecasting method; machine learning; non-parametric regression; photovoltaic power generation; smooth trend; tensor product splines

Citation: Matsumoto, T.; Yamada, Y. Comprehensive and Comparative Analysis of GAM-Based PV Power Forecasting Models Using Multidimensional Tensor Product Splines against Machine Learning Techniques. *Energies* **2021**, *14*, 7146. https://doi.org/10.3390/en14217146

Academic Editor: Fernando Sánchez Lasheras

Received: 9 October 2021
Accepted: 25 October 2021
Published: 1 November 2021

Publisher's Note: MDPI stays neutral with regard to jurisdictional claims in published maps and institutional affiliations.

1. Introduction

PV power generation forecasting is indispensable for electric utilities. Based on the forecast of demand and renewable energy generation, power producers and retailers submit their generation and procurement plans for the next day or hour to the system operator. Forecast errors in PV generation can lead to losses in the form of imbalance charges (penalties imposed for supply–demand mismatch). Therefore, accurate forecasting of PV power generation has become an essential issue for the economical business operation of electric utilities. In particular, in recent years, many countries around the world have adopted PV power generation as a clean energy source to address global warming, and the need for PV power generation forecasting has been increasing each year. For example, in Japan, the feed-in tariff (FIT) system has provided incentives for the introduction of PV power generation, and many small-scale businesses have recently entered the PV power generation business [1]. This trend of increasing (or diversifying) the number of players is similar in other countries, although there are minor differences in the systems. Against this background, there is a need for a forecasting method that is not only highly accurate but also easy to implement and interpret by a wide range of practitioners, including small businesses.

For PV power forecasting, many previous studies have proposed various forecasting methods, which are very diverse in terms of the forecasting variables used, time granularity and forecast horizon, and algorithms. There are also various survey studies [2–8]. This study focuses on publicly available weather forecast information and its application to PV forecasting methods for more general electric utilities, but even if we focus on such a target, many machine learning (ML)-based models have been proposed, such as artificial neural networks (ANN) [9] and support vector regression (SVR) [10], and ML has become a mainstream approach.

Recent ML-based forecasting methods that focus on using publicly available weather data include Maitanova et al. [11], which applies a special architecture of an artificial recurrent neural network (RNN) called long short-term memory (LSTM). In the context of comparing forecast accuracy in ML methods, there have been many empirical studies, especially in the last few years. For example, Mohammed and Aung [12] compared seven ML methods, such as k-nearest neighbor (kNN) [13], decision tree, gradient boosting, and random forests (RF) [14]. Additionally, Das et al. [15] proposed a PV power prediction using SVR and compared it with ANN. Rosato et al. [16] proposed three techniques based on neural and fuzzy neural networks. Khandakar et al. [17] proposed an ANN-based prediction model to explore effective feature selection techniques. Nespoli et al. [18] also dealt with ANN-based forecasting methods, where a comparison was made between the historical dataset alone and a hybrid approach combining weather forecast. Abdel-Nasser and Mahmoud [19] proposed a model using LSTM and compared it with multiple linear regression and NN, among others. These previous studies have empirically shown that ML methods are superior in terms of forecast error reduction, but most forecasting methods using ML have challenges, such as high computational load and difficulty in interpretation, which is a well-known drawback in general.

In electric utility practice, the ease of interpretation is often the most important factor in building consensus and ensuring the reliability of the model [20]. Ease of implementation and computational tractability are also important factors, as well as accuracy. It is emphasized that these points have often been overlooked in recent research. Motivated by these practical needs, this study proposes a forecasting model based on the generalized additive model (GAM) [21], a statistical approach, rather than ML. In particular, we demonstrate that GAM-based PV forecasting models are easy to handle for a wide range of electric utilities, including new entrants, and compare our methodology with ML-based PV forecasting models.

The GAM-based PV forecasting models in this study generalize our previous studies [1,22] for the case of multidimensional tensor product spline functions. Note that these previous studies effectively modeled the smooth trend inherent in the seasonal (and in seasonal and hourly) [22] direction of PV power generation using a univariate or a bivariate tensor product spline function. In addition, note that another study using GAM for PV forecasting [23] aimed to improve the algorithm for capturing nonlinear dependencies in ensemble learning, one of the ML methods, but is slightly different from our interest. We would like to pursue ease of implementation and interpretation for practitioners; that is, our focus is more on application methods rather than on algorithm development.

Another issue of interest in this study is a "comprehensive and comparative analysis." Although previous studies [1,22] have shown the effectiveness of GAM-based forecasting models in the context of interpretability and robustness, it has not been sufficiently verified, in the context of comparison between multiple models, how much better the forecasting accuracy of the GAM-based model is. In this study, we demonstrate the accuracy and practicability of GAM-based PV forecasting models in comparison with other regression models and several ML models. In addition, this study deals with data for both types of PV generation; that is, area-wide PV power generation in the grid area and an individual PV panel. In particular, for the latter dataset, we extend the models of previous studies [1,22] by utilizing a three-dimensional (3D) tensor product spline function. To the best of our knowledge, this is the first attempt to apply 3D tensor product spline functions not only for PV forecasting, but also for energy time series forecasting.

In the empirical analysis, we verify the reliability of the models from the structural aspect by visualizing the smooth trends estimated using tensor product spline functions for the proposed GAM-based PV forecasting models and provide reasonable interpretations of the estimated trends. Then, we conduct a comprehensive and comparative analysis with

ML methods, such as kNN, ANN, SVR, and RF, to verify the accuracy of the forecasts. Overall, we conclude that the GAM-based model with multidimensional tensor product spline functions is superior in terms of interpretability, robustness, low computational load, and prediction accuracy. However, depending on the sample period ("in-sample period" for model estimation or "out-of-sample period" for accuracy validation.) and prediction error indices (MAE or RMSE), ML methods outperform in some cases, and additional empirical analysis leads to interesting insights into the conditions under which GAM-based models have advantages.

This paper is organized as follows: Section 2 introduces the GAM method and builds a forecasting model using tensor product spline functions. Section 3 provides an overview of the ML methods that this study compares. Section 4 presents an empirical analysis using actual data, provides an interpretation of the estimated multidimensional smooth trends, and analyzes the forecast errors from various aspects. Finally, Section 5 concludes the paper.

2. PV Power Forecasting Models based on GAM

In the following sections, we first construct an area-wide PV power generation forecast model, and then construct a forecasting model for individual PV panels.

2.1. Area-Wide PV Power Generation Forecasting Model

To forecast area-wide power generation, it is necessary to consider its yearly increasing trend; that is, the installed capacity of PV power generation is increasing year by year. To this end, our forecasting model is constructed by first adjusting the yearly trend of increasing capacity to derive a unit power generation. The total procedure is described as follows (note that the main variables used in these models can be found in the nomenclature at the end of this paper and the intuitive relationship between data preparation and the flow of model estimation is shown in Figure 1):

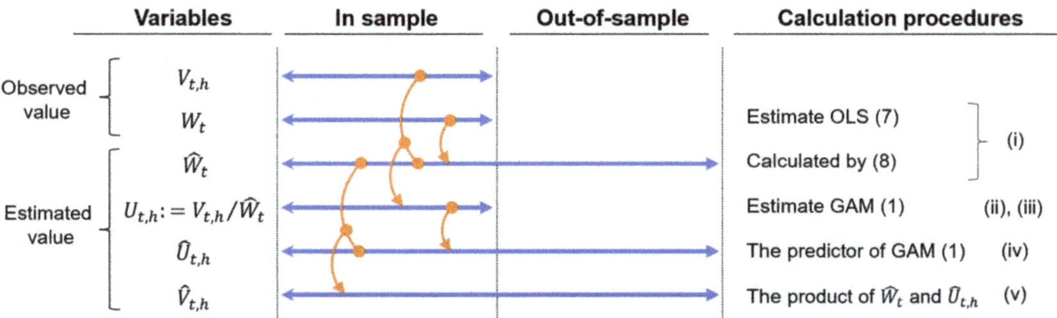

Figure 1. Estimation procedures and data periods in the area-wide PV power generation forecasting model. Note: In our empirical analysis, we use the in-sample period from 1 April 2016, to 31 December 2017, and the out-of-sample period from 1 January to 31 December 2018 (as described in Section 4.1). When using the ML methods introduced in Section 3, the only difference is that each ML model is adopted instead of GAM (1) in steps (iii) and (iv), and the other steps are the same.

(i)   Using the actual areawide PV power generation capacity W_t (observed no more frequently than monthly), the daily increasing trend is estimated using a linear model. As a result, the daily forecast value of PV power generation capacity is obtained as $\widehat{W}_t$ (see Appendix A).

(ii)   The unit power generation $U_{t,\,h}$ is obtained by dividing the measured hourly area-wide PV power generation $V_{t,h}$ by $\widehat{W}_t$ (i.e., $U_{t,\,h} := V_{t,h}/\widehat{W}_t$).

(iii)   For $U_{t,\,h}$ a GAM is built with calendar information, general weather conditions forecast, and maximum/minimum temperature forecast as explanatory variables.

(iv) The out-of-sample forecast unit power generation $\hat{U}_{t,\,h}$ is obtained by substituting the explanatory variables observed in the estimated GAM forecast formula (predictor part of the GAM).

(v) The out-of-sample forecast power generation $\hat{V}_{t,\,h}$ is obtained as the product of the forecast PV power capacity $\hat{W}_t$ and the forecast unit power generation $\hat{U}_{t,\,h}$.

In the following, we apply GAM for unit power generation $U_{t,\,h}$ by using temperature and general (descriptive) weather forecasts. (Note that another possible method is to use solar radiation forecasting, but this method is not used in our area-wide PV forecast model because the solar radiation data are not available in many countries [24], and it is difficult to handle local fluctuations that depend on the observation points).

$$
\begin{aligned}
U_{t,\,h} = {} & u_{sunny}(t,h) I_{sunny,t,h} + u_{cloudy}(t,h) I_{cloudy,t,h} + u_{rainy}(t,h) I_{rainy,t,h} \\
& + u_{snowy}(t,h) I_{snowy,t,h} + u_{tmax}(t,h) \epsilon_{tmax,t} + u_{tmin}(t,h) \epsilon_{tmin,t} \\
& + \eta_{t,\,h}
\end{aligned}
\tag{1}
$$

where $u_.(t,h)$ is the tensor product spline function to be estimated by applying GAM, which is the 2D time trend that smoothly connects in the direction of both the date t and hour h (see Appendix B for an overview of the tensor product spline function and the smoothing mechanism). Note that the "snowy" term is only included in the model for snowy areas, which in this study are referred to Hokkaido, Tohoku, and Hokuriku, out of the nine target areas. The tensor product spline function provides smoothing conditions in two orthogonal directions (see [25] and Appendix B). This ensures robustness and makes it possible to incorporate multiple explanatory variables even with small sample sizes.

Note that $\epsilon_{tmax,\,t}$ and $\epsilon_{tmin,\,t}$ denote the maximum and minimum "temperature forecast deviations", respectively, and they are obtained by applying GAM as follows:

$$
Tmax_t = f_{tmax}(t) + \epsilon_{tmax,\,t}, \, Tmin_t = f_{tmin}(t) + \epsilon_{tmin,\,t}
\tag{2}
$$

We estimate the yearly cyclical trends as the spline functions $f_.(Seasonal_t)$ in (2) and $u_.(Seasonal_t, h)$ in (1), using yearly cyclical dummy variables $Seasonal_t \, (= 1, \ldots, 365 \, (or \, 366))$. In this study, the starting point of the cyclical dummy variables is January 1, days are allocated in order from 1 to 365 (366 for leap years). To make the notation more concise, we denote $f_.(Seasonal_t)$ and $u_.(Seasonal_t, h)$ as $f_.(t)$ and $u_.(t,h)$.

Note that we select the cyclic cubic spline function [25] in the direction of $Seasonal_t$ for $f_.(Seasonal_t)$ and $u_.(Seasonal_t, h)$. In this way, each spline function is given as a function that is smoothly continuous (the value and the first and second derivative values are all connected) not only throughout the domain of definition, but also at the beginning and end of the domain of definition. (See [26] for a detailed description and formulas of "basis functions" to apply to similar models with annual periodicity.)

2.2. Individual PV Power Generation Forecasting Model

In this section, we develop a model for forecasting the power generation of individual PV panels. In this study, we assume that solar radiation forecasts are available in the same region where the individual PV panels are located, and use this information in the model (note that in Japan, 5 km mesh solar radiation forecasts by the Japan Meteorological Agency (JMA) are widely distributed through the Japan Meteorological Business Support Center [27]). First, we propose the following forecast model "M3" for PV power generation $V_{t,\,h}$ at date t time h.

$$
\text{M3}: \qquad\qquad V_{t,\,h} = v(t,\,h,\,R_{t,\,h}) + \eta_{t,\,h}
\tag{3}
$$

where $R_{t,\,h}$ is the forecast solar radiation, $\eta_{t,\,h}$ is the residual term, $v(\cdot)$ is the 3D tensor product spline function estimated by GAM (3), and $Seasonal_t$ is denoted by t.

Next, we construct three alternative models, M0–M2, for comparison to make sure that the nonlinear conditions in 3D directions in M3 contribute to the improvement of explanatory power and robustness.

$$\text{M0}: \qquad V_{t,\,h} = \beta R_{t,\,h} + \alpha + \eta_{t,\,h} \tag{4}$$

$$\text{M1}: \qquad V_{t,\,h} = \sum_{M,H} \left(\beta^{(M,H)} R_{t,\,h} + \alpha^{(M,H)} \right) I_{t,h}^{(M,\,H)} + \eta_{t,\,h} \tag{5}$$

$$\text{M2}: \qquad V_{t,\,h} = \beta(t,\,h) R_{t,\,h} + \alpha(t,\,h) + \eta_{t,\,h} \tag{6}$$

where βs and αs correspond to the regression coefficients and intercepts, respectively, when each model is viewed as a linear regression equation for solar radiation. Note that these values are constant for the entire period in M0, constant under a specific month and time (M, H) in M1 (where $I_{t,h}^{(M,\,H)}$ is a dummy variable that is set to 1 if the target time (t, h) of the sample corresponds to (M, H) and 0 otherwise), and variables depending on the date and hour in M2. In fact, $\beta(t,\,h)$ and $\alpha(t,\,h)$ are defined by 2D tensor product spline functions (where t denotes *Seasonal$_t$*) with smoothing conditions in the date/hour direction.

In other words, M1 is synonymous with constructing a linear regression model for each hour of the month. M2 is the same type of model as the method described in Section 2.1 in which 2D tensor product spline functions express the regression coefficient and constant terms change smoothly in the date and hour directions. M2 is a more granular model than M1 in that the estimated parameters vary from day to day, and M3 is a more refined model in incorporating nonlinearity in the direction of solar radiation to M2.

3. Machine Learning Methods to Be Compared

To validate the accuracy of the GAM-based forecasting model proposed in the previous section, we also perform forecasts using multiple ML methods. Next, we introduce four popular ML algorithms, kNN, ANN, SVR, and RF, to perform similar forecasts to the previous section using the "caret" package [28] (short for classification and regression training). We also compare the forecast accuracy of our proposed methods using GAMs with these ML techniques.

A brief explanation of these four ML methods is given below.

- *k-nearest neighbor (kNN):* kNN [13] is one of the simplest yet effective ML algorithms [29]. kNN can be used for classification and regression problems. The main idea is to use the proximity of features to predict the value of new data points. When used for classification problems, the classification of an object is determined by the votes of its neighboring groups of objects (i.e., the most common class in the k nearest neighbor groups is assigned to the object). kNN regression, on the other hand, uses the average of the values of the k nearest neighbors, or the inverse distance weighted average of the k nearest neighbors as the expected result. The kNN algorithm measures the distance between the numerical target parameters and a set of parameters in the dataset, usually the Euclidean distance (which is also used by caret's "knn"). Other distances, such as the Manhattan distance, can also be used. kNN methods have the challenge of being sensitive to the local structure of the data.
- *Artificial Neural Networks (ANN):* ANN [9] is "a mathematical model or computational model based on biological neural networks; in other words, it is an emulation of a biological neural system" [30]. The perceptron is the starting point for the neural-network formation procedure. Simply put, the input is received by the perceptron, where it is multiplied by a series of weights and then passed to the activation function of choice (linear, logistic, hyperbolic tangent, or ReLU) to produce the output. A neural network consists of a multilayer perceptron model, which consists of a cascade of perceptron layers: an input layer, a hidden layer, and an output layer. Data are received in the input layer, and the final output is generated in the output layer. The

hidden layer, as is commonly known, is located between the input and output layers, where transient computations occur.

- ***Support vector machine (SVM)/support vector regression (SVR):*** SVM is considered "one of the most robust and accurate methods among all well-known algorithms" [31]. When used for classification problems, SVMs learn the boundary that most boldly separates a given training sample (maximizing the margin, which is the distance between the boundary and the data). The unique feature of SVM is that it can be combined with the kernel method [32], which is a method for nonlinear data analysis. That is, by using a method that maps data to a finite (or infinite) dimensional feature space using a kernel function and performs linear separation on that feature space, it is possible to apply this method to nonlinear classification problems. When SVM is used for regression, known as support vector regression (SVR), as originally proposed in [10], some properties of the SVM classifier are inherited. That is, the problem is solved in such a way that the error and the weights (regression coefficients) of the mapping functions are minimized simultaneously (see [33] for the formula). This prevents overlearning in a manner similar to ridge regression [34]. SVR has a structure similar to that of kernel ridge regression (KRR), but it is unique in that a loss function called "(linear) ε-insensitive loss functions" (see, e.g., "Figure 1" of [35]) is used to evaluate the prediction error. In this respect, it differs from KRR, which has squared error loss as its loss function [36].
- ***Random Forest (RF):*** RF [14] is an ensemble model that combines several prediction models called "decision trees." It is called "forest" because it consists of a large number of decision trees, and "random" because the decision trees (classification trees or regression trees) are constructed using k (which is given in advance) sorts of randomly chosen explanatory variables instead of all explanatory variables. In random forest regression, when new data are given, each generated regression tree predicts the output of the individual prediction, and they are averaged to output the final prediction [14]. The random forest regressor has advantages in that it can solve complex problems on a variety of datasets using different functions and find and unbiased estimate the generalization error; however, it can be overfitted for some datasets and add noisy classification/regression tasks [37].

The caret package, which is used in this study, has been developed to facilitate the use of various ML algorithms [38], which is a very useful tool because it allows the user to manipulate the creation of forecast models, tuning of hyperparameters, and forecasting using the created models. The parameters to be tuned for each of the four methods in the caret package are presented in Appendix C. In caret, the default number of hyperparameters to be tuned (how many ways to evaluate for each hyperparameter) is four, but this number is set to 10 for more precise tuning and to avoid underestimating the prediction accuracy of ML methods as much as possible. We also make all the explanatory variables (including periods) used in the ML models perfectly consistent with those in the GAM.

4. Empirical Analysis

In this section, we present an empirical analysis using observation data from Japan and perform a comprehensive and comparative study between our proposed methods and the ML algorithms explained in the previous section.

4.1. Area-Wide PV Power Generation Forecasting Model

First, we demonstrate the empirical accuracy of the area-wide PV power generation forecasting models constructed in Section 2.1. We estimate each model using in-sample period data from 1 April 2016, to 31 December 2017, and we verify the forecast errors using out-of-sample data from 1 January 1 to 31 December 2018, in nine different power areas. For forecasting area-wide PV power generation, the following observed data are used:

- PV power generation volume $V_{t,h}$ (MW): published by nine electricity power companies (e.g., data for the Tokyo area was downloaded from [39])

- PV power capacity W_t (MW): month-end results published by the Ministry of Economy, Trade and Industry [40]
- Weather condition dummy $I_{.,t,h}$, max (min) temperature $Tmax_t$ ($Tmin_t$) (°C): forecast values (of one major city in each of the nine areas) announced by the JMA on the previous morning [41]

4.1.1. Estimated Trend

Figure 2 shows the estimated trends (2D tensor product splines in GAM(1)) for the Tokyo area (See Appendix D for estimated trends in all nine areas). It can be confirmed that power generation is greater in the order of sunny, cloudy, and rainy weather. The estimated trend of sunny weather declines in the summer, which reflects the technical fact that PV power generation decreases in efficiency during summer due to high temperatures. The significant decline in the rainy trend in winter is consistent with extreme darkening because the weather tends to change into sleets or snow. While the maximum temperature contributes to increasing power generation, the minimum temperature contributes to decreasing power generation, and this could be interpreted as under a fixed maximum temperature. The lower the minimum temperature (usually recorded around early dawn), the larger the solar radiation during the day (to raise the temperature). For this reason, there is a negative correlation between the minimum temperature and PV power generation.

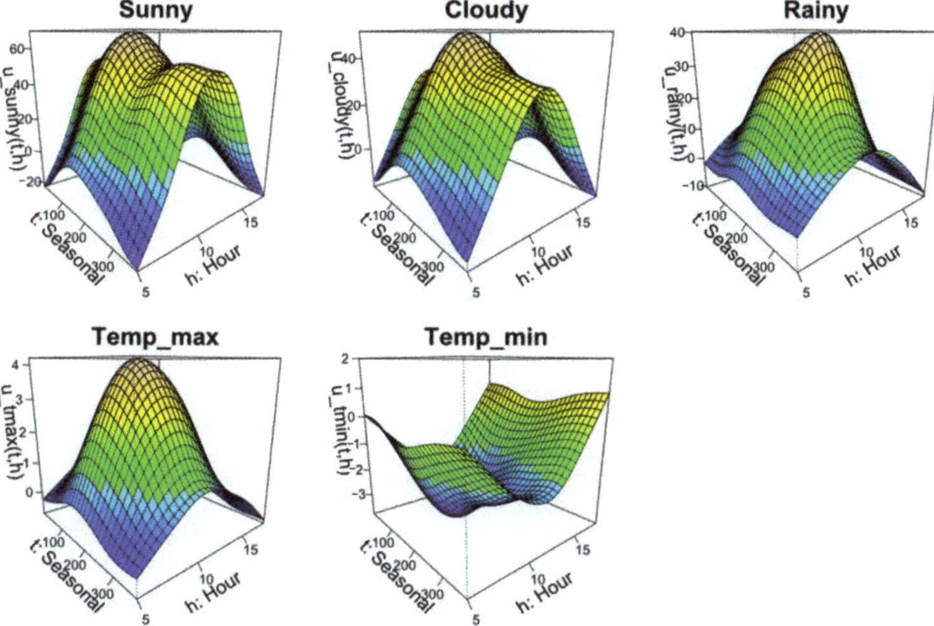

Figure 2. Estimated trends for areawide PV power generation model (example of Tokyo area). Note: The unit of vertical axis for "Sunny," "Cloudy" and "Rainy" is (%), and that for "Temp_max" and "Temp_min" is (%/°C). "Seasonal" denotes a yearly cyclical dummy variable (see Section 2.1), and "Hour" denotes the time (o'clock).

4.1.2. Comparison of Forecast Accuracy

In this section, in order to compare and verify the forecasting accuracy of the four ML methods, Table 1 shows the R-squared (RSQ), mean absolute error (MAE), and root mean square error (RMSE) for each of the nine areas by in-sample and out-of-sample periods, respectively (see e.g., [42] for out-of-sample R-squared statistic). The computation time (in seconds) required for model estimation is also shown. Note that this simulation was run on a system with an Intel Core i9 CPU running at 3.10 GHz with 32 GB RAM (Windows 10), and this is also the case for the computation time in the empirical result in Section 4.2.2 shown later.

Table 1. Forecasting accuracy and computation time of each method for PV power generation for nine areas.

Metrix	Model	In Sample									Out-of-Sample								
		1	2	3	4	5	6	7	8	9	1	2	3	4	5	6	7	8	9
RSQ	GAM	0.793	0.776	0.821	0.835	0.761	0.796	0.840	0.801	0.805	**0.718**	**0.754**	**0.792**	**0.790**	**0.716**	**0.766**	**0.807**	**0.759**	**0.805**
	kNN	0.900	0.910	0.916	0.923	0.889	0.910	0.927	0.915	0.914	0.620	0.621	0.712	0.716	0.644	0.684	0.735	0.664	0.705
	ANN	0.741	0.740	0.763	0.799	0.726	0.750	0.812	0.757	0.770	0.626	0.681	0.709	0.742	0.666	0.715	0.767	0.725	0.772
	SMR	0.835	0.822	0.859	0.867	0.798	0.821	0.869	0.830	0.837	0.629	0.542	0.722	0.665	0.676	0.741	0.704	0.655	0.713
	RF	**0.979**	**0.973**	**0.979**	**0.978**	**0.968**	**0.975**	**0.979**	**0.976**	**0.976**	0.672	0.723	0.775	0.776	0.691	0.741	0.782	0.741	0.776
MAE	GAM	0.227	0.264	0.233	0.217	0.275	0.242	0.220	0.240	0.253	**0.298**	**0.295**	**0.238**	**0.226**	**0.323**	**0.309**	**0.233**	**0.250**	**0.243**
	kNN	0.165	0.174	0.168	0.156	0.191	0.169	0.153	0.158	0.176	0.356	0.364	0.290	0.284	0.363	0.342	0.281	0.309	0.306
	ANN	0.272	0.303	0.295	0.261	0.313	0.292	0.254	0.286	0.291	0.365	0.350	0.308	0.278	0.376	0.341	0.271	0.298	0.278
	SMR	0.189	0.219	0.195	0.179	0.230	0.207	0.182	0.197	0.212	0.329	0.356	0.264	0.282	0.332	0.318	0.272	0.297	0.280
	RF	**0.071**	**0.091**	**0.082**	**0.081**	**0.101**	**0.086**	**0.079**	**0.084**	**0.090**	0.310	0.304	0.244	0.234	0.329	0.317	0.243	0.257	0.258
RMSE	GAM	0.318	0.373	0.336	0.311	0.386	0.353	0.314	0.345	0.368	**0.415**	**0.422**	**0.348**	**0.345**	**0.450**	**0.430**	**0.336**	**0.379**	**0.363**
	kNN	0.222	0.236	0.231	0.214	0.264	0.235	0.212	0.226	0.246	0.481	0.505	0.404	0.401	0.503	0.477	0.394	0.449	0.446
	ANN	0.356	0.401	0.386	0.344	0.413	0.391	0.340	0.381	0.400	0.482	0.470	0.409	0.381	0.488	0.454	0.368	0.408	0.392
	SMR	0.284	0.332	0.298	0.281	0.355	0.330	0.284	0.319	0.338	0.474	0.551	0.399	0.437	0.480	0.438	0.415	0.453	0.440
	RF	**0.102**	**0.128**	**0.115**	**0.114**	**0.142**	**0.123**	**0.113**	**0.119**	**0.129**	0.447	0.446	0.361	0.355	0.469	0.445	0.359	0.391	0.389
Time	GAM	12.1	21.4	6.1	6.0	42.5	6.6	5.9	5.8	5.7									
	kNN	6.0	1.4	1.1	1.7	1.7	1.1	1.8	2.1	1.2									
	ANN	55.0	53.8	50.2	52.3	55.4	55.0	53.9	53.8	52.3									
	SMR	73.1	71.3	68.4	72.7	62.1	77.3	79.0	71.9	73.1									
	RF	313.3	426.5	211.6	358.7	152.6	144.1	132.5	327.5	139.9									

Note: Column numbers correspond to the following areas: 1. Hokkaido, 2. Tohoku, 3. Tokyo, 4. Chubu, 5. Hokuriku, 6. Kansai, 7. Chugoku, 8. Shikoku, 9. Kyushu. Among the five different models with the same area, sample, and accuracy index, a gradient is applied so that the model with the best accuracy is in red. The best values (maximum for RSQ and minimum for MAE and RMSE) are shown in bold.

As the table shows, GAM has the best accuracy indices (RSQ, MAE, and RMSE) among the five models in the out-of-sample period for all area cases. In terms of computation time, GAM is the second shortest after kNN, which is more than an order of magnitude shorter than that of the other three ML methods. Note that the computation time for each ML method includes the time required for tuning the hyperparameters by cross-validation (see Appendix C). However, GAM also performs similar calculations in that it uses cross-validation to find the optimal smoothing parameters (see Appendix B). In addition, it should be noted that we performed parallel computation (which is allowed for the caret package) on 28 cores in this study; the time required for tuning the four ML methods is approximately 1/28th of the original time for a simple calculation [43].

Next, in order to make a comparison between methods for both in-sample and out-of-sample forecast errors easier to understand, these are shown in Figure 3 (MAE) and Figure 4 (RMSE) as scatter plots by nine areas. The dashed lines in each graph represent straight lines where the values of the vertical and horizontal axes are equal (i.e., a 45-degree line). As the graphs show, GAM has the smallest out-of-sample forecast error in all cases, and the model is relatively robust in that it does not deviate significantly from the dashed line (i.e., it has the same level of accuracy in the out-of-sample as in-sample). The same is true for ANN in that it does not deviate from the 45-degree line, but it is still inferior to GAM in terms of out-of-sample prediction accuracy (in any case) because of the relatively poor fit of the original model (large in-sample prediction error).

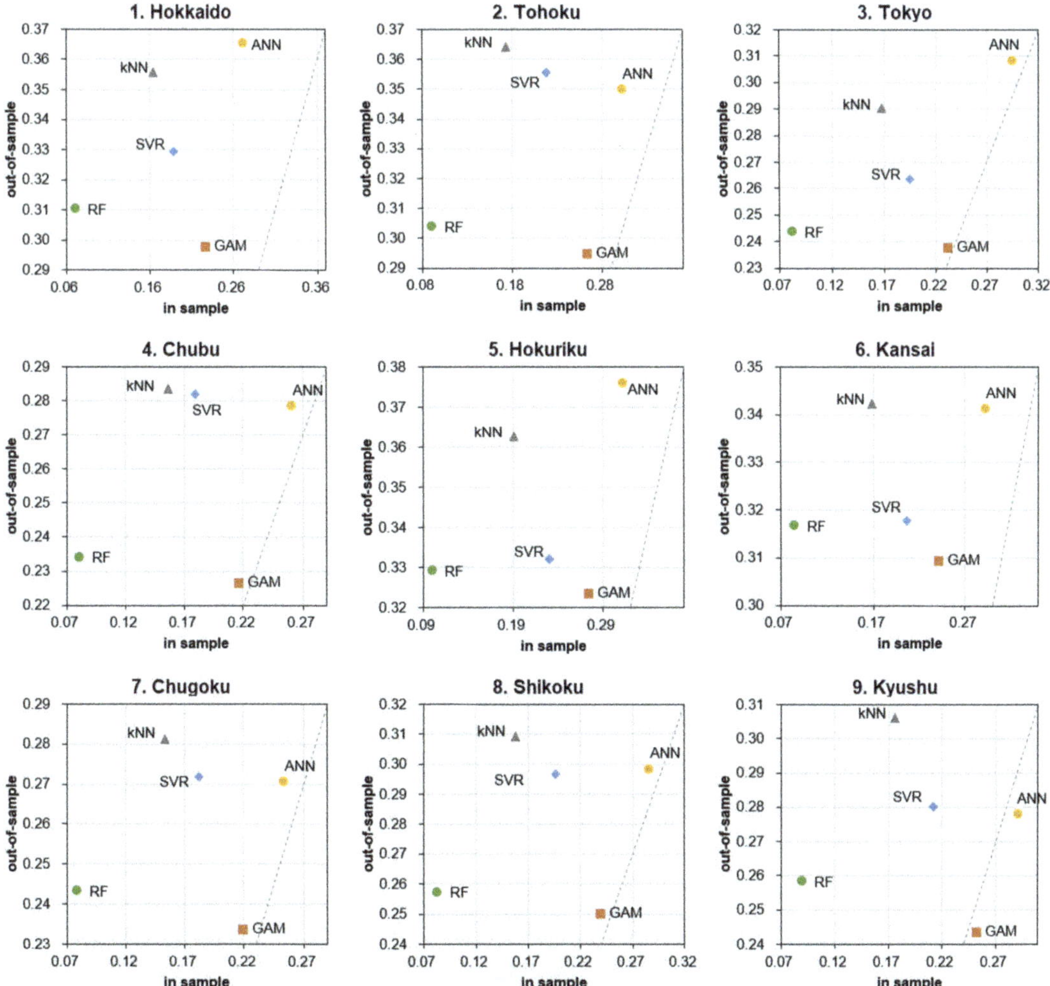

Figure 3. MAE of each method on PV power generation in nine areas.

Characteristically, the in-sample forecast errors of RF are minimal in all cases but the out-of-sample errors are larger than GAM. As explained in Section 3, this result reflects the fact that RF regression is prone to overlearning (the fitting image of RF regression is intuitive in the graph in "Section 2" of [44]). However, note that RF is relatively more accurate than the other ML methods even in terms of out-of-sample prediction error; if the sample size in the in-sample period had been sufficiently large, it is possible that the out-of-sample forecast error would have been even smaller. Additionally, although RF showed the longest computation time, it may be possible to reduce the computation time significantly by using a GPU environment instead of a CPU, since parallel computation is possible for the calculation of each decision tree, but such an investigation is left for future work (the same consideration also applies to the empirical results in Section 4.2.2).

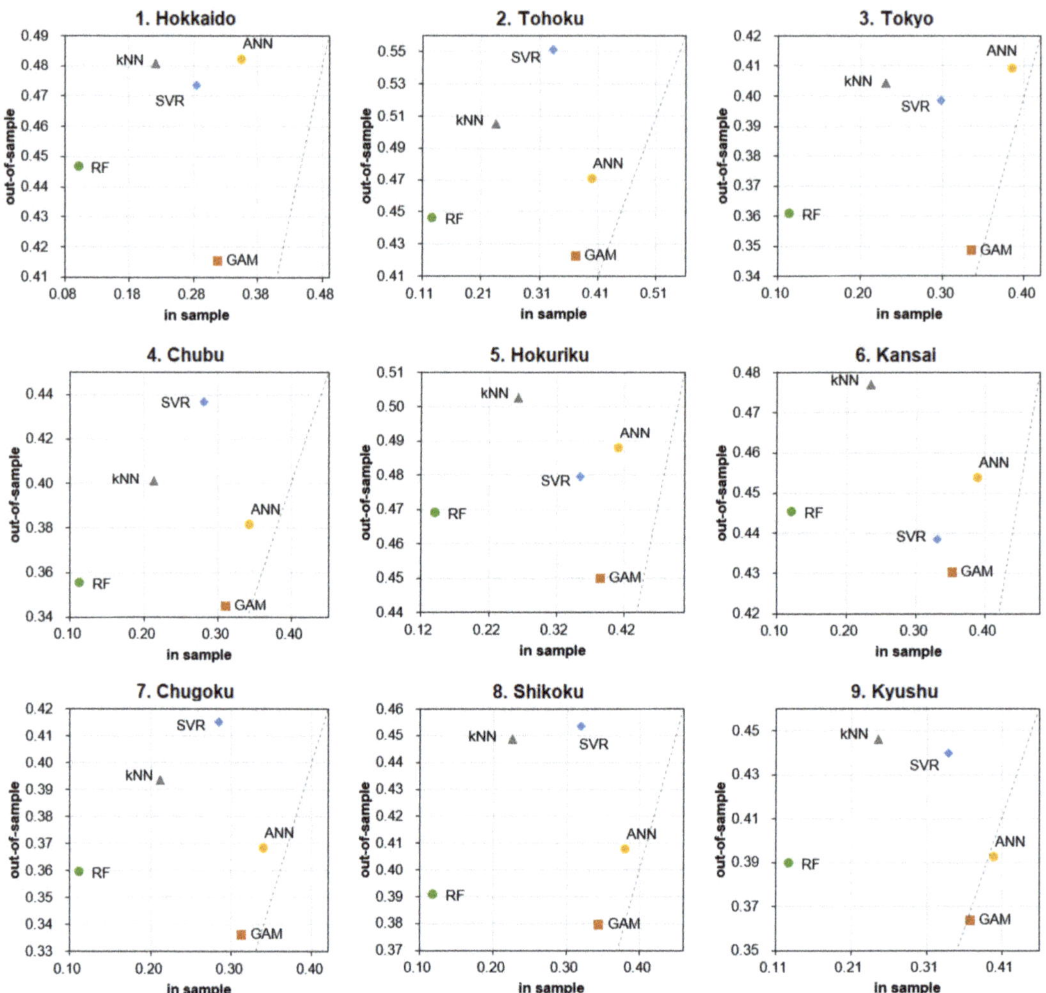

Figure 4. RMSE of each method on PV power generation in nine areas.

4.2. Individual PV Power Generation Forecasting Model

In this section, we present an empirical analysis of the individual PV power generation models constructed in Section 2.2. We estimate each model using in-sample period data from 1 January 2013, to 31 December 2017, and we verify the forecast errors using out-of-sample data from 1 January to 31 December 2018. Note that only when we conduct additional empirical analysis in the second half of Section 4.2.2 will we conduct forecast error analysis when varying the in-sample or out-of-sample period, which will be defined again at that time (note that they are also summarized in Figure 5 in advance). For the forecast of individual PV power generation, the following observed data are used:

Table 2. Forecasting accuracy and computation time of each method for individual PV power generation.

Model	Time	In Sample			Out-of-Sample		
		RSQ	MAE	RMSE	RSQ	MAE	RMSE
GAM-M0	0.01	0.844	0.245	0.363	0.850	0.247	0.360
GAM-M1	4.86	0.912	0.163	0.273	0.923	0.153	0.258
GAM-M2	0.41	0.910	0.167	0.275	0.924	0.153	0.256
GAM-M3	0.40	0.912	0.162	0.272	**0.926**	0.149	**0.253**
kNN	7.52	0.918	0.151	0.262	0.925	0.146	0.254
ANN	212.70	0.909	0.171	0.277	0.923	0.158	0.258
SVR	418.73	0.910	0.151	0.276	0.924	**0.140**	0.258
RF	328.24	**0.975**	**0.084**	**0.146**	0.919	0.151	0.264

Note: Among the five different models with the same area, sample, and accuracy index, a gradient is applied so that the model with the best accuracy is in red. The best values (maximum for RSQ and minimum for MAE and RMSE) are shown in bold. This is "Default validation case" in Figure 5.

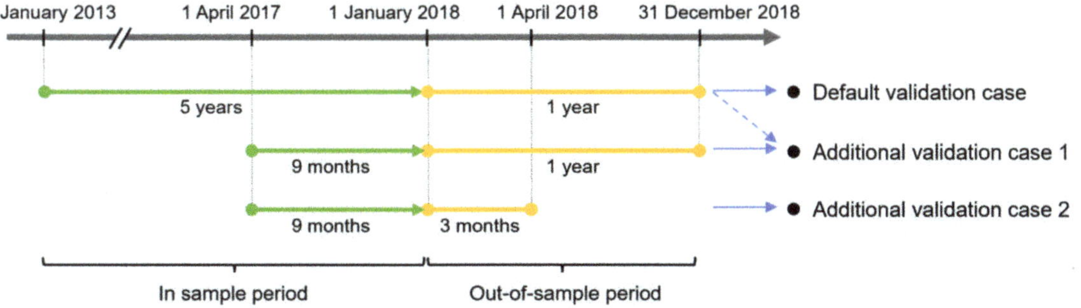

Figure 5. The relationship between in-sample and out-of-sample periods of each validation case for the individual PV power generation forecasting model. Note: "Default validation case" corresponds to Figures 6 and 7 and Table 1; "Additional validation case 1" corresponds to Figures 8 and A2; "Additional validation case 2" corresponds to Figure 9.

- PV power generation volume $V_{t,h}$ (MW): measured value of the household's solar power system (with the permission of the owner, we use the data of a private roof-mounted power system in Hiroshima city, Japan).
- Solar radiation $R_{t,h}$ (MJ/m^2): Measured solar radiation in Hiroshima City as published by the JMA [45].

For solar radiation $R_{t,h}$, this study uses actual measured values that are available free of charge for the purpose of comparison among models, considering that there is no essential difference in whether to use forecast values or measured values in the comparison between models. Incidentally, weather forecasts of the JMA have been reported to be effective in forecasting electricity time series up to about one week ahead [46], and it may be possible to compare models under different forecast horizons, but such analysis is a future issue.

4.2.1. Estimated Trend

Figure 6 shows the trend estimated for M3's 3D tensor product spline function $v(Seasonal_t, h, R_{t,h})$ in GAM (3) by hour h. The surface on the 2D coordinates of the date $Seasonal_t$ and solar radiation $R_{t,h}$ at each hour changes gradually with the hour. It should be noted that the slope of the PV generation with respect to the solar radiation tends to decrease (and even diminish with solar radiation) from spring to early summer, when the solar radiation is large at each time. This reflects the technical characteristics of PV panels, where the power generation efficiency decreases as the temperature (solar radiation) increases.

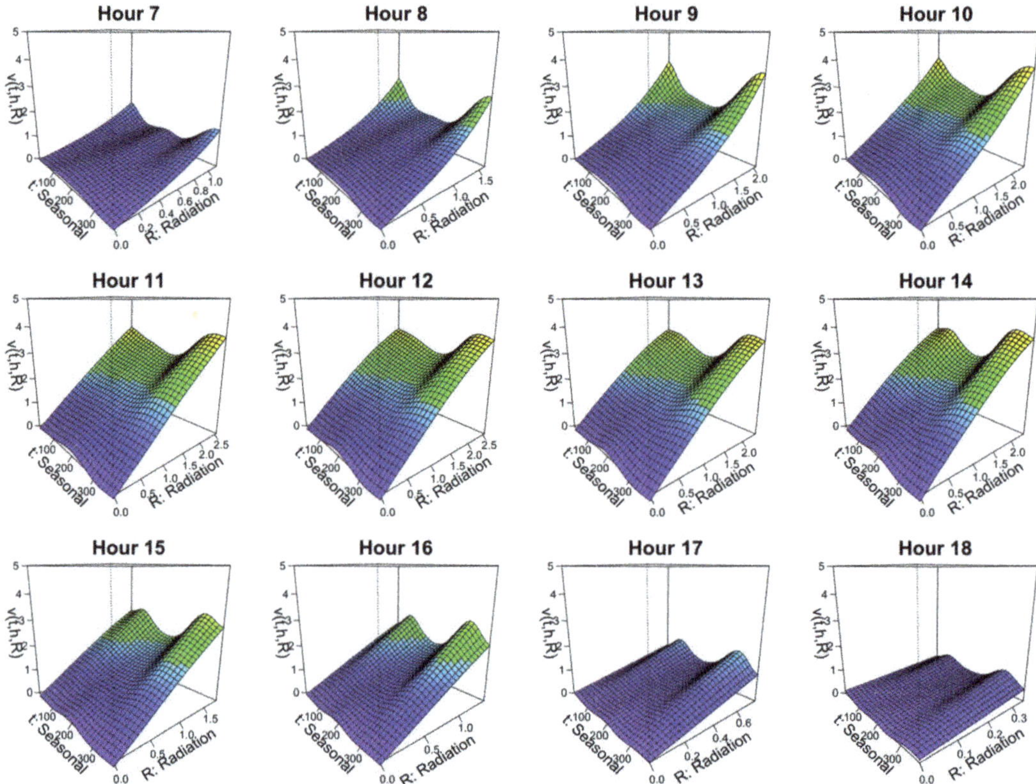

Figure 6. Trend estimation results for the M3 model (estimated from 5 years of in-sample data). Note: The units of vertical axis and "Radiation" axis are (MW) and (MJ/m^2), respectively. "Seasonal" denotes a yearly cyclical dummy variable (see Section 2.1).

4.2.2. Comparison of Forecast Accuracy

Table 2 shows the results of the comparison of the forecast error and computation time for each model. The results are discussed from two perspectives: the comparison between statistical models (M0-M3) and the comparison between GAM (M3) and ML methods, which are shown in the following two subsections.

Comparison of Forecast Accuracy among Statistical Models

In this section, we analyze the results in terms of comparison between the statistical models (M0 to M3). Since this study is the first attempt to apply 3D tensor product spline functions to energy time series data forecasting, it would be interesting to examine the effect of adding smoothing (nonlinear) conditions in multiple directions on the accuracy.

First, as seen in Table 2, the overall forecast accuracy is generally the highest for M3 for both in-sample and out-of-sample, while M0, which assumes a constant linear regression equation for the whole year, is significantly less accurate than the other models. The MAE and RMSE of M1 are better than M2 (comparable to M3) in the in-sample, but worse than M2 in the out-of-sample. This can be interpreted as M1 building a different linear model for each month and time, which makes it less robust than M2 and M3, where smoothing conditions are imposed in the date direction.

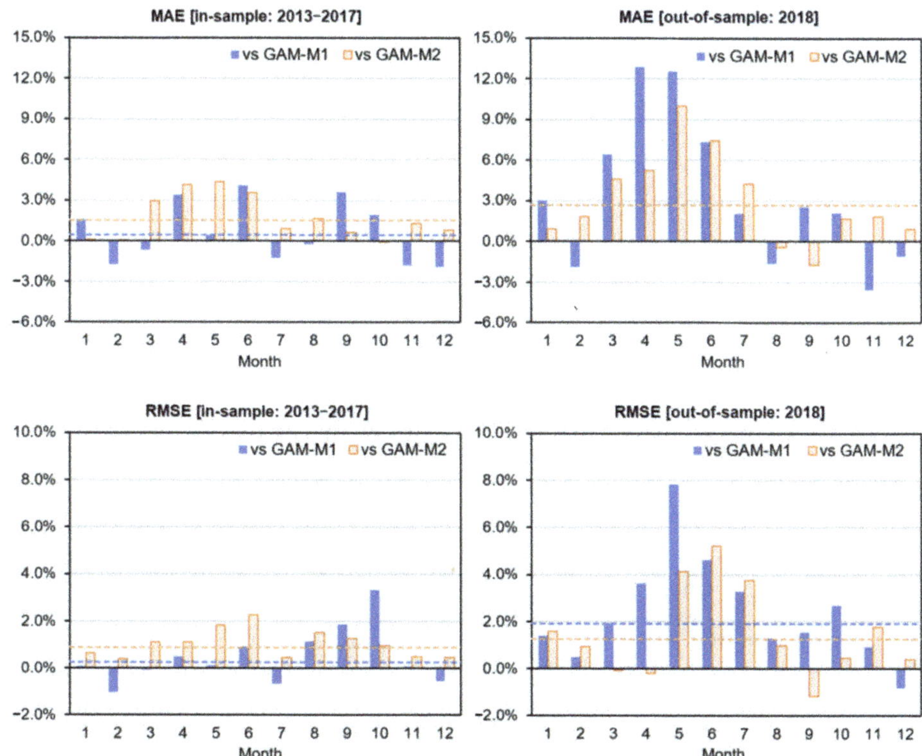

Figure 7. Monthly relative forecast error (MAE and RMSE) of M1 and M2 with respect to M3. Note: The dashed line is the relative increment of the forecast error to M3 over the period (the out-of-sample MAEs overlap the two because they are equal). This is "Default validation case" in Figure 5.

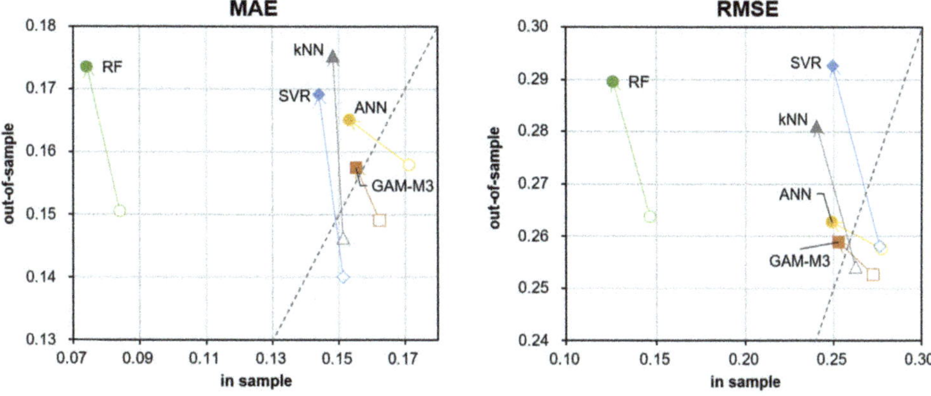

Figure 8. MAE and RMSE of each method on individual PV power generation (change in forecast error by method when the in-sample period is shortened). Note: White dots represent the 5-year in-sample period (corresponding to values in Table 2), and color-filled dots represent the 9-month in-sample period. Out-samples were both 2018 (i.e., this is "Additional validation case 1" in Figure 5).

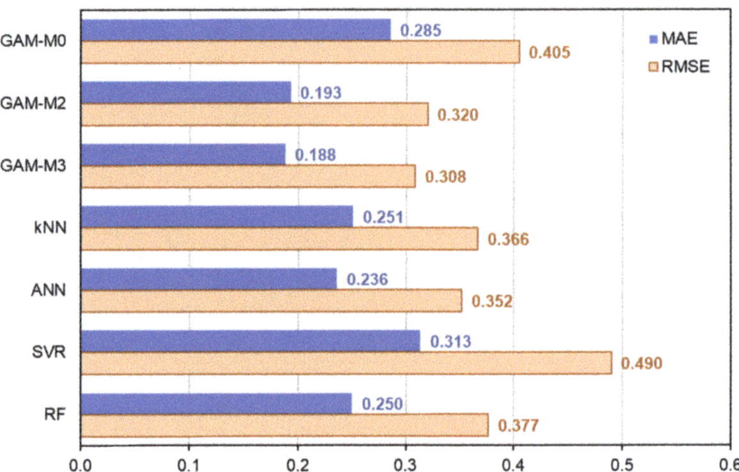

Figure 9. Comparison of forecast errors for the next three months when the model is estimated from nine months of data. Note: The in-sample period is 9 months, from 1 April to 31 December 2017; the out-of-sample period is 3 months, from 1 January to 31 March 2018 (i.e., this is "Additional validation case 2" in Figure 5). The reason for the absence of GAM-M1 is that model estimation is not possible because of the lack of in-sample data for the same month.

Next, to obtain a more detailed understanding of the effect of incorporating the nonlinearity between PV generation and solar radiation in the M3 model, we compare the prediction errors between models on a monthly basis. Figure 7 plots the relative error increments for the MAE (or RMSE) of both M1 and M2 relative to the MAE (or RMSE) of M3 by month (e.g., MAE, $MAE_{M1\ or\ M2}/MAE_{M3} - 1$). It can be seen that the relative errors (MAE and RMSE) of M1 and M2, both of which are linear models with respect to solar radiation, are particularly large during the spring and early summer months in the out-of-sample period. This is consistent with the fact that the slope of M3 diminished in relation to the solar radiation during the same period when the solar radiation increased, as is confirmed in Figure 6. This means that the nonlinearity that exists between PV generation and solar radiation during the same period was modeled relatively robustly in M3. In addition, although the forecast errors of M1 did not differ significantly from M3 during the in-sample period, the error increased for the out-of-sample period. This result also suggests that M1, which does not have a smoothing condition, had an undesirable (excessive) model fitting by month.

Comparison of Forecasting Accuracy between GAM and ML Methods

Next, in this section, we compare the prediction accuracies of GAM and ML. As seen in the previous section, M3 was the model with the highest forecast accuracy among the GAMs (including linear models), so we omitted the M0–M2 models and dealt only with the M3 model (in this section, the term "GAM" refers solely to M3).

As can be seen in Table 2, in the comparison between GAM and the four ML methods, RF has the best fit in the in-sample, which is the same result as that of the area-wide PV generation model in Section 4.1.2. On the other hand, in terms of out-of-sample forecast errors, SVR is the best for MAE, and GAM is the best for RSQ and RMSE; the reason why SVR is the best for MAE is presumably because the objective function of SVR is close to MAE minimization. As mentioned in Section 3, SVR uses linear ε-insensitive loss functions (as shown in "Figure 1" of [35]), but because ε is relatively small, SVR can be said to approximately minimize the MAE.

In any case, it is true that GAM is a highly superior model overall in terms of computation time and ease of interpretation with intuitive visualization. However, it should be noted that when compared with the forecast accuracy results of the area-wide PV power generation model examined in Section 4.1.2, the results in this section seem to indicate that the superiority of GAM may not be so clear (at least, it is inferior to SVR and kNN in MAE).

One of the reasons for this may be the following: the area-wide PV generation model had a large number of incorporated explanatory variables, even though the period of the data used was relatively short (about a year and a half). That is, in the GAM, the "model structure"—which is based on rational human reasoning—was able to be taught in advance such that the sensitivities of the various explanatory variables (weather conditions and temperatures) each have a daily smooth yearly cyclical trend, along with a smooth trend in the orthogonal time direction. On the other hand, the individual PV model is a relatively simple model that only estimates a smooth trend in the three directions, which means that the ML methods (without prior knowledge of the existence of the multi-dimensional smooth trends) were able to estimate it reasonably effectively. In other words, when the modeling practitioner recognizes or detects the existence of a global model structure that is difficult to learn from the data alone, a statistical model, such as GAM, has an advantage in that it is relatively easy to describe it in the formulation to facilitate accurate forecasts.

To make this consideration more credible, in the following, we will assume a case where the in-sample period is short (a missing period is intentionally created), and an experiment to see how the accuracy of each forecasting method changes in that case. Below, we separately calculate the case where the in-sample period is from 1 April to 31 December 2017 (9 months), and plot in Figure 8 how the forecast error for out-of-sample (2018) changes from the original case where the in-sample period is 1 January 2013 to 31 December 2017 (5 years); the latter is already calculated in Table 2. In other words, the newly estimated model here contains missing data from January to March.

As can be seen from the graphs, when the in-sample period is shortened, the in-sample forecast error (MAE or RMSE) becomes smaller, but the out-of-sample forecast error becomes larger, which is common to all forecasting methods. This result is consistent with the intuition that the shorter the period, the better the fit of each model, but the less robust it will tend to be. More interestingly, while kNN, SVR, and RF significantly deteriorated the out-of-sample accuracy (when missing periods were included), GAM and ANN did not, and were relatively robust in trend completion for missing values. As a result, GAM had the smallest MAE and RMSE for the 9-month case. The reason for the robustness of the ANN suggests that some trend completion may have occurred in the hidden layer. On the other hand, kNN, SVR, and RF are unsuitable for trend estimation (especially extrapolation) that complements missing periods.

In the following, in order to make the performance comparison of trend completion clearer, using the estimated model based on nine months of in-sample data from 1 April to 31 December 2017, we compare the forecast errors measured from the following three months of out-of-sample data (1 January to 31 March 2018) in Figure 9. Note that in practice, insufficient historical data is common, especially for new entrants. This result clearly shows that the prediction errors of kNN, SVR, and RF by extrapolation are relatively large. In particular, the extremely poor prediction error of SVR may be due to the radial basis function (RBF) kernel used, which is also called a local kernel and is suitable for interpolation but not for extrapolation (see "Figure 8" in [47]). It has also been proposed that SVR methods can be improved by combining them with polynomial kernels (global kernels) [47], but the improvement of ML algorithms is beyond the scope of this study.

For reference, Figure A2 in Appendix E shows the trend of the 3D tensor product spline function estimated from nine months of in-sample data from 1 April to 31 December 2017. It can be seen that even though there is no data for January–March, the shape is almost the same as in Figure 6, which uses data for five full years, and smooth trend estimation in the solar radiation and hour directions is achieved. This result also confirms the robustness of our GAM-based model.

5. Conclusions

This study proposed and validated a GAM-based model with multidimensional tensor product splines to support the forecasting of PV power generation in practice. In summary, our contribution lies in the following points:

- We constructed different GAM-based forecasting models for area-wide PV power generation and individual PV power generation, and demonstrated the effectiveness of the models by visualizing the estimated trends and providing reasonable interpretations.
- For the individual PV power generation model, we constructed a new forecasting model using 3D tensor product splines and demonstrated its effectiveness. We quantitatively demonstrated that the robustness and forecasting accuracy of the model increased when smoothing (nonlinear) conditions were incorporated in three directions by comparing it with linear models.
- By comparing the proposed GAM-based models with other popular ML methods, such as kNN, ANN, SVR, and RF for each PV power model, it was shown that the GAM-based models have advantages in terms of computational speed and forecast error minimization. Specifically, we have shown that the GAM-based model is highly effective for global nonlinear trend completion.

In general, ML has been reported to be superior to statistical models in terms of predictability. However, this study showed that our forecasting approach using GAM may have several advantages over ML methods, such as interpretability, robustness, computational load, and forecasting accuracy. Moreover, when the existence of a smooth (periodic) trend is inferred in advance, the GAM may capture the structure and provide better forecasting accuracy. For example, in this study, the 2D tensor product spline model was formulated in advance that the coefficients of each variable of weather and temperature should have smoothly connected trends in the seasonal and time directions, respectively. The 3D model was described as having a smooth trend in the directions of seasonal, time, and solar radiation. In addition, the cyclic cubic spline function was used to incorporate the condition that the seasonal trends are connected at the start and end points of the yearly cycle.

A statistical model, such as GAM, has various advantages in practical use, such as ease of reflecting the model designer's prior knowledge, understanding the results intuitively, and explaining them to others. In particular, it has the advantage that the recognized global model structure can be formulated (taught) in advance, which makes it easier to build a reasonable and robust model compared to ML methods that recognize patterns from data only. Therefore, the descriptiveness of the model (with high interpretability) is an important factor that potentially contributes to an improvement in accuracy. In conclusion, it may be fair to say that our GAM-based models with multi-dimensional tensor product splines provide a promising forecast approach for practitioners that require model tractability, reliability, and forecasting accuracy in the PV business.

Author Contributions: Conceptualization, Y.Y.; methodology, T.M.; software, T.M.; validation, T.M.; data curation, T.M.; writing—original draft preparation, T.M.; writing—review and editing, Y.Y.; visualization, T.M.; supervision, Y.Y.; project administration, Y.Y.; funding acquisition, T.M. and Y.Y. All authors have read and agreed to the published version of the manuscript.

Funding: This work was funded by a Grant-in-Aid for Scientific Research (A) 20H00285, Grant-in-Aid for Challenging Research (Exploratory) 19K22024, and Grant-in-Aid for Young Scientists 21K14374 from the Japan Society for the Promotion of Science (JSPS).

Institutional Review Board Statement: Not applicable.

Informed Consent Statement: Not applicable.

Data Availability Statement: Please contact the authors.

Conflicts of Interest: The authors declare no conflict of interest.

Nomenclature

$V_{t,h}$	measured PV power generation volume at date t, hour h
W_t	installed PV power capacity at date t, hour h
$U_{t,h}$	unit power generation at date t, hour h
$I_{.,t,h}$	dummy variables, which are 1 if the forecast general weather condition at date t hour h is the same as the suffix's weather condition, or 0 otherwise
$Seasonal_t$	yearly cyclical dummy variables $(= 1, \ldots, 365\ (or\ 366))$
$Tmax_t, Tmin_t$	previous day's maximum or minimum temperature forecast at date t
$\epsilon_{tmax,t}, \epsilon_{tmin,t}$	maximum or minimum temperature forecast deviation at date t (observed temperature forecast minus its trend)
$u_.(t,h)$	2D tensor product spline functions estimated by the GAM of the area-wide PV power forecast model (t denote $Seasonal_t$)
$f_.(t)$	univariate spline functions estimated by the GAM of the temperature trend model (t denote $Seasonal_t$)
$\eta_{t,h}$	residual terms with the average of 0
$R_{t,h}$	forecast solar radiation at date t, hour h
β, α	coefficients and constant terms for the individual PV power generation models when each model is viewed as a linear regression equation for solar radiation (constant for M0 and M1, or variables defined by 2D tensor product spline function for M2)
$v(t, h, R_{t,h})$	3D tensor product spline functions estimated by the GAM of the individual PV power forecast model (t denote $Seasonal_t$)

Appendix A. Installed Capacity Trend Estimation for Area PV Generation Forecasting

The area-wide PV power capacity W_t is modeled by the following ordinary least squares regression (OLS):

$$W_t = w_1 Period_t + w_2 + \eta_t. \tag{A1}$$

where w_1 and w_2 are the coefficient and intercept, respectively, estimated by the OLS (A1), $Period_t$ is the (annualized) daily dummy variable representing the number of years that have passed, and η_t is the residual term. Using this equation, the forecast capacity $\hat{W}_t$ can be obtained as follows.

$$\hat{W}_t = w_1 Period_t + w_2. \tag{A2}$$

Note that the original observed capacity W_t is monthly data with some missing values, but the forecast value $\hat{W}_t$ can be obtained as daily granularity data because we use the daily dummy $Period_t$.

Appendix B. Smoothing Spline Functions

The univariate smoothing spline function is estimated as the function h that minimizes the penalized residual sum of squares (PRSS), given by

$$\mathrm{PRSS} = \sum_{n=1}^{N} \{y_n - h(x_n)\}^2 + J(h), \quad \text{where} \quad J(h) = \lambda \int \{h''(x)\}^2 dx. \tag{A3}$$

In (A3), the first term measures the approximation of the data, and the second term (penalty term) $J(h)$ adds penalties according to the magnitude of the curvature of the function. In this study, we construct the GAM using the R package "mgcv" [48] to obtain a series of smoothing spline functions, where the smoothing parameter λ is calculated using a general cross-validation criterion.

When estimating the 2D (3D) tensor product spline function $h(x,z)$ ($h(x,z,v)$), the following penalty term $J_2(h)$ ($J_3(h)$) is included in the PRSS that should be minimized [25]:

$$J_2(h) = \int_{x,z} \left[\lambda_x \left(\frac{\partial^2 h}{\partial x^2} \right)^2 + \lambda_z \left(\frac{\partial^2 h}{\partial z^2} \right)^2 \right] dxdz. \tag{A4}$$

$$J_3(h) = \int_{x,z,v} \left[\lambda_x \left(\frac{\partial^2 h}{\partial x^2} \right)^2 + \lambda_z \left(\frac{\partial^2 h}{\partial z^2} \right)^2 + \lambda_v \left(\frac{\partial^2 h}{\partial v^2} \right)^2 \right] dxdzdv. \tag{A5}$$

Thus, the tensor product spline function can incorporate independent smoothing conditions for each variable (direction). Previous studies that applied 2D tensor product splines to different energy time series data include [49,50], where the tensor product spline functions are used for pricing weather derivatives for risk hedging rather than for forecasting models.

Appendix C. Hyperparameters to be Tuned for the Caret Package

The four ML methods in this study are automatically tuned by the caret package, and the hyperparameters to be tuned for each method are listed in Table A1 [51]. For details of the parameters, please refer to the references of each package. (Note that in the caret package, epsilon in the "ε-insensitive loss functions" of SVM (SVR) is not tuned, and the default value of 0.1 is used.)

Table A1. Hyperparameters to be tuned in the four ML methods used in this study.

Model	Method Value	Package		Tuning Parameters
kNN	knn	caret [28]	k	Number of neighbors considered
ANN	nnet	nnet [52]	decay	The parameter for weight decay
			size	Number of units in the hidden layer
SVM (SVR)	svmRadial	kernlab [53]	sigma	The inverse kernel width used by the Gaussian kernel
			C	The cost regularization parameter, which controls the smoothness
RF	rf	randomForest [54]	mtry	Number of variables randomly sampled as candidates at each split

In the caret package, the parameter "tuneLength" (set to 4 by default) allows the user to select the number of tunings (how many different scenarios of each hyperparameter are compared and verified) for the target hyperparameters. In this study, by setting this parameter to 10, we tuned 10 scenarios for "knn" and "rf," and 100 scenarios for "nnet" and "svmRadial". For the evaluation method of the tuning, we adopted caret's default method of 10-fold cross-validation (i.e., cross-validation was performed by dividing the training data into 10 equal parts).

Appendix D. Estimated Trends for Areawide PV Power Generation Model

As shown in Figure A1, the estimated trends of the 2D tensor product spline functions for each of the nine areas are generally similar in shape, although there are some differences among the areas (the interpretation of the trend shapes described in Section 4.1.1 is common to all areas). If we look at the details, we can see that in Hokkaido, an area of high latitude, power generation is relatively level throughout the seasons (e.g., the decline of the rainy trend in winters is relatively small), and the seasonal trend of snow is extracted relatively clearly. The reason why Tohoku and Hokuriku, which are other snowfall areas, do not have such seasonal snowy trends, is perhaps because the sample sizes of snowy weather of these areas were not large enough.

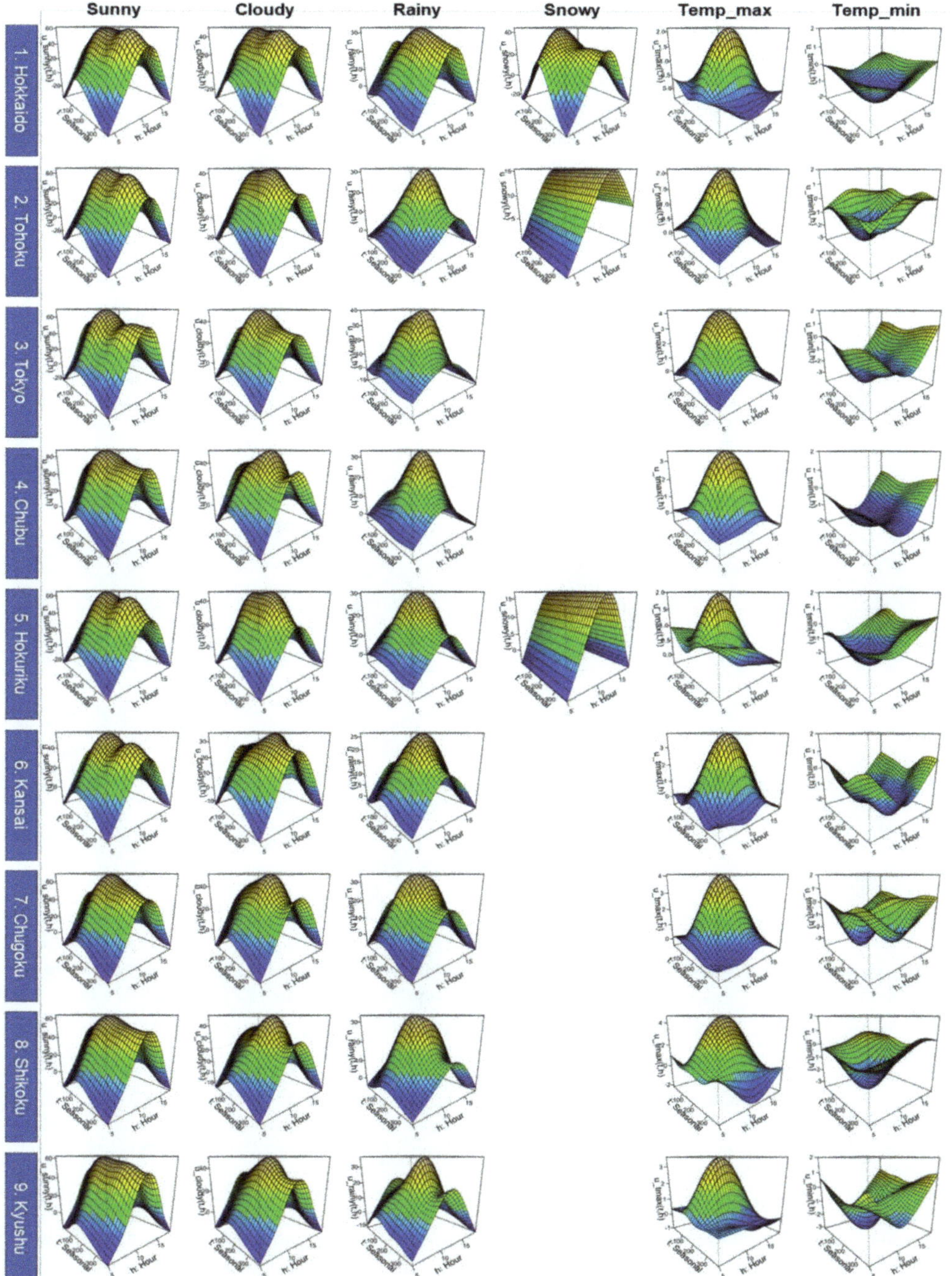

Figure A1. Estimated trends for areawide PV power generation model (all nine areas). Note: The unit of vertical axis for "Sunny," "Cloudy," "Rainy" and "Snowy" is (%), and that for "Temp_max" and "Temp_min" is (%/°C). "Seasonal" denotes a yearly cyclical dummy variable (see Section 2.1), and "Hour" denotes the time (o'clock).

Appendix E. Trend Estimation Results for the M3 Model (Estimated from 9 Months of In-Sample Data)

The estimation results of the 3D tensor product spline function of the individual PV power generation model for the in-sample period from 1 April to 31 December 2017 are shown in Figure A2. All graphs have almost the same shape as the estimation results when the in-sample period is five years from 2013 to 2017 (Figure 6), indicating that the 3D tensor product spline function is highly robust.

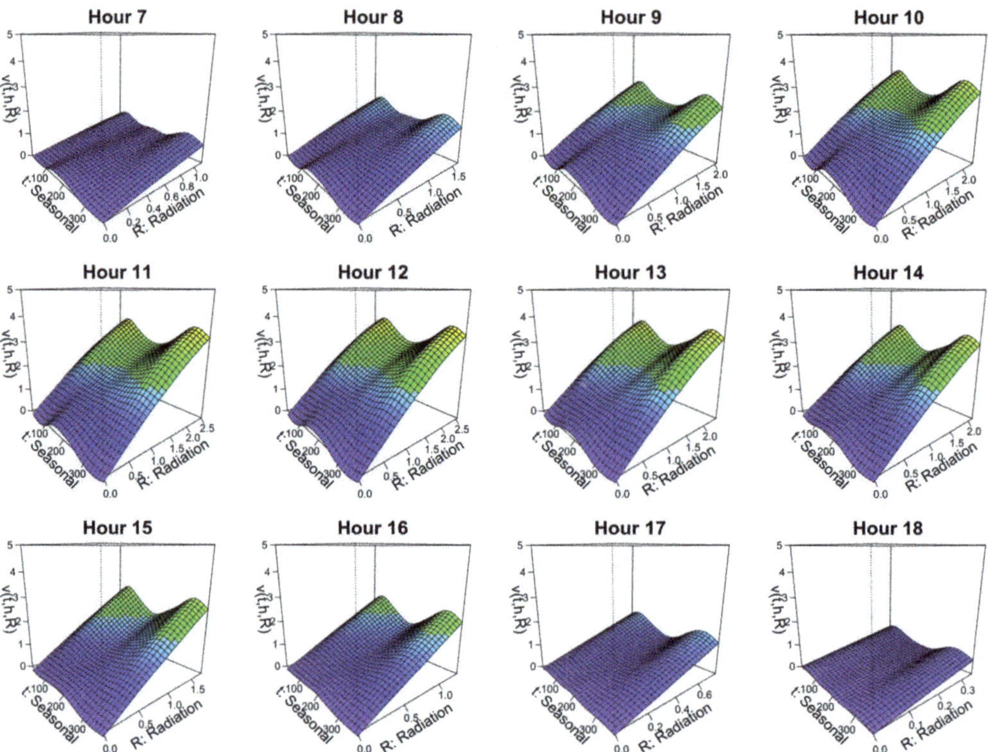

Figure A2. Trend estimation results for the M3 model (estimated from 9 months of in-sample data). Note: The units of vertical axis and "Radiation" axis are (MW) and (MJ/m^2), respectively. "Seasonal" denotes a yearly cyclical dummy variable (see Section 2.1).

References

1. Matsumoto, T.; Yamada, Y. Prediction method for solar power business based on forecasted general weather conditions and periodic trends by weather. *Trans. Mater. Res. Soc. Jpn.* **2019**, *62*, 1–22. (In Japanese) [CrossRef]
2. Antonanzas, J.; Osorio, N.; Escobar, R.; Urraca, R.; Martinez-de-Pison, F.J.; Antonanzas-Torres, F. Review of photovoltaic power forecasting. *Sol. Energy* **2016**, *136*, 78–111. [CrossRef]
3. Pelland, S.; Remund, J.; Kleissl, J.; Oozeki, T. *Photovoltaic and Solar Forecasting: State of the Art*; Report PVPS T14-10; International Energy Agency: Paris, France, 2013.
4. Kato, T. Development of forecasting method of photovoltaic power output. *J. IEEJ* **2017**, *137*, 101–104. (In Japanese) [CrossRef]
5. Raza, M.Q.; Nadarajah, M.N.; Ekanayake, C. On recent advances in PV output power forecast. *Sol. Energy* **2016**, *136*, 125–144. [CrossRef]
6. Mosavi, A.; Salimi, M.; Faizollahzadeh Ardabili, S.; Rabczuk, T.; Shamshirband, S.; Varkonyi-Koczy, A.R. State of the art of machine learning models in energy systems, a systematic review. *Energies* **2019**, *12*, 1301. [CrossRef]
7. Das, U.K.; Tey, K.S.; Seyedmahmoudian, M.; Mekhilef, S.; Idris, M.Y.I.; Van Deventer, W.; Horan, B.; Stojcevski, A. Forecasting of photovoltaic power generation and model optimization: A review. *Renew. Sustain. Energy Rev.* **2018**, *81*, 912–928. [CrossRef]

8. Sobri, S.; Koohi-Kamali, S.; Rahim, N.A. Solar photovoltaic generation forecasting methods: A review. *Energy Convers. Manag.* **2018**, *156*, 459–497. [CrossRef]
9. Ripley, B.D. *Pattern Recognition and Neural Networks*; Cambridge University Press: Cambridge, UK, 2007.
10. Drucker, H.; Burges, C.C.; Kaufman, L.; Smola, A.J.; Vapnik, V.N. Support vector regression machines. In *Advances in Neural Information Processing Systems 9, NIPS 1997*; MIT Press: Cambridge, MA, USA, 1996; pp. 155–161.
11. Maitanova, N.; Telle, J.S.; Hanke, B.; Grottke, M.; Schmidt, T.; Maydell, K.V.; Agert, C. A machine learning approach to low-cost photovoltaic power prediction based on publicly available weather reports. *Energies* **2020**, *13*, 735. [CrossRef]
12. Mohammed, A.A.; Aung, Z. Ensemble learning approach for probabilistic forecasting of solar power generation. *Energies* **2016**, *9*, 1017. [CrossRef]
13. Altman, N.S. An introduction to kernel and nearest-neighbor nonparametric regression. *Am. Stat.* **1992**, *46*, 175–185.
14. Breiman, L. Random forests. *Mach. Learn.* **2001**, *45*, 5–32. [CrossRef]
15. Das, U.K.; Tey, K.S.; Seyedmahmoudian, M.; Idris, M.Y.I.; Mekhilef, S.; Horan, B.; Stojcevski, A. SVR-based model to forecast PV power generation under different weather conditions. *Energies* **2017**, *10*, 876. [CrossRef]
16. Rosato, A.; Altilio, R.; Araneo, R.; Panella, M. Prediction in photovoltaic power by neural networks. *Energies* **2017**, *10*, 1003. [CrossRef]
17. Khandakar, A.; Chowdhury, M.E.H.; Khoda Kazi, M.; Benhmed, K.; Touati, F.; Al-Hitmi, M.; Gonzales, J.S. Machine learning based photovoltaics (PV) power prediction using different environmental parameters of Qatar. *Energies* **2019**, *12*, 2782. [CrossRef]
18. Nespoli, A.; Ogliari, E.; Leva, S.; Massi Pavan, A.; Mellit, A.; Lughi, V.; Dolara, A. Day-ahead photovoltaic forecasting: A comparison of the most effective techniques. *Energies* **2019**, *12*, 1621. [CrossRef]
19. Abdel-Nasser, M.; Mahmoud, K. Accurate photovoltaic power forecasting models using deep LSTM-RNN. *Neural Comput. Appl.* **2019**, *31*, 2727–2740. [CrossRef]
20. Matsumoto, T. Forecast Based Risk Management for Electricity Trading Market. Ph.D. Thesis, University of Tsukuba, Tokyo, Japan, 2020.
21. Hastie, T.; Tibshirani, R. *Generalized Additive Models*; Chapman and Hall: London, UK, 1990.
22. Matsumoto, T.; Yamada, Y. Construction of forecast model for power demand and PV power generation using tensor product spline function. In *IOP Conference Series: Earth and Environmental Science*; IOP Publishing: Bristol, UK, 2021; Volume 812, p. 012001.
23. Sundararajan, A.; Ollis, B. Regression and generalized additive model to enhance the performance of photovoltaic power ensemble predictors. *IEEE Access* **2021**, *9*, 111899–111914. [CrossRef]
24. Gadiwala, M.S.; Usman, A.; Akhtar, M.; Jamil, K. Empirical models for the estimation of global solar radiation with sunshine hours on horizontal surface in various cities of Pakistan. *Pak. J. Meteorol.* **2013**, *9*, 43–55.
25. Wood, S.N. *Generalized Additive Models: An Introduction with R*; Chapman and Hall: London, UK, 2017.
26. Matsumoto, T.; Yamada, Y. Customized yet Standardized temperature derivatives: A non-parametric approach with suitable basis selection for ensuring robustness. *Energies* **2021**, *14*, 3351. [CrossRef]
27. Japan Meteorological Business Support Center. Available online: http://www.jmbsc.or.jp/jp/online/file/f-online10200.html (accessed on 26 October 2021).
28. Kuhn, M.; Wing, J.; Weston, S.; Williams, A.; Keefer, C.; Engelhardt, A.; Cooper, T.; Mayer, Z.; Kenkel, B.; Benesty, M.; et al. Package Caret. Available online: https://cran.r-project.org/web/packages/caret/caret.pdf (accessed on 26 October 2021).
29. Cherkassky, V.; Mulier, F.M. *Learning from Data: Concepts, Theory, and Methods*; John Wiley & Sons: Hoboken, NJ, USA, 2007.
30. Singh, Y.; Chauhan, A.S. Neural networks in data mining. *J. Theor. Appl. Inf. Technol.* **2009**, *5*, 37–42. [CrossRef]
31. Wu, X.; Kumar, V.; Quinlan, J.R.; Ghosh, J.; Yang, Q.; Motoda, H.; McLachlan, G.; Ng, A.; Liu, B.; Yu, P.; et al. Top 10 algorithms in data mining. *Knowl. Inf. Syst.* **2008**, *14*, 1–37. [CrossRef]
32. Aizerman, M. Theoretical foundations of the potential function method in pattern recognition learning. *Autom. Remote Control* **1964**, *25*, 821–837.
33. Friedman, J.; Hastie, T.; Tibshirani, R. *The Elements of Statistical Learning*; Springer Series in Statistics: New York, NY, USA, 2001; Volume 1, Available online: https://web.stanford.edu/~{}hastie/ElemStatLearn/printings/ESLII_print12_toc.pdf (accessed on 26 October 2021).
34. Marquardt, D.W.; Snee, R.D. Ridge regression in practice. *Am. Stat.* **1975**, *29*, 3–20.
35. Smola, A.J.; Schölkopf, B. A tutorial on support vector regression. *Stat. Comput.* **2004**, *14*, 199–222. [CrossRef]
36. Scikit-learn. Kernel Ridge Regression. 2018. Available online: http://scikit-learn.org/stable/modules/kernel_ridge.html (accessed on 26 October 2021).
37. Dutta, A.; Dureja, A.; Abrol, S.; Dureja, A. Prediction of ticket prices for public transport using linear regression and random forest regression methods: A practical approach using machine learning. In *International Conference on Recent Developments in Science, Engineering and Technology*; Springer: Singapore, 2019; pp. 140–150.
38. Kuhn, M. Building predictive models in R using the caret package. *J. Stat. Softw.* **2008**, *28*, 1–26. [CrossRef]
39. Tokyo Electric Power Co. "Announcement of Area Supply and Demand Results". Available online: https://www.tepco.co.jp/forecast/html/area_data-j.html (accessed on 26 October 2021).
40. Ministry of Economy, Trade and Industry. "Feed-In Tariff (FIT) Program Information Release Website". Available online: https://www.fit-portal.go.jp/PublicInfoSummary (accessed on 26 October 2021).
41. Changes in Weather Forecast. Available online: http://weather-transition.gger.jp/ (accessed on 26 October 2021).

42. Campbell, J.Y.; Thompson, S.B. Predicting excess stock returns out of sample: Can anything beat the historical average? *Rev. Financ. Stud.* **2008**, *21*, 1509–1531. [CrossRef]
43. Kuhn, M. Parallel Processing. Available online: https://topepo.github.io/caret/parallel-processing.html (accessed on 27 September 2021).
44. Kaggle, mlcourse.ai. Open Machine Learning Course by OpenDataScience. Available online: https://www.kaggle.com/kashnitsky/topic-5-ensembles-part-2-random-forest (accessed on 26 October 2021).
45. Japan Meteorological Agency "Historical Weather Data Download". Available online: https://www.data.jma.go.jp/gmd/risk/obsdl/ (accessed on 26 October 2021).
46. Takuji, M.; Misao, E. One-week-ahead electricity price forecasting using weather forecasts, and its application to arbitrage in the forward market: An empirical study of the Japan electric power exchange. *J. Energy Mark.* **2021**, *14*, 1–26.
47. Smits, G.F.; Jordaan, E.M. Improved SVM regression using mixtures of kernels. In Proceedings of the 2002 International Joint Conference on Neural Networks, Honolulu, HI, USA, 12–17 May 2002; IJCNN'02 Cat. No. 02CH37290. IEEE: New York, NY, USA, 2002; Volume 3, pp. 2785–2790.
48. Wood, S.N. 2019 Package "mgcv". Available online: https://cran.r-project.org/web/packages/mgcv/mgcv.pdf (accessed on 15 October 2020).
49. Matsumoto, T.; Yamada, Y. Simultaneous hedging strategy for price and volume risks in electricity businesses using energy and weather derivatives. *Energy Econ.* **2021**, *95*, 105101. [CrossRef]
50. Matsumoto, T.; Yamada, Y. Cross hedging using prediction error weather derivatives for loss of solar output prediction errors in electricity market. *Asia-Pac. Financ. Mark.* **2018**, *26*, 211–227. [CrossRef]
51. Kuhn, M. The Caret Package. *J. Stat. Softw.* **2009**, *28*. Available online: http://citeseerx.ist.psu.edu/viewdoc/download?doi=10.1.1.216.2142&rep=rep1&type=pdf (accessed on 26 October 2021).
52. Ripley, B.; Venables, W.; Ripley, M.B. Package 'nnet'. R Package Version. 2016. Available online: https://cran.r-project.org/web/packages/nnet/nnet.pdf (accessed on 26 October 2021).
53. Karatzoglou, A.; Smola, A.; Hornik, K.; Karatzoglou, M.A. Package 'Kernlab'. CRAN R Project. 2019. Available online: https://cran.r-project.org/web/packages/kernlab/kernlab.pdf (accessed on 26 October 2021).
54. Breiman, L.; Cutler, A. *Package 'RandomForest'*; University of California, Berkeley: Berkeley, CA, USA, 2018; Available online: https://cran.r-project.org/web/packages/randomForest/randomForest.pdf (accessed on 26 October 2021).

Article

Going for Derivatives or Forwards? Minimizing Cashflow Fluctuations of Electricity Transactions on Power Markets

Yuji Yamada [1,*] **and Takuji Matsumoto** [2]

1 Faculty of Business Sciences, University of Tsukuba, Tokyo 112-0012, Japan
2 Socio-Economic Research Center, Central Research Institute of Electric Power Industry, Tokyo 100-8126, Japan; mtakuji1122@gmail.com
* Correspondence: yuji@gssm.otsuka.tsukuba.ac.jp

Abstract: In a competitive electricity market, both electricity retailers and generators predict future prices and volumes and execute electricity delivery contracts through power exchange. In such circumstances, they may suffer from uncertainties caused by fluctuations in spot prices and future demand due to their high volatility. In this study, we develop a unified approach using derivatives and forwards on the spot electricity price and weather data to mitigate the cashflow fluctuation for power utilities. We aim to clarify the applicability of our proposed methods and provide a new and useful perspective on hedging schemes involving various electricity utilities, such as power retailers, solar photovoltaic (PV) generators, and thermal generators. Moreover, we analyze the risk of risk takers (such as the insurance companies in this study) in the derivatives market. In addition, we perform empirical simulations to measure out-of-sample hedging effects on their cashflow management using actual data in Japan.

Keywords: cashflow management of electricity businesses; electricity derivatives and forwards; retailers and power producers; solar power and thermal energy; optimal hedging using nonparametric techniques; empirical simulations

Citation: Yamada, Y.; Matsumoto, T. Going for Derivatives or Forwards? Minimizing Cashflow Fluctuations of Electricity Transactions on Power Markets. *Energies* **2021**, *14*, 7311. https://doi.org/10.3390/en14217311

Academic Editor: Dimitrios Asteriou

Received: 20 October 2021
Accepted: 28 October 2021
Published: 4 November 2021

Publisher's Note: MDPI stays neutral with regard to jurisdictional claims in published maps and institutional affiliations.

1. Introduction

In electricity markets, the transactions of electricity delivery contracts between power retailers and generators are based on predictions of demand and supply that reflect the actual consumption of the end-users as well as the renewable power generation in the future. For example, the demand volume for power retailers largely depends on the future temperature, whereas the power output from solar photovoltaic (PV) and other renewable energy generation fluctuates over time according to the future weather conditions. In addition, energy prices, such as oil and natural gas, affect the electricity price as well as the supply and demand predictions, and so the spot electricity price is quite volatile in a competitive power exchange market. In such a situation, power retailers and generators suffer from the risk of simultaneous price and volume fluctuations, leading to large volatility in their cashflows, and adequate strategies for reducing the cashflow fluctuations are required for power utilities. Therefore, financial instruments, including derivatives and forwards on spot electricity prices and weather indexes, are considered effective tools [1,2].

There are several previous studies on electricity derivatives and weather derivatives, and various methods that have been proposed, especially in the context of pricing. For electricity derivatives, there is a relatively wide variety of studies on the pricing of option-type derivatives (e.g., [3–6]), which are systematically reviewed in [1]. Other characteristic-related works include, for example, Oum et al. [7], who proposed an expected utility-based approach for constructing electricity derivatives with arbitrary nonlinear payoff functions. Recently, there have been pricing methods for electricity derivatives with various granularities and payoffs, such as "cap/floor futures", where the underlying asset is the hourly intraday electricity price, traded on a weekly basis [8], and "day-ahead cap futures" with the day-ahead price as the underlying asset, traded on a daily basis [9].

As for weather derivatives, various studies have been carried out, mainly on pricing methods. There are a wide range of indices that can be used as underlying assets for weather derivatives, such as temperature [10–20], wind [21–24], solar radiation [25], and rainfall [26]; thus, the applicability of weather derivatives has been demonstrated by many researchers. Recently, the research on the investigation of hedging effects has gained attention as well. As an example of such previous studies, Bhattacharya et al. [27] have illustrated the hedging effect of weather derivatives (using heating degree days (HDDs) and cooling degree days (CDDs)) on the profit fluctuations of a solar PV generator using a data-driven approach.

Instead of applying standard derivatives, unique derivatives based on nonparametric regression techniques have been proposed to further improve the hedging effectiveness [28–34]. The approach of those studies is to estimate the nonlinear functions of the optimal payoffs and/or the optimal contract volumes of the derivatives using generalized additive models (GAMs [35,36]). That is, those studies focus on, for example, the fact that price volatility can lead to losses for retail businesses that sell electricity at fixed prices [31] and clarify the importance and effectiveness of strategies to effectively suppress fluctuations in cash flows. Among them, our recent study [33] has demonstrated that derivatives based on temperature and solar radiation are highly effective in hedging the risk of revenue fluctuations for electricity retailers and solar PV generators, and a more recent study [34] has focused on the methodological refinement of the choice of spline basis functions.

In this study, we systematically organize the theoretical aspects of our previous studies [33,34] to develop a unified approach using electricity and weather derivatives/forwards and demonstrate a comprehensive analysis of various types of players. We aim to not only to clarify the applicability of our proposed methods, but also to provide a new and useful perspective on derivative trading schemes involving different electricity utilities and insurance companies. In our empirical analysis, we assume three types of players—electricity retailers, solar PV generators, and thermal power generators—and measure the hedging effects on their cashflow management using electricity and weather derivatives (as well as forward contracts). What is unique about this study is that we deal with "forwards" with linear payoffs as well as "derivatives" with nonlinear payoffs for three different types of electricity businesses and compare the hedging effects (and hedging errors) of both types of hedge instruments from various perspectives. In addition, we apply the methodology of previous studies on daily granular derivative contracts [33,34] to derivatives with hourly granularity payoffs and show that empirical hedging effects are sufficiently high using out-of-sample data despite the high volatility of hourly volume and price data. In this way, this study provides valuable insights into the applicability of our method for high-granularity hedging transactions for distributed power sources and peer-to-peer electricity markets, which are expected to increase soon. More specifically, with the massive introduction of distributed power sources, the number of electricity traders will diversify, and very small businesses and (in some cases) individuals may engage in electricity trading. For such traders, the need to control fine-grained cashflow fluctuation risk is expected to be particularly large, and the hedging method in this study is expected to provide an effective solution to such a need.

Furthermore, the new perspective provided by this study is not limited to the hedging effect of electric utilities as hedgers, but also applies to the underwriting (residual) risk of their counter parties, including insurance companies, as risk takers. That is, our previous studies [33,34] focus on the perspective of improving the hedging effect of the electric utilities (hedgers), while the reality of transactions from the risk taker's perspective (i.e., whether there is a risk taker as a matter of reality, or what circumstances the risks are more likely to be accepted) remains an open question. In this study, we explicitly introduce counter parties of derivative and forward transactions, such as insurance companies, who can profit if a commission is purchased for every transaction. Moreover, their risks may be averaged out by executing derivative contracts with power retailers and generators simultaneously. This is because their cash flow directions may be different, or opposite for the electricity purchase, and the payoffs of derivatives may be canceled out.

Based on the above discussions, this study reveals an interesting empirical result that insurance companies (risk takers) can significantly reduce risk by simultaneously executing individual electricity/weather forwards/derivatives with both generators and retailers, compared to the aggregate risk for separate transactions (i.e., when risk underwriting transactions with both parties are performed independently, or when risks are underwritten by different insurance companies). In other words, by taking a comprehensive view of the entire electricity trading market, including the sellers and buyers of electricity (actuals) and derivatives, and the intermediaries of derivative contracts, this study presents a solution to the problem of improving the efficiency of the entire market trading scheme, including the reality of risk underwriting transactions, which has not been solved by previous studies. Thus, this study provides an ambitious approach to obtain beneficial suggestions on not only the applicability of the methodology but also the further extension of the market model for the practical use of the methodology.

This paper is organized as follows: in Section 2, we introduce the minimum variance hedging problems of cashflow fluctuations for three types of electricity utility players and describe the overview for the market, including the derivative transactions; in Section 3, we construct hedging schemes based on GAMs for given observation data and describe their estimation and test procedures in detail; in Section 4, we perform empirical hedging simulations based on actual data and estimate the optimal payoff functions/coefficients of derivatives/forwards, and conduct an extensive empirical analysis including the hedging effects and accuracy; in Section 5, we illustrate the empirical risk reduction for insurance companies through the simultaneous transactions of derivatives; finally, in Section 6 we provide a comprehensive discussion based on the results of our analysis.

2. Minimum Variance Hedging Problems of Cashflow Fluctuations

In liberalized electricity markets, it is common for electricity retailing companies to purchase spot electricity through the central power exchange and deliver it to their consumers (or demanders). On the other hand, power generation companies place sales orders on power exchange and produce electricity based on the executed volume. In this situation, their profit or loss may depend on the cashflows defined by the product of spot electricity price and volume. In this section, we introduce the minimum variance hedging problems to mitigate cashflow fluctuations for power retailers and generators.

2.1. Minimum Variance Hedging Problem for Power Retailers

Assume that there is a central power exchange that allows power retailers to procure spot electricity every day at every hour. Each power retailer predicts the future demand for end users (i.e., consumers) and places a buy order on the power exchange. Let S_t be the spot electricity price that delivers a fixed amount of electricity at time t for a certain time interval. The cashflow of this transaction is determined by the product of the executed volume V_t and the spot price S_t. Since the retailer needs to equalize the demand and supply every moment, the volume V_t is required to match the electricity demand of end-users; that is, $V_t \equiv V_t^{demand}$, where V_t^{demand} stands for the total demand and the notation $a \equiv b$ is used to denote that a is defined to be b. Note that S_t and V_t are both volatile, and the cashflow determined by the product is extremely volatile; that is, the cashflow volatility, denoted by its variance, $\text{Var}[V_t S_t]$, is supposed to be quite large. In this study, we aim to reduce the risk of the cashflow fluctuation by using financial instruments such as derivatives or forwards.

We will now formulate the problem. Assume that there exist underlying assets or indexes observed at time $t \in \{t_0, \ldots, t_1\}$, where $t_0, \ldots, t_1$ are contract periods of interest corresponding to electricity delivery. Note that potential candidates for such variables are weather indexes for weather derivatives or spot price S_t for electricity derivatives. Let W_t be the value of weather indexes observed at time t. Note that W_t may be a multidimensional vector; that is, multivariate weather derivatives/forwards can be defined using vector notation. In addition, we suppose that these derivative contracts are cash settlement contracts without risk premiums; that is, the introduction of derivative or forward contracts will not change the expected total cashflow (or, equivalently, the mean value of total cashflow). A general formulation of variance minimization is given as follows:

Find optimal derivative contracts on S_t and W_t to minimize

$$\mathrm{Var}\left[V_t S_t - payoff(S_t, W_t)\right]$$
$$\text{s.t. } \overline{payoff(S_t, W_t)} = 0 \tag{1}$$

where $\mathrm{Var}[\cdot]$ stands for variance and $\bar{a}$ is the mean value (or expected value) of a.

In (1), $payoff(S_t, W_t)$ is defined by the payoff functions of the underlying variables, (S_t, W_t), which may depend on time t. Moreover, because the volume V_t reflects consumer demand, which largely depends on temperature, we select $W_t \equiv T_t$, where T_t is the value of temperature at time t in the demand area. In this study, we focus on synthesizing separate payoff functions only; that is, $payoff(S_t, W_t)$ is the sum of single variate functions satisfying

$$payoff(S_t, W_t) \, f(S_t) + g(W_t) \tag{2}$$

In the case of forward contracts, the payoff functions are supposed to be linear on S_t and W_t but we assume that the coefficients depend on time t as follows:

$$payoff(S_t, W_t)\delta(t)\left\{S_t - F_t^S\right\} + \gamma(t)\left\{W_t - F_t^W\right\} \tag{3}$$

where $\delta(t)$ and $\gamma(t)$ are the numbers of forward contracts, and F_t^S and F_t^W are the forward prices of spot electricity and weather indexes, respectively. Note that forward prices need to be specified for computing forward cashflows, but as far as hedge errors are concerned, as in our analysis, it is not necessarily to specify the forward prices explicitly. In our formulation using GAM, the forward prices are incorporated in the time trend term, which will be estimated separately.

In this study, we construct optimal payoff functions or optimal positions of forward contracts based on the historical data of variables in (1) using statistical estimation techniques. To this end, we split the data period into in-sample parameter estimation period and out-of-sample performance evaluation period; that is, the entire data period $t \in \{0, \ldots, t_1\}$ will be split into $t \in \{0, \ldots, t_0 - 1\}$ and $t \in \{t_0, \ldots, t_1\}$, respectively. Note that when statistical estimation techniques are applied for problem (1), $\mathrm{Var}[\cdot]$ and the overline notation (e.g., $\overline{S_t}$) may be interpreted as sample variance and mean, respectively.

2.2. Minimum Variance Hedging Problem for Solar PV Generators

The minimum variance hedging problem (1) defined in the previous subsection is for power retailers, but in fact it can be said that it is the hedging problem of a load aggregator who procures the total demand on behalf of a group of power retailers in the same area. There, individual retailers place buy orders to the load aggregator which compiles all orders to execute them in the power exchange market. In this case, the prediction errors of consumer demand for individual retailers may be averaged out such that the gap between the ordered volume and actual consumption decreases. Otherwise, retailers may suffer from the imbalance risk as well, and we may need other instruments for the hedge such as prediction error derivatives [31].

A similar argument may be applied for a group of solar power generators, where the percentage of solar power generation is increasing rapidly but the power output largely depends on solar radiation with uncertainty. Here, we consider an aggregator of solar PV generators in the same area and assume that the aggregator complies with all the sales orders from individual PV generators. Then, the total prediction error may be averaged out and the aggregator may focus on the risks of cashflow fluctuation. For the solar PV aggregator, the cashflow at each period is defined by the product of spot price S_t and the total volume (corresponding to the total PV output), and a similar hedging problem may be formulated using (1), in which the volume V_t is now defined by the total PV output as $V_t \equiv V_t^{solar}$. In this case, an appropriate variable associated with V_t^{solar} is the solar radiation; thus, we select $W_t \equiv R_t$, where R_t is the value of the solar radiation index at time t in the same area.

2.3. Minimum Variance Hedging Problem for Thermal Power Generators

Although we used the same notation to define the cashflows for the load aggregator and the solar PV aggregator, the directions of cashflows are opposite. That is, for the load aggregator, $S_t V_t$ provides the procurement cost corresponding to the cash outflow, whereas for the PV aggregator, it provides the sales revenue corresponding to the cash inflow. In fact, the load and the PV aggregators may become counterparties to each other; that is, the load aggregator can purchase the PV output from the PV aggregator and deliver it to end-users through the power transmission and distribution company.

However, direct transactions between retailers and solar PV generators are generally difficult because demand and supply volumes are volatile and change over time. Hence, we need to regulate the supply generations of thermal power in the electricity market. In this study, we introduce thermal generators and consider their hedging problem.

To simplify the discussion, assume that there is a supply aggregator that compiles all the generation stacks from thermal generators. We define the minimum variance hedging problem for the supply aggregator using (1), where V_t represents the total supply volume of thermal generators; that is, $V_t \equiv V_t^{thermal}$. In the electricity market, the volume of thermal generators $V_t^{thermal}$ should be balanced to match consumers' demand minus the renewable energy output. Although renewable energy power includes other resources such as wind and biomass, we only focus on the effect of solar power. This is because, in the Japanese electricity market tested in this study, the ratio of solar power introduction is much higher than other renewable power resources, except for hydro energy. In the minimum variance hedging problem for thermal generators, we set $V_t \equiv V_t^{thermal}$ in (1) and selected W_t as the temperature and solar radiation; that is, $W_t \equiv [T_t, \ R_t]^T$. The payoffs for the derivatives and forwards are defined as

$$payoff(S_t, W_t) = f(S_t) + g(T_t) + h(R_t) \tag{4}$$

and

$$payoff(S_t, W_t) = \delta(t)\left\{S_t - F_t^S\right\} + \gamma^T(t)\left\{W_t - F_t^W\right\} \tag{5}$$

respectively, where h is another payoff function of R_t, and F_t^W and $\gamma(t)$ in (3) are now column vectors.

2.4. Electricity Transaction Market Including Derivatives

In summary, we consider the following three types of hedging problems:

1. Power retailers' hedge. $V_t \equiv V_t^{demand}$: Total demand, $W_t \equiv T_t$: Temperature;
2. Solar PV generators' hedge. $V_t \equiv V_t^{solar}$: Total PV generation, $W_t \equiv R_t$: Solar radiation;
3. Thermal generators' hedge. $V_t \equiv V_t^{thermal}$: Total thermal power generation, $W_t \equiv [T_t, \ R_t]^T$: Temperature and solar radiation.

Figure 1 depicts the electricity and derivatives transactions considered in this study. We assume that there is a derivatives market that enables power generators and retailers to execute derivative contracts with arbitrary payoff functions. The power generators and retailers first solve the hedging problems with an appropriate choice of variables, as shown in items 1–3. Then, find a counterparty who agrees with the transaction to execute the derivative contracts. Potential candidates of such counterparties are insurance companies, and we assume that insurance companies can execute the electricity and weather derivative transactions with any payoff functions. Note that, since the net payoffs of derivatives are supposed to be zero on average, the insurance companies can make a profit if a small commission or transaction fee is purchased for each transaction. Furthermore, since the insurance companies can make transactions with power retailers and generators simultaneously, who may cause cashflows in opposite directions, the insurance companies' risks may be reduced (see Section 5).

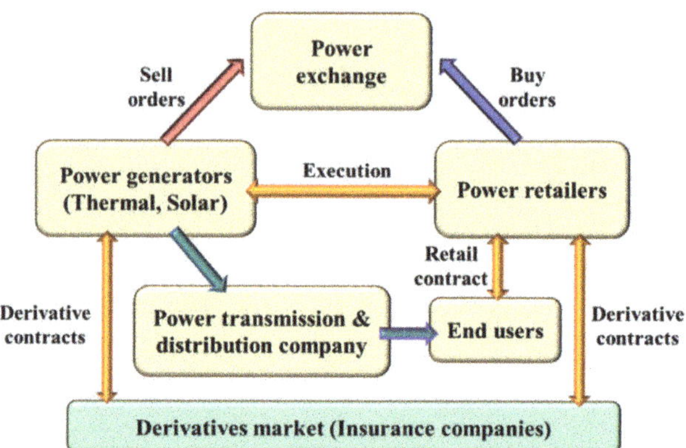

Figure 1. Transaction model of electricity and derivatives.

In the case of the forward contracts in (3), it may be possible to introduce market makers (e.g., insurance companies or financial institutions) who provide fair bid and ask prices and accept sell and buy orders from power generators and retailers. Note that market makers can make a profit from the bid–ask spread, whose sizes are limited by regulations in the market. If the numbers of short and long positions are the same for the same product, the market makers do not have any risks. Therefore, the balance between the long and short positions is important for estimating the market maker's risk. In this study, we will not analyze such balance risk between long and short positions for forward contracts, but only demonstrate the risk reduction of insurance companies by simultaneously making transactions of derivative contracts with power retailers and generators. A detailed analysis of the balance risk for forward transactions will be left for future work.

Note that, since retailers will typically have already made electricity delivery contracts for any consumed volume over longer durations, the cashflow risk with high volatility in some hours would be a serious concern, as discussed in [9]. In addition, solar power generators may be concerned about a significant drop or surplus in output because of the weather, and so may suffer from unexpected high or low prices as well as volume fluctuation for the power output. Therefore, a selective hedge against the price and volume fluctuation in particular hours would be desirable. Considering the above, we will represent hourly and daily periods using different indexes and define the variables accordingly after this section. In that case, the subscript t will be used for a day index whereas m will be used for an hour index. For example, the spot price on day t at hour m will be denoted as $S_{t,m}$ in Section 3 and thereafter.

3. Estimation and Test Procedures

In this section, we will explain the statistical estimation models to solve our hedging problems. Since the basic idea is already explained in our previous literature [28,29,33], we will briefly summarize our hedging models.

3.1. Variables Used for Hedging Problems

We will express the variables using the day and hour indexes. Let $t \in \{0, \ldots, t_0 - 1\}$ be the observation data period on a daily basis and $m \in \{0, \ldots, 23\}$ be an hour index.

Then, the spot price delivering 1 kWh of electricity from hour m (to hour $m + 1$) on day t is denoted by $S_{t,m}$. Furthermore, the volume is defined $V_{t,m}$, but depending on the situation we may specify the category of the volume using a superscript such as $V_{t,m}^{demand}$, $V_{t,m}^{solar}$ and $V_{t,m}^{thermal}$, respectively, for the total volume of retailers (i.e., consumers' demand), the total PV generation, and the total thermal generation. In this study, we construct hedging models for each m and demonstrate the hedge effects.

For the weather index data $W_{t,m}$, we used hourly temperature and solar radiation data, denoted by $T_{t,m}$ and $R_{t,m}$. Note that the choice of weather index data is different for power retailers, solar PV generators, and thermal generators, and is given by $W_{t,m} = W_{t,m}^{retail} \equiv T_{t,m}, W_{t,m} = W_{t,m}^{solar} \equiv R_{t,m}$, and $W_{t,m} = W_{t,m}^{thermal} \equiv [T_{t,m}, R_{t,m}]^T$, respectively. In addition, when weather data are available at multiple points in one region, we compute a local demand weighted average for temperature and an installed capacity of local PV weighted average for radiation, respectively, and create the temperature and radiation indexes.

3.2. Minimum Variance Hedging Using Derivatives

Consider the minimum variance hedging problem with the payoff functions of the derivatives in (2). To find the optimal payoff functions, we apply GAM for each $m \in \{0, \ldots, 23\}$ as follows:

$$V_{t,m}S_{t,m} = f_m(S_{t,m}) + g_m(W_{t,m}) + Calendar_m(t) + \epsilon_{t,m} \tag{6}$$

where f_m and g_m are smoothing spline functions to be estimated in GAM and $\epsilon_{t,m}$ is a residual satisfying zero mean condition, $\overline{\epsilon_{t,m}} = 0$.

In (6), $Calendar_m(t)$ contains day of week, long-term, and seasonal trends as

$$Calendar_m(t) = \beta_1 Mon_t + \cdots + \beta_6 Sat_t + \beta_7 Holidays_t + Seasonal(t) + Longterm(t) \tag{7}$$

where $Mon_t, \ldots, Sat_t$, and $Holidays_t$ are day of week and holiday dummy variables that take $Mon_t = 1$ if the day of t is Monday or $Mon_t = 0$ otherwise, and so on. $Seasonal(t)$ denotes a yearly cyclical smoothing spline function and reflects the seasonal trend in $V_{t,m}S_{t,m}$, whereas $Longterm(t)$ is a smoothing spline function (e.g., a cubic spline function) of the day variable t. These functions can be estimated using the day dummy variables. Note that the coefficients and spline functions in (7) are different by hour m, but we omit specifying this dependence for brevity. In addition, because the solar power may be independent of the day of the week and holidays, we assume that $\beta_i \equiv 0, \forall i = 1, \ldots, 7$ for solar PV generations.

For each m, GAMs can be estimated by minimizing the following penalized residual sum of squares (PRSS):

$$\text{PRSS}: \sum_{t=1}^{N} \{\epsilon_{t,m}\}^2 + J(\lambda), \quad \lambda = [\lambda_1, \ldots, \lambda_j]^T \in \Re^j \tag{8}$$

where N is the number of observations for each variable. In (8), the first term is the sum of squares for residuals, and the second term provides the smoothness constraint on spline functions with smoothing parameter vector $\lambda \in \Re^j$, where j is the number of smoothing spline functions in GAMs, and the larger the λ_i, the smoother the ith spline function. The smoothing parameter vector λ needs to be fixed a priori, but an optimal λ may be searched based on the generalized cross-validation criteria, as shown in [35].

Then, the PRSS is minimized over smoothing spline functions and coefficients given λ to construct the GAMs.

From (6), we have

$$\text{Var}\left[V_{t,m}S_{t,m} - (f_m(S_{t,m}) + g_m(W_{t,m}) + Calendar_m(t))\right] = \text{Var}[\epsilon_{t,m}] \tag{9}$$

Hence, minimizing the sample variance of $\epsilon_{t,m}$ with smoothing conditions may be considered as PRSS minimization. Note that we can add the constraints $\overline{f_m(S_{t,m})} = 0$ and $\overline{g_m(W_{t,m})} = 0$ when solving the PRSS so that $\overline{payoff(S_{t,m}, W_{t,m})} = 0$ is satisfied. In this case, we have $\overline{V_{t,m}S_{t,m} - Calendar_m(t)} = 0$, and $Calendar_m(t)$ may be considered as the time trend, such as day, seasonal, and long-term contained in $V_{t,m}S_{t,m}$. In practice, the deterministic term $Calendar_m(t)$ may be replicated by buying a bound that pays off the same amount of $Calendar_m(t)$ at the settlement period. Consequently, we conclude that the minimum variance hedging problem (1) with (2) can be formulated using GAM (6).

3.3. Minimum Variance Hedging Using Forwards

In the previous subsection, we explained that the optimal payoff functions of electricity and weather derivatives may be found by applying GAM. Here, we show that the minimum variance hedging problem (1) with (3), in which the payoff is defined by time-dependent forward positions, may also be formulated using GAM.

Consider the following GAM with cross variables, $S_{t,m}$ and $W_{t,m}$:

$$V_{t,m}S_{t,m} = \delta_m(t)S_{t,m} + \gamma_m(t)W_{t,m} + Calendar_m(t) + \epsilon_{t,m} \tag{10}$$

where δ_m and γ_m are smoothing spline functions to be estimated and $\epsilon_{t,m}$ is a residual satisfying zero mean condition, $\overline{\epsilon_{t,m}} = 0$. The smoothing spline functions, $\delta_m(t)$ and $\gamma_m(t)$, are given by a yearly cyclical smoothing spline function like $Seasonal(t)$ in (7). Note that in the case of the solar PV generators hedging problem, a long-term trend (like $Longterm(t)$) may be added.

In GAM (10), forward prices, $F^S_{t,m}$ and $F^W_{t,m}$, are not specified explicitly, but we can show that $F^S_{t,m}$ and $F^W_{t,m}$ may be extracted from $Calendar_m(t)$ by decomposing as

$$Calendar_m(t) \equiv -\delta_m(t)F^S_{t,m} - \gamma_m(t)F^W_{t,m} + d_m(t) \tag{11}$$

where $F^S_{t,m}$ and $F^W_{t,m}$ are forward prices satisfying $\overline{\delta_m(t)\left\{S_t - F^S_{t,m}\right\}} = 0$ and $\overline{\gamma_m(t)\left\{W_t - F^W_{t,m}\right\}} = 0$, and $d_m(t)$ is an additional term that may be calculated by (11) after $F^S_{t,m}$ and $F^W_{t,m}$ are found. The calculation of $F^S_{t,m}$ and $F^W_{t,m}$ requires solving additional regression problems, but as far as hedge errors are concerned, we do not have to explicitly specify $F^S_{t,m}$ and $F^W_{t,m}$. Then, we see that minimizing the sample variance of $\epsilon_{t,m}$ with smoothing conditions may be considered as the minimum variance hedging problem (1) with (3), that is,

$$\mathrm{Var}\left[V_{t,m}S_{t,m} - \left(\delta(t)\left\{S_{t,m} - F^S_{t,m}\right\} + \gamma(t)\left\{W_{t,m} - F^W_{t,m}\right\} + d_m(t)\right)\right] = \mathrm{Var}[\epsilon_{t,m}] \tag{12}$$

3.4. Empirical Test Procedure

As explained in the end of Section 2.1, our empirical test consists of parameter estimation and performance verification based on in-sample and out-of-sample data, respectively. Assume that the entire data period is given by $t \in \{1, \ldots, t_1\}$ in which the hourly data are also available. Our empirical test procedure is as follows:

Step 1. Given observation data of $V_{t,m}$, $S_{t,m}$ and $W_{t,m}$, split the data period into $t \in \{1, \ldots, t_0 - 1\}$ and $t \in \{t_0, \ldots, t_1\}$;

Step 2. For each hourly period m, apply GAM (6) (or GAM (10)) to find optimal smooth functions, f_m and g_m (or δ_m and γ_m), and calendar trend function, $Calendar_m$;

Step 3. For the optimal smooth functions and $Calendar_m$ obtained in Step 2, compute the out-of-sample hedge errors by

$$\epsilon^{out}_{t,m} V_{t,m}S_{t,m} - (f_m(S_{t,m}) + g_m(W_{t,m}) + Calendar_m(t)), \ t \in \{t_0, \ldots, t_1\}$$

$$\text{or} \ \ \epsilon^{out}_{t,m} V_{t,m}S_{t,m} - (\delta_m(t)S_{t,m} + \gamma_m(t)W_{t,m} + Calendar_m(t)), \ t \in \{t_0, \ldots, t_1\} \tag{13}$$

Step 4. For the out-of-sample data of $t \in \{t_0, \ldots, t_1\}$, evaluate the out-of-sample hedge performance using the following variance reduction rate (VRR):

$$\frac{\text{Var}\left[\epsilon_{t,m}^{out}\right]}{\text{Var}[V_{t,m}S_{t,m}]} \tag{14}$$

and the normalized mean absolute error (NMAE),

$$\frac{\overline{\left|\epsilon_{t,m}^{out}\right|}}{\overline{|V_{t,m}S_{t,m}|}} \tag{15}$$

4. Empirical Hedge Simulations

In this section, we conduct empirical simulations of our hedging problems and demonstrate hedge performance using Japanese electricity market and meteorological data. (In this study, we estimate GAMs using R 4.0.5 (https://www.R-project.org/, accessed on 27 October 2021) and the package mgcv [37] (https://cran.r-project.org/web/packages/mgcv/index.html, accessed on 27 October 2021) to obtain the series of smoothing spline functions, wherein the smoothing parameter is calculated by the generalized cross-validation criterion. All figures are plotted using MATLAB 2021a (MathWorks, Inc., Natick, MA, USA).)

4.1. Data

We use the electricity price, volume, and weather data observed in the Tokyo area, Japan. The data period is chosen from 1 April 2016 (when the Japanese electricity market was fully liberalized) to 31 December 2019, in which we set the first three years (from 1 April 2016 to 31 March 2019) as the in-sample estimation period and the remaining 275 days (from 1 April 2019 to 31 December 2019) was reserved for the out-of-sample performance evaluation.

The following is the list of data used in our analysis:

1. Electricity price $S_{t,m}$ [Yen/kWh]: JEPX spot price in Tokyo area (JEPX Tokyo area price) delivering 1 kWh of electricity from hours m to $m + 1$ (downloaded from http://www.jepx.org/market/index.html, accessed on 27 October 2021); since the length of delivery is 30 min for JEPX spot prices, we compute the average of two consecutive prices per hour; for example, we compute the average of 1:00–1:30 p.m. and 1:30–2:00 p.m. delivery prices for the 1:00–2:00 p.m. price;

2. Volume $V_{t,m}$ [kWh]: Hourly realized demand and supply data in the Tokyo area, including the total demand ($V_{t,m}^{demand}$), the total solar power generation ($V_{t,m}^{solar}$), and the total thermal power generation ($V_{t,m}^{thermal}$) between hours m and $m + 1$ on day t (downloaded from https://www.tepco.co.jp/forecast/html/area_data-j.html, accessed on 27 October 2021);

3. Temperature $T_{t,m}$ [°C]: We use hourly realized temperature data on day t in the Tokyo area (downloaded from https://www.data.jma.go.jp/gmd/risk/obsdl/, accessed on 27 October 2021). A temperature index is constructed using the electricity consumption-based weighted average of nine observation points (we used the yearly local electricity consumption data in Tokyo area as of the end of March 2016, obtained from https://www.tepco.co.jp/corporateinfo/illustrated/business/business-scale-area-j.html, accessed on 27 October 2021);

4. Solar radiation $R_{t,m}$ [MJ/m^2]: We use hourly realized solar radiation data on day t in the Tokyo area (downloaded from https://www.data.jma.go.jp/gmd/risk/obsdl/, accessed on 27 October 2021). A solar radiation index is constructed using an installed capacity of local PV weighted average of seven observation points (we used the installation capacity data in Tokyo area as of the end of March 2019 (correspond-

ing to the end of in-sample period), obtained from https://www.fit-portal.go.jp/PublicInfoSummary, accessed on 27 October 2021).

Figure 2 shows the daily electricity price in the entire period, where the blue line is the fluctuation of the daily spot price (i.e., the average of 30 min prices per day) and the red line is the 60 days moving average. Figure 3 provides the volume data of solar PV and thermal generations in total, as well as the total supply (which is the same as the total demand) in the Tokyo area. Figure 4 shows the temperature index in the Tokyo area, where the average temperature for 24 h per day and its 60 days moving average are plotted as the blue and red lines, respectively. Similarly, Figure 5 provides the solar radiation index in the Tokyo area. Note that the temperature and radiation indexes are constructed by taking the weighted averages of several observation points by local electricity consumption and installation capacities of local PV generation in Tokyo, respectively. Furthermore, note that these figures are plotted daily by taking averages, but we construct hedging models based on hourly data, as explained in the previous section.

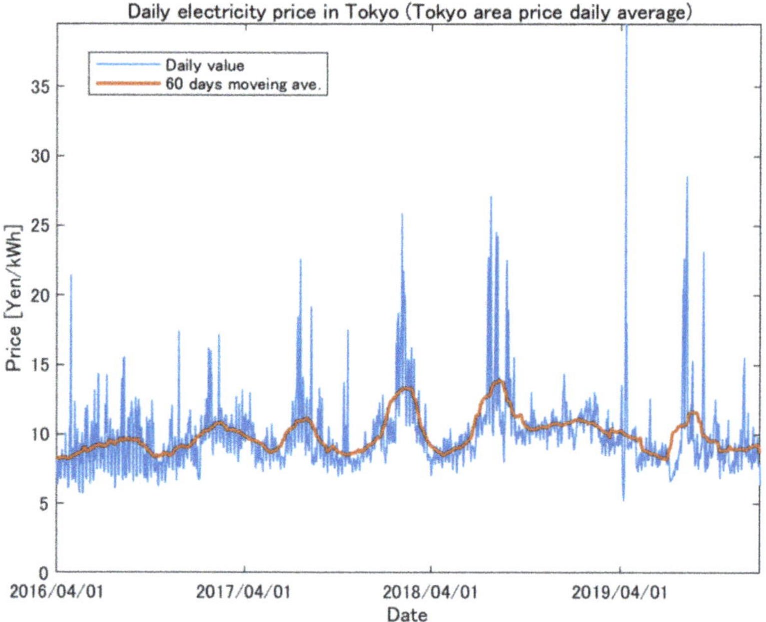

Figure 2. Daily average price in Tokyo and its 60 days moving average in the period of 1 April 2016 to 31 December 2019.

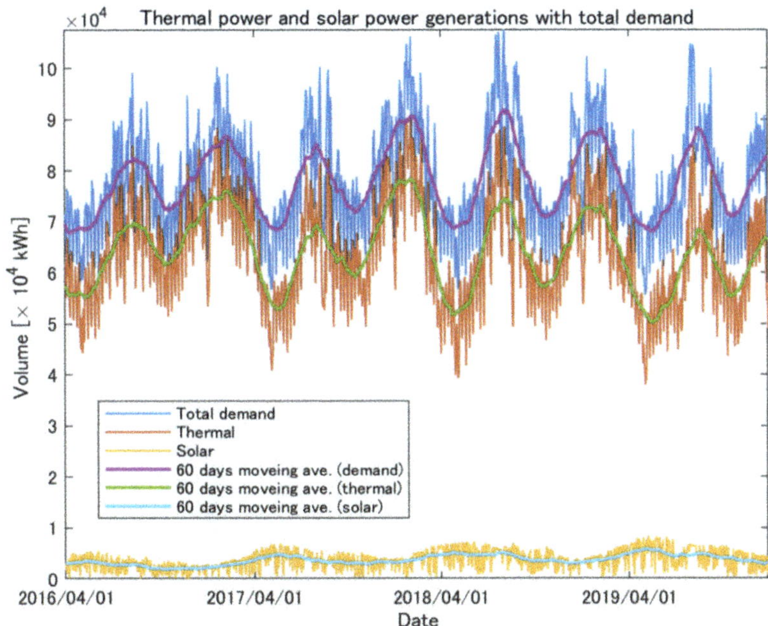

Figure 3. Daily fluctuations of thermal power and solar power generations, total demand, and their 60 days moving averages in the period of 1 April 2016 to 31 December 2019.

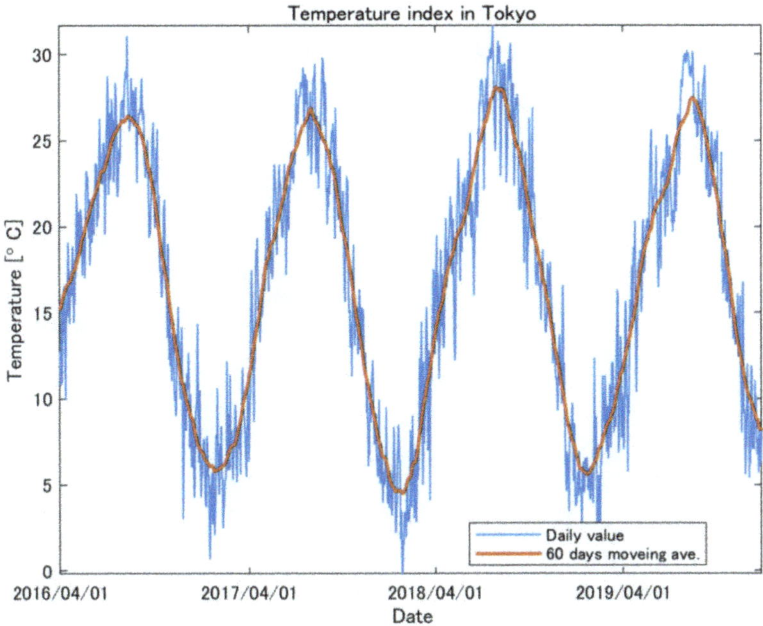

Figure 4. Daily average of temperature index in Tokyo and its 60 days moving average in the period of 1 April 2016 to 31 December 2019.

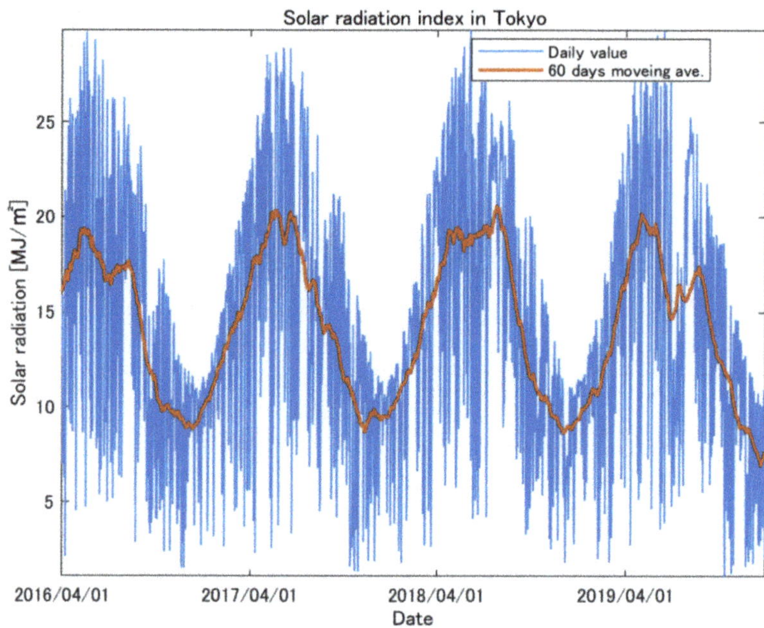

Figure 5. Daily average of solar radiation index in Tokyo and its 60 days moving average in the period of 1 April 2016 to 31 December 2019.

4.2. Estimation Result for Power Retailers' Hedges

First, we solved the minimum variance hedging problem for power retailers (or equivalently, the hedging problem of the load retailer) by applying GAM (6) with $V_{t,m} \equiv V_{t,m}^{demand}$ and $W_{t,m} \equiv T_{t,m}$. We estimated the optimal spline functions and other required parameters in (6) based on the in-sample data. Then, we computed the out-of-sample hedge errors based on Equation (13) to evaluate the hedge performance in terms of VRR and NMAE in (14) and (15), respectively.

Panels (a) and (b) of Figure 6 represent the payoff functions estimated by applying GAM (6), where the payoff functions of electricity derivatives f_m for $m = 2, 6, 10, 14, 18, 22$ are plotted in panel (a) among 24 estimated functions and those of temperature derivatives, g_m, are shown in panel (b). These payoff functions satisfy $\overline{f_m(S_{t,m})} = 0$ and $\overline{g_m(W_{t,m})} = 0$ given the parameter estimation period and may provide negative values of the payoffs. We see that the payoff functions for electricity derivatives increase monotonically, whereas those of temperature derivatives increase with a larger temperature and a smaller temperature for both sides. The latter is interpreted as the effects of temperature on electricity demand. For example, the payoff function at 2 p.m. increases rapidly when the temperature is higher than 25 °C, reflecting the electricity consumption in summer for the usage of air conditioners. In addition, in the morning and the evening (e.g., 10 a.m. and 6 p.m.), the payoff functions increase rapidly when the temperature is below 10 °C mainly from the electricity consumption in winter.

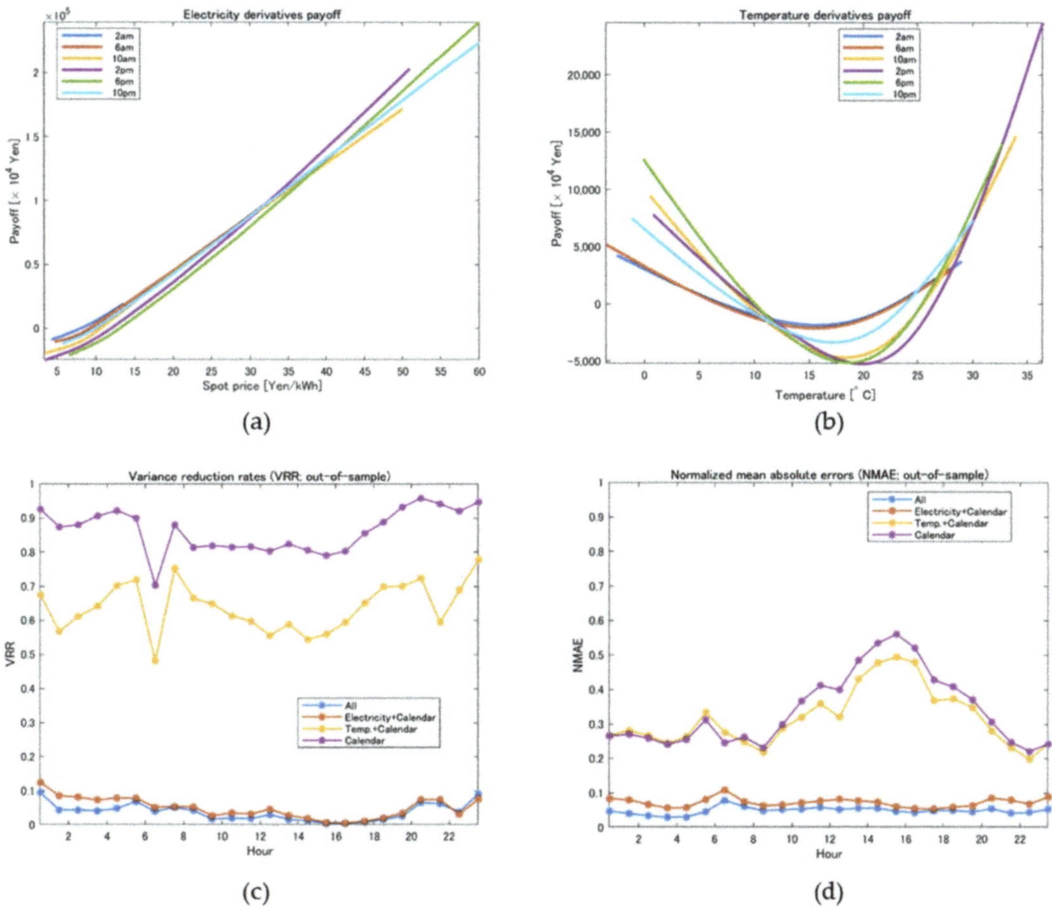

Figure 6. Results of minimum variance hedging with derivatives for electricity retailers' cash flows, $V_{t,m}^{demand} S_{t,m}$, based on empirical data: (**a**) optimal payoff functions of electricity derivatives; (**b**) optimal payoff functions of temperature derivatives; (**c**) out-of-sample VRR for each hour; (**d**) out-of-sample NMAE for each hour.

Panels (c) and (d) represent the out-of-sample hedge performance of our methodology, which provide VRRs and NMAEs computed by Equations (14) and (15), respectively. Note that both blue lines at the bottom of the figures are those obtained by using GAM (6) for different values of $m = 0, 1, \ldots, 23$, whereas other lines are obtained by applying GAMs with f_m (electricity derivative) and $Calendar_m$ only, g_m (temperature derivative) and $Calendar_m$ only, and $Calendar_m$ only, respectively. These lines are plotted as red, yellow, and purple lines, respectively, in panels (c) and (d). Comparing the purple and red lines, we see that the hedge performance is improved significantly by incorporating electricity derivatives. Then, the VRRs and NMAEs are further improved by adding temperature derivatives.

Furthermore, we solved the minimum variance hedging problem (1) with (3) to find the optimal coefficients of forward contracts, δ_m and γ_m, by applying GAM (10) based on in-sample data. Panels (a) and (b) of Figure 7 provide the estimation results, where the estimated values of δ_m and γ_m are plotted, providing the coefficients of electricity forwards and temperature forwards, respectively. The dates of the in-sample and out-of-sample periods are assigned on the horizontal axes instead of the day and cyclical dummy variables used for the estimation of GAM (10). Since we assumed that the in-sample data period was until 31 March 2019, the estimated functions after 1 April 2019 provide the predicted values of the coefficients. We see that the coefficients of electricity forwards have two peaks in a year, which reflect the demand peaks in summer and winter.

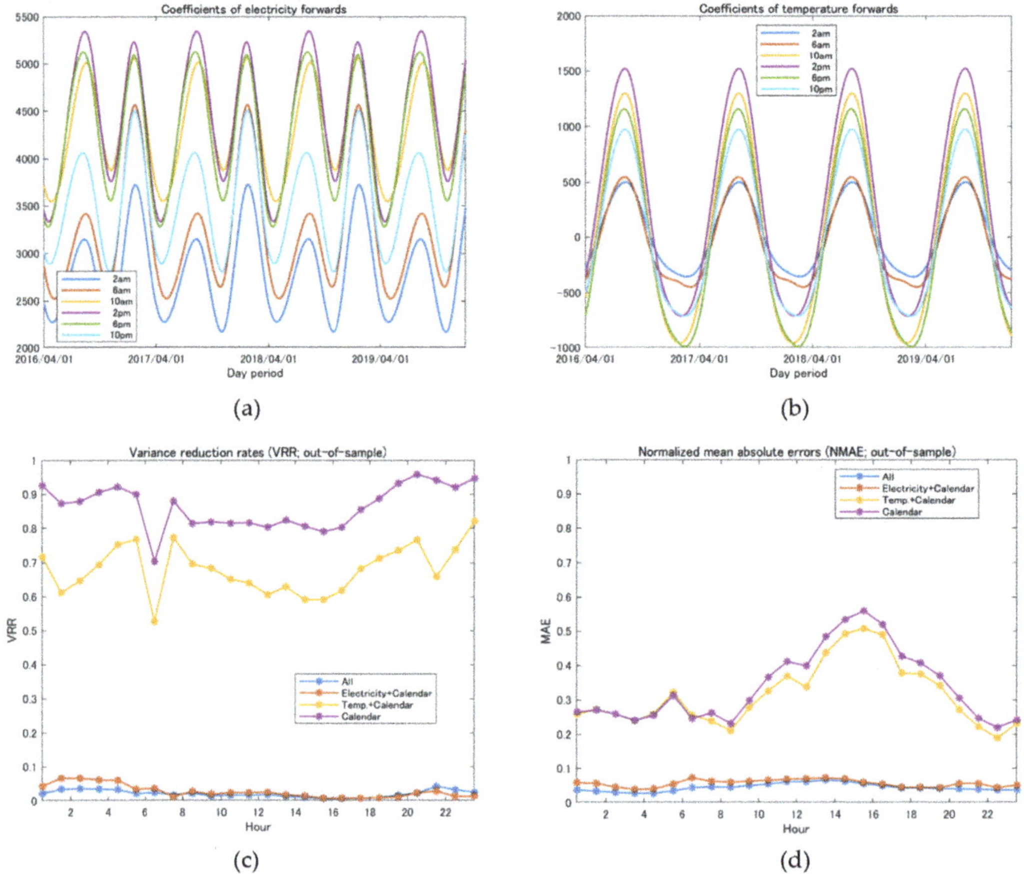

Figure 7. Results of minimum variance hedging with forwards for electricity retailers' cash flows, $V_{t,m}^{demand} S_{t,m}$, based on empirical data: (**a**) coefficients of electricity forwards; (**b**) coefficients of temperature forwards; (**c**) out-of-sample VRR for each hour; (**d**) out-of-sample NMAE for each hour.

The coefficients of temperatures reflect the effect of temperature on demand. For example, in summer the demand has a positive correlation with temperature, whereas in winter the correlation becomes negative so that the demand increases as the temperature decreases. Panels (c) and (d) show out-of-sample VRRs and NMAEs. Like panels (c) and (d) of Figure 6, we see that the hedge performance is improved significantly by incorporating electricity forwards, which is further improved by adding temperature forwards.

4.3. Estimation Results for Solar PV Generators' Hedges

Next, we demonstrate our empirical results for hedging problems with solar PV generations. To the end, we applied GAMs (6) and (10) to solve the minimum variance hedging problems with $V_{t,m} \equiv V_{t,m}^{solar}$ and $W_{t,m} \equiv R_{t,m}$, as with the previous subsection, but the hour index m is restricted to the range of $m = 8, \ldots, 15$ and the solar PV generations from 8 a.m. to 4 p.m. are considered. We estimated the optimal spline functions and other required parameters in (6) and (10) based on in-sample data and computed out-of-sample hedge errors.

Figures 8 and 9 present the empirical results. Panels (a) and (b) in Figure 8 represent the payoff functions estimated by applying GAM (6), where the payoff functions of electricity derivatives and radiation derivatives, f_m and g_m, for $m = 8, \ldots, 15$ are plotted, respectively. These payoff functions satisfy $\overline{f_m(S_{t,m})} = 0$ and $= 0$ given the parameter estimation period and may provide negative values of the payoffs. We see that both the payoff functions for electricity and radiation derivatives increase monotonically, incorporating the effects of electricity price and solar PV generation on the cashflow. Panels (c) and (d) in Figure 8 provide out-of-sample VRRs and NMAEs, respectively, which were computed by applying Equations (14) and (15) based on out-of-sample data. Similar to panels (c) and (d) in Figure 6, the blue lines denote VRRs and NMAEs obtained using all the terms in GAM (6), whereas other lines were obtained with f_m (electricity derivative), $Calendar_m$ only; g_m (radiation derivative), $Calendar_m$ only; and $Calendar_m$ only. Although both VRRs and NMAEs were not improved significantly by electricity derivatives compared to Figure 6, we see that the combinations of electricity and radiation derivatives largely improved VRRs and NMAEs. Thus, we conclude that radiation derivatives are effective for hedging problems.

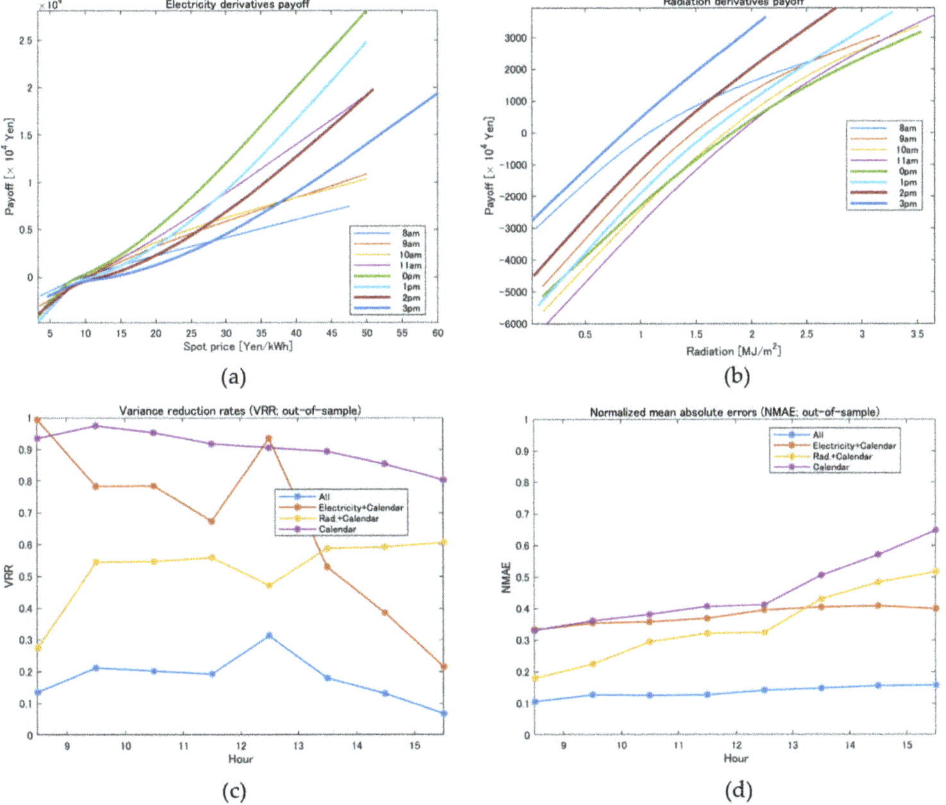

Figure 8. Results of minimum variance hedging with derivatives for solar PV generators' cash flows, $V_{t,m}^{solar} S_{t,m}$, based on empirical data: (**a**) optimal payoff functions of electricity derivatives; (**b**) optimal payoff functions of solar radiation derivatives; (**c**) out-of-sample VRR for each hour; (**d**) out-of-sample NMAE for each hour.

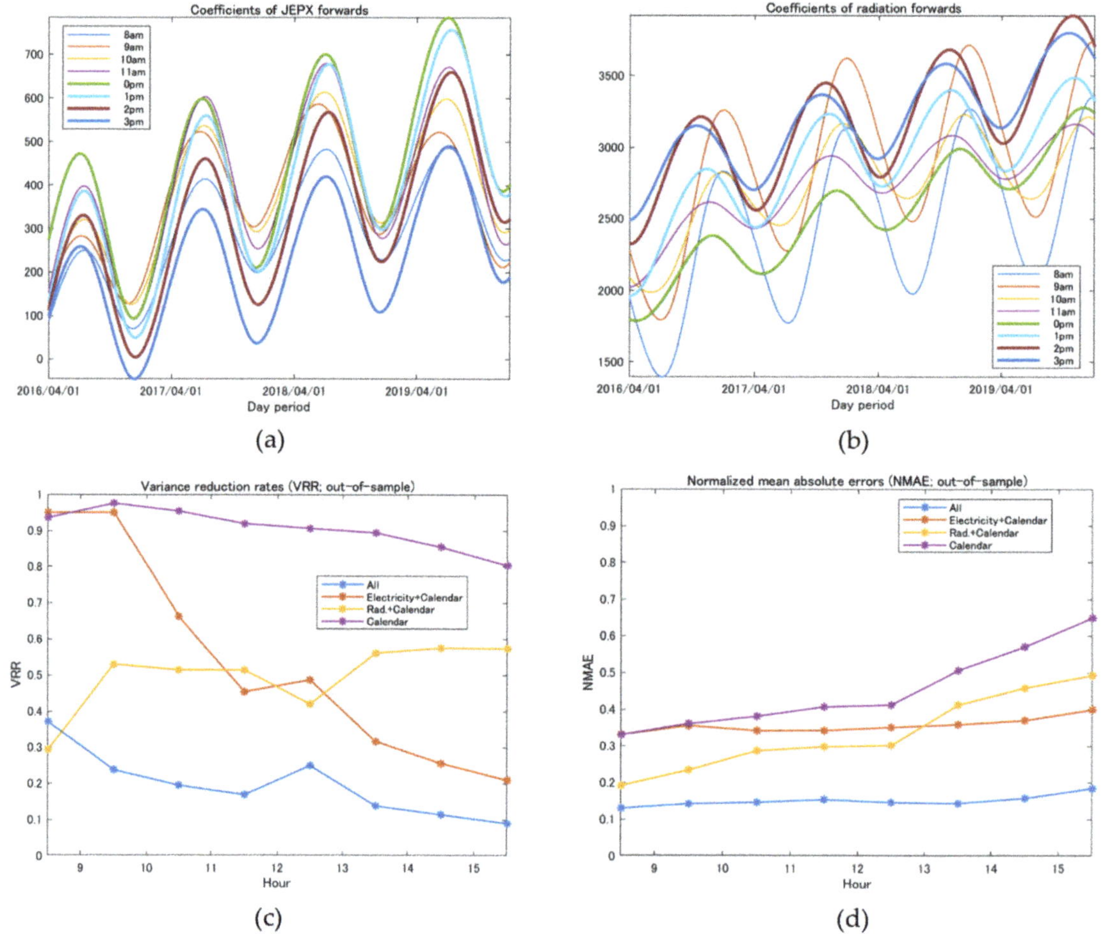

Figure 9. Results of minimum variance hedging with forwards for solar PV generators' cash flows, $V_{t,m}^{solar} S_{t,m}$, based on empirical data: (**a**) coefficients of electricity forwards; (**b**) coefficients of solar radiation forwards; (**c**) out-of-sample VRR for each hour; (**d**) out-of-sample NMAE for each hour.

Panels (a) and (b) of Figure 9 provide the estimated values of δ_m and γ_m corresponding to the coefficients of electricity forwards and radiation forwards, respectively. Like panels (a) and (b) of Figure 7, the day dummy variables are replaced by the dates of the in-sample and out-of-sample periods. In these figures, we see that both coefficients have increasing trends, which incorporate the increase in total PV generation in the Tokyo area. Furthermore, the periodicity of these coefficients reflects the seasonality of solar radiation in a year. Panels (c) and (d) show out-of-sample VRRs and NMAEs, like Figure 8. From these figures, we see that the hedge performance was improved significantly by incorporating both electricity and radiation forwards.

4.4. Estimation Results for Thermal Generators' Hedges

Finally, we present our empirical simulation results for hedging problems with thermal generations. To this end, we applied the following GAMs with $V_{t,m} \equiv V_{t,m}^{thermal}$ and $W_{t,m} \equiv [T_{t,m}, R_{t,m}]^T$, respectively, for constructing derivatives and forwards:

$$V_{t,m} S_{t,m} = f_m(S_{t,m}) + g_m(T_t) + h_m(R_t) + Calendar_m(t) + \epsilon_{t,m} \tag{16}$$

$$V_{t,m} S_{t,m} = \delta_m(t) S_{t,m} + \gamma_m^{temp}(t) T_{t,m} + \gamma_m^{rad}(t) R_{t,m} + Calendar_m(t) + \epsilon_{t,m} \tag{17}$$

where f_m, g_m and h_m in (16) are smoothing spline functions, δ_m, γ_m^{temp} and γ_m^{rad} in (17) are cyclic spline functions, and $\epsilon_{t,m}$ is a residual term satisfying $\overline{\epsilon_{t,m}} = 0$. Note that we used a separate notation h_m for a function of R_t in (16) to emphasize that g_m and h_m are individual single variate functions. We estimated optimal spline functions and other required parameters in (16) and (17) based on in-sample data and computed out-of-sample hedge errors like those in the previous subsections.

Panels (a) and (b) in Figure 10 represent the estimated payoff functions, f_m and g_m, for electricity derivatives and temperature derivatives, respectively. Like other payoff functions, these functions satisfy $\overline{f_m(S_{t,m})} = 0$ and $\overline{g_m(W_{t,m})} = 0$ given the parameter estimation period and may provide negative payoffs. We see that the shapes of both payoff functions are like those in Figure 8 but have different scales in the y-axis. This is because the volume covered by thermal generation was approximately 80% on average with respect to the total demand for the period of our analysis. Panels (c) and (d) in Figure 10 provide out-of-sample VRRs and NMAEs, respectively, which were computed by applying Equations (14) and (15) based on out-of-sample data. In this test, radiation derivatives were included for $m = 8, \ldots, 15$ only, and we estimated the payoff functions of radiation derivatives, as shown in panel (e). In these figures, note that VRRs and NMAEs, including radiation derivatives, are plotted using blue lines, although they are almost hidden by the red lines corresponding to VRRs and NMAEs without radiation derivatives. To emphasize the difference between them, we further plotted VRRs and NMAEs with and without radiation derivatives, as shown in panel (f). Then, we can observe that the radiation derivatives contribute to the improvement of out-of-sample hedge performance.

Panels (a) and (b) of Figure 11 provide the estimated values for the coefficients of electricity forwards and temperature forwards, respectively. In the hedging problems with forwards, radiation terms were included for $m = 8, \ldots, 15$ and their coefficients were computed, as shown in panel (e). Like the previous figures, the day dummy variables were replaced by the dates of in-sample and out-of-sample periods. Furthermore, panels (c) and (d) show out-of-sample VRRs and NMAEs. In these figures, the blue lines provide VRRs and NMEs with radiation derivatives; however, they are almost completely hidden. Then, we further investigated VRRs and NMAEs with and without radiation forwards, as shown in panel (f). However, it turned out that the contribution of radiation forwards to the improvement of hedge effect was weak and unstable compared to the case of radiation derivatives, at least in the out-of-sample simulations.

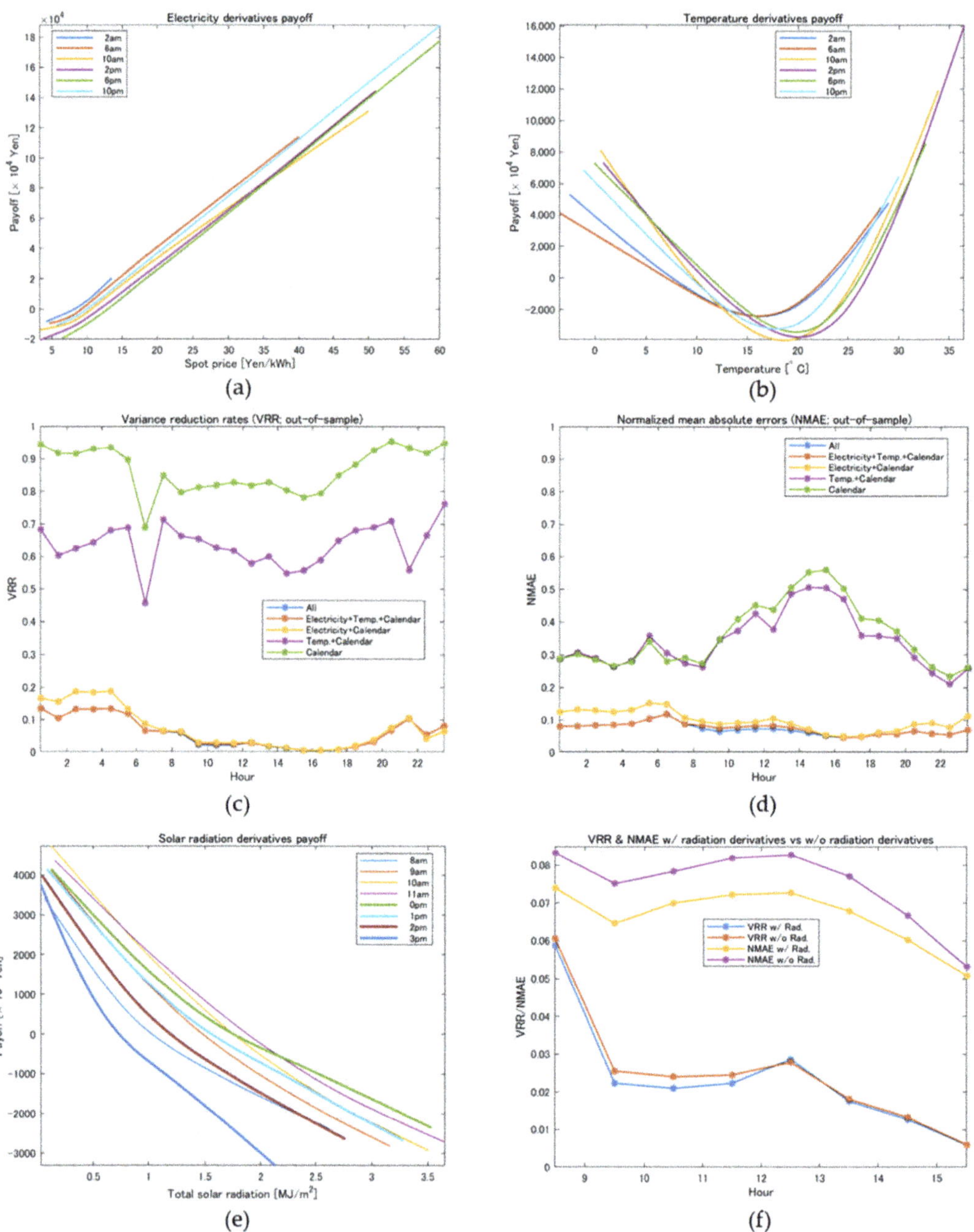

Figure 10. Results of minimum variance hedging with derivatives for thermal generators' cash flows, $V_{t,m}^{thermal} S_{t,m}$: (**a**) optimal payoff functions of electricity derivatives; (**b**) optimal payoff functions of temperature derivatives; (**c**) out-of-sample VRR for each hour; (**d**) out-of-sample NMAE for each hour; (**e**) optimal payoff functions of radiation derivatives; (**f**) out-of-sample VRR & NMAE with or without radiation derivatives for each hour.

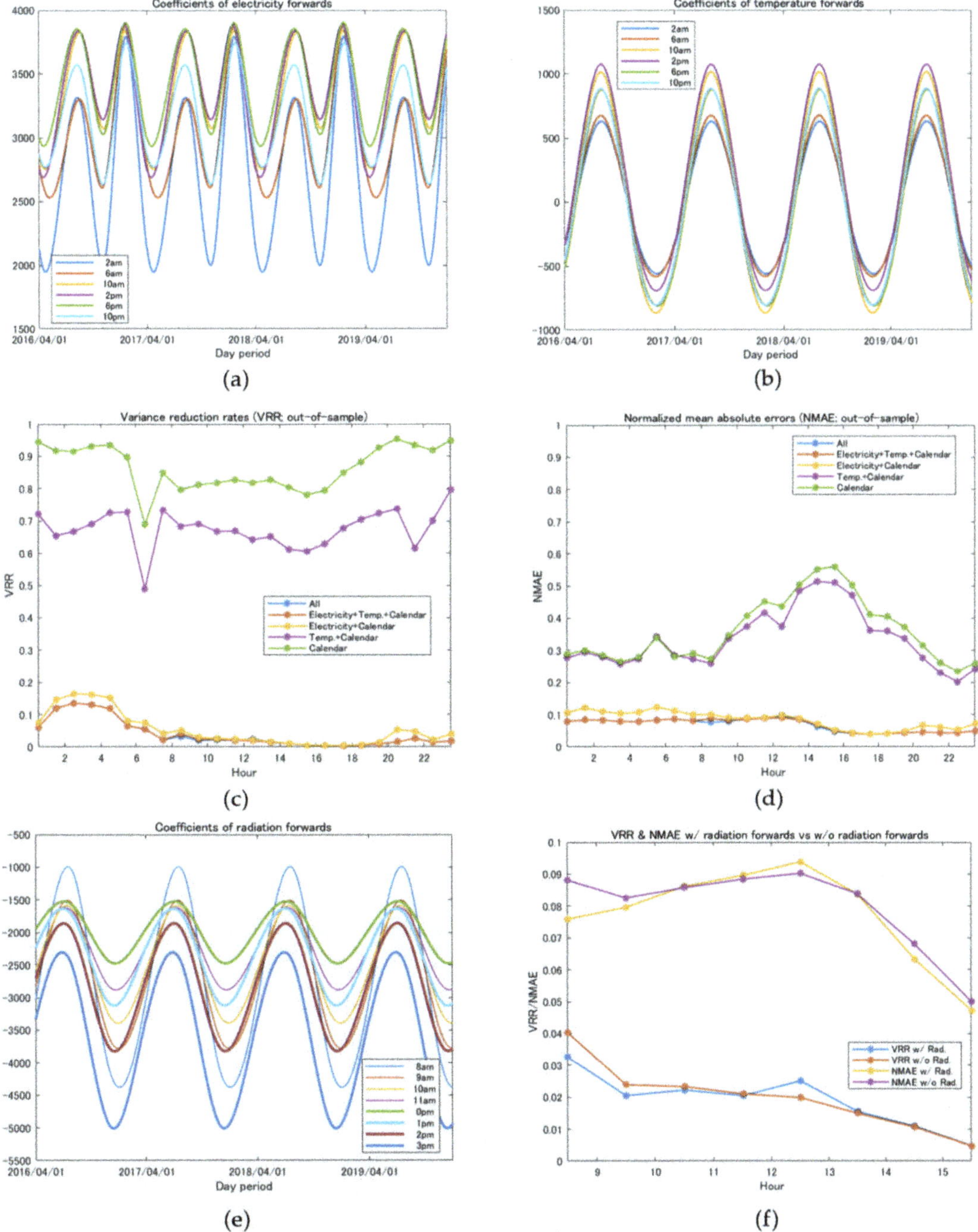

Figure 11. Results of minimum variance hedging with forwards for thermal generators' cash flows, $V^{thermal}_{t,m} S_{t,m}$: (**a**) optimal coefficients functions of electricity forwards; (**b**) optimal coefficients of temperature forwards; (**c**) out-of-sample VRR for each hour; (**d**) out-of-sample NMAE for each hour; (**e**) optimal coefficients of radiation forwards; (**f**) out-of-sample VRR & NMAE with or without radiation forwards for each hour.

To compare the cross-sectional hedge performance, we computed the averages of hourly VRRs and NMAEs in the out-of-sample period, corresponding to the averages of "All" (the blue lines) in panels (c) and (d) of Figures 6–11, respectively. Table 1 provides the averages of VRRs and NMAEs of minimum variance hedging problems for retailers, solar PV generators, and thermal generators, where the averages are taken for $m = 9, \ldots, 16$ in the case of solar PV generators. If compared between minimum variance hedging problems using derivatives and those using forwards for the same electricity utility players (i.e., retailers, solar PV generators, or thermal generators), we see that retailers and thermal generators achieve both better VRRs and NMAEs using forwards, as emphasized by bold letters in Table 1. On the other hand, in the case of solar PV generators, minimum variance hedging using derivatives provides a better hedge performance.

Table 1. Averages of hourly VRRs and NMAEs in the out-of-sample simulations.

	Retailers	Solar PV	Thermal
Average VRR (Derivatives)	0.0378	**0.1796**	0.0603
Average NMAE (Derivatives)	0.0493	**0.1366**	0.0716
Average VRR (Forwards)	0.0218	0.1958	**0.0401**
Average NMAE (Forwards)	**0.0452**	0.1506	**0.0676**

5. Reduction of Risks for Insurance Companies

In this study, we have assumed that the counter parties for derivative contracts are insurance companies (see Figure 1). Then, as explained in Section 2, the risks of insurance companies can be averaged out by executing derivatives or forward contracts with players in different positions, such as power retailers and generators. In this section, we illustrate that the risks of insurance companies can be reduced by executing derivative contracts with such players simultaneously.

5.1. Basic Idea

Assume that there is a derivative contract in the market offered by an insurance company whose payoff at t is denoted by X_t and satisfies $\overline{X_t} = 0$. Then, the insurance company's expected cashflow from the derivative is given by $-\overline{X_t} = 0$, and the insurance company can make a positive profit by receiving a commission from a buyer if the risk of cashflow fluctuation is small. However, there is a possibility that large cashflow fluctuations lead to a significant loss to insurance companies, and so the insurance company needs to evaluate the risk a priori; one measure of such risk is given by its variance.

We further assume that there exists another derivative contract offered by another insurance company, whose payoff at time t is Y_t and satisfies $\overline{Y_t} = 0$. Then, the aggregate risk in the market from X_t and Y_t may be given by the sum of variances, $\mathrm{Var}[X_t] + \mathrm{Var}[Y_t]$. Instead of considering aggregate risk, one may introduce the risk of aggregate cashflow, $X_t + Y_t$, defined by $\mathrm{Var}[X_t + Y_t]$. This may be a situation of evaluating the risk of an insurance company that is willing to offer both derivatives with payoffs, X_t and Y_t, and the following quantity provides a relative effectiveness of such position compared to the aggregate risk in the market:

$$\frac{\mathrm{Var}[X_t + Y_t]}{\mathrm{Var}[X_t] + \mathrm{Var}[Y_t]} \tag{18}$$

If X_t and Y_t are independent in (18), we see that $\mathrm{Var}[X_t + Y_t] = \mathrm{Var}[X_t] + \mathrm{Var}[Y_t]$ and that the quantity in (18) equals 1. On the other hand, if X_t and Y_t are negatively correlated, then $\mathrm{Var}[X_t + Y_t] < \mathrm{Var}[X_t] + \mathrm{Var}[Y_t]$ holds and the quantity in (18) becomes less than 1, which leads to a reduction in variance by combining two cashflows. In this sense, the quantity in (18) measures the variance reduction effect of the two cashflows for the insurance company.

In general, assuming that $X_t^{(1)}, \ldots, X_t^{(n)}$ are cashflows from n derivative contracts executed with power retailers and generators, the insurance company's VRR may be defined as follows:

$$\text{Insurance company's VRR} \; : \; \frac{\text{Var}\left[X_t^{(1)} + \cdots + X_t^{(n)}\right]}{\text{Var}\left[X_t^{(1)}\right] + \cdots + \text{Var}\left[X_t^{(n)}\right]} \tag{19}$$

Furthermore, we define the insurance company's NMAE as

$$\text{Insurance company's NMAE} : \; \frac{\left|X_t^{(1)} + \cdots + X_t^{(n)}\right|}{\left|X_t^{(1)}\right| + \cdots + \left|X_t^{(n)}\right|} \tag{20}$$

Note that the sum of cashflows in (20) is not an error, but we use the same terminology as the previous definitions to avoid a redundant definition. Although we can introduce insurance companies' VRR and NMAE for cashflows from forward contracts as well, here we focus on the cashflows from derivative contracts only; that is, we consider cashflows of derivatives obtained by solving minimum variance hedging problems for power retailers and generators.

5.2. Evaluation of Insurance Company's VRRs and NMAEs Using Empirical Data

Now, we evaluate insurance companies' VRRs and NMAEs using empirical data. Let f_m^{retail}, f_m^{solar}, and $f_m^{thermal}$ be the payoff functions of the electricity derivatives obtained by applying GAM (6) with $V_{t,m} \equiv V_{t,m}^{demand}$ and $W_{t,m} \equiv T_{t,m}$ for power retailers, GAM (6) with $V_{t,m} \equiv V_{t,m}^{solar}$ and $W_{t,m} \equiv R_{t,m}$ for solar PV generators, and $V_{t,m} \equiv V_{t,m}^{thermal}$ and $W_{t,m} \equiv [T_{t,m}, R_{t,m}]^T$ for thermal generators, where these payoff functions are estimated using in-sample data, as shown in panel (a) of Figure 6, Figure 8, and Figure 10, respectively.

Assume that all transactions of electricity derivatives in hedging problems are executed with the same insurance company. Since the direction of cashflow for exchanging the electricity delivery contract through power exchange is opposite between power retailers and generators, the insurance company is supposed to pay $f_m^{retail}(S_{t,m})$ to retailers and receive $f_m^{solar}(S_{t,m})$ and $f_m^{thermal}(S_{t,m})$ from power generators. Therefore, the aggregate cashflow (i.e., cash-out from the insurance company) is given as

$$f_m^{retail}(S_{t,m}) - \left(f_m^{solar}(S_{t,m}) + f_m^{thermal}(S_{t,m})\right) \tag{21}$$

Panels (a) and (b) in Figure 12 show the cashflows from the payoffs of electricity derivatives, where the blue line is the payoff of derivatives that the retailer receives and the red line is the sum of payoffs for generators (i.e., the solar power generator and the thermal generator). Panel (a) represents cashflows corresponding to the electricity delivery of 10–11 a.m. and panel (b) cashflows for 2–3 p.m. In these figures, the x-axis denotes the dates of the in-sample and out-of-sample periods, in which the in-sample period is until 31 March 2019. The yellow lines provide the aggregate cashflows of (21).

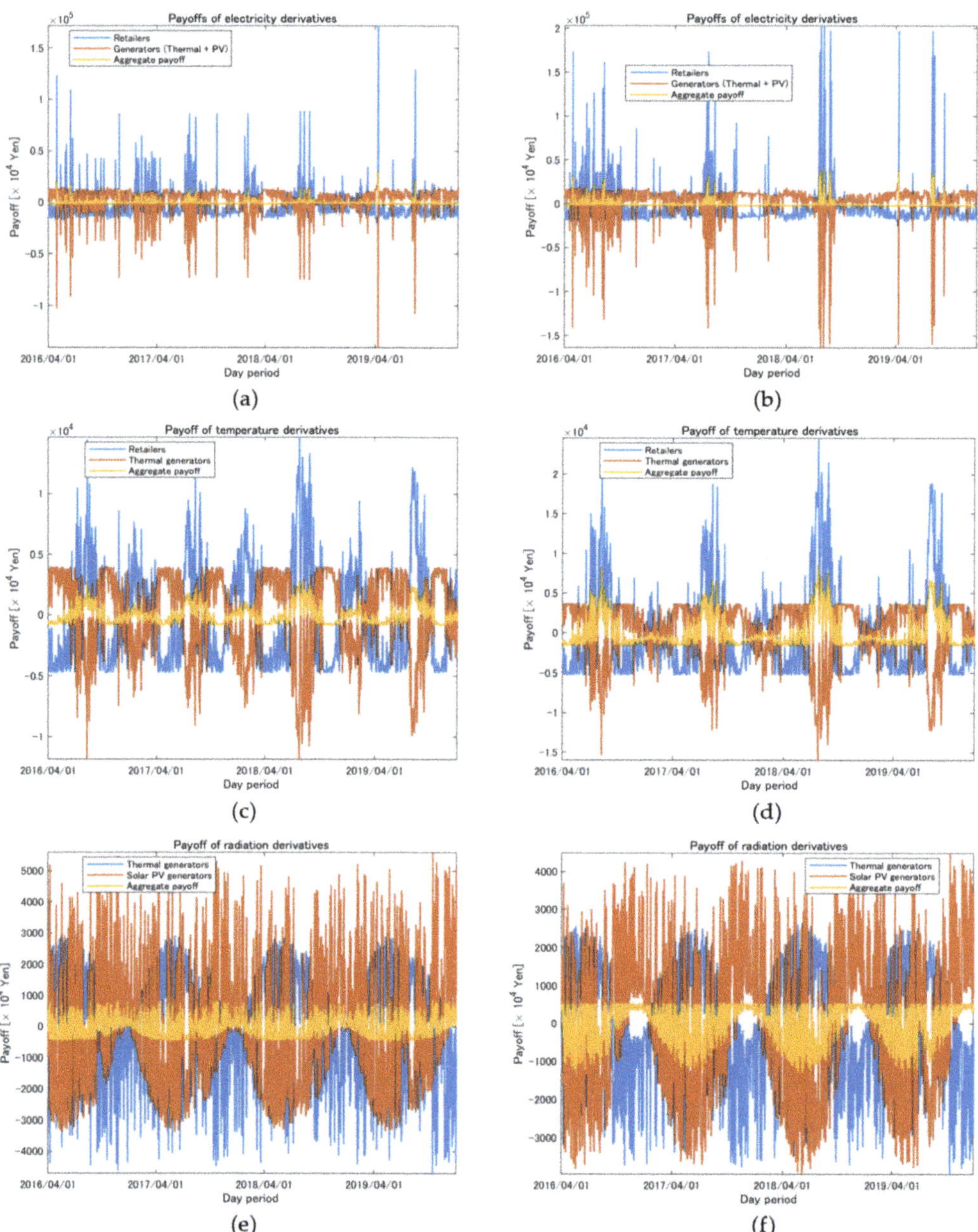

Figure 12. Cash flows (CFs) from derivatives payoffs: (**a**) payoff of retailers, the sum of payoffs for thermal generators and solar PV generators, and their aggregate payoff from electricity derivatives for 10–11 a.m.; (**b**) payoff of retailers, the sum of payoffs for thermal generators and solar PV generators, and their aggregate payoff from electricity derivatives for 2–3 p.m.; (**c**) payoffs of retailers and thermal generators and their aggregate payoff from temperature derivatives for 10–11 a.m.; (**d**) payoffs of retailers and thermal generators and their aggregate payoff from temperature derivatives for 2–3 p.m.; (**e**) payoffs of thermal generators and solar PV generators and their aggregate payoff from radiation derivatives for 10–11 a.m.; (**f**) payoffs of thermal generators and solar PV generators and their aggregate payoff from radiation derivatives for 2–3 p.m.

Similarly, panels (c) and (d) show the cashflows from payoffs of temperature derivatives, and panels (e) and (f) those of radiation derivatives. In these cases, the aggregate cashflows are given by

$$g_m^{retail}(T_{t,m}) - g_m^{thermal}(T_{t,m}) \tag{22}$$

for temperature derivatives, and

$$- h_m^{thermal}(R_{t,m}) - g_m^{solar}(R_{t,m}) \tag{23}$$

for radiation derivatives, where the optimal payoff functions in (22) and (23) are obtained by applying GAM (6) with appropriate variables. The superscripts of these functions denote the problems we have solved. For example, g_m^{retail} is obtained by solving the minimum variance hedging problem for retailers and $g_m^{retail}(T_{t,m})$ provides the retailer's payoff of temperature derivatives. The minus signs in front of payoff functions for power generators indicate that the direction of cashflows defined by payoff functions is opposite from that for retailers. For example, the payoff that the thermal generator receives is defined by $-g_m^{thermal}(T_{t,m})$.

Panels (a) and (b) of Figure 13 provide insurance companies' VRRs and NMAEs of electricity derivatives, respectively, for each $m = 0, \ldots, 23$, where the blue lines denote those obtained by using in-sample data, the red lines denote those obtained using out-of-sample data, and the yellow lines indicate the entire period data. Since the aggregate cashflow is given by (21), the insurance company's VRRs and NMAEs are computed by replacing $X_t^{(1)} \equiv f_m^{retail}(S_{t,m}), X_t^{(2)} \equiv -f_m^{solar}(S_{t,m})$, and $X_t^{(3)} \equiv -f_m^{thermal}(S_{t,m})$ in (19) and (20), respectively. From these figures, we see that both VRRs and NMAEs are small in the case of electricity derivative transactions, and the variance is reduced significantly by combining cashflows from derivatives executed with retailers and generators.

Panels (c)–(f) provide insurance companies' VRRs and NMAEs of temperature and radiation derivatives, respectively, like panels (a) and (b) of electricity derivatives. In the case of temperature derivatives, the aggregate cashflow of (22) consists of $X_t^{(1)} \equiv g_m^{retail}(T_{t,m})$ and $X_t^{(2)} \equiv -g_m^{thermal}(T_{t,m})$, whereas that of (23) consists of $X_t^{(1)} \equiv -h_m^{thermal}(R_{t,m})$ and $X_t^{(2)} \equiv -g_m^{solar}(R_{t,m})$ in the case of radiation derivatives, respectively. Then, the VRRs and NMAEs are computed based on (19) and (20). From these figures, we see that the risk reduction effect is reasonably significant, although it is not as large as that of electricity derivatives.

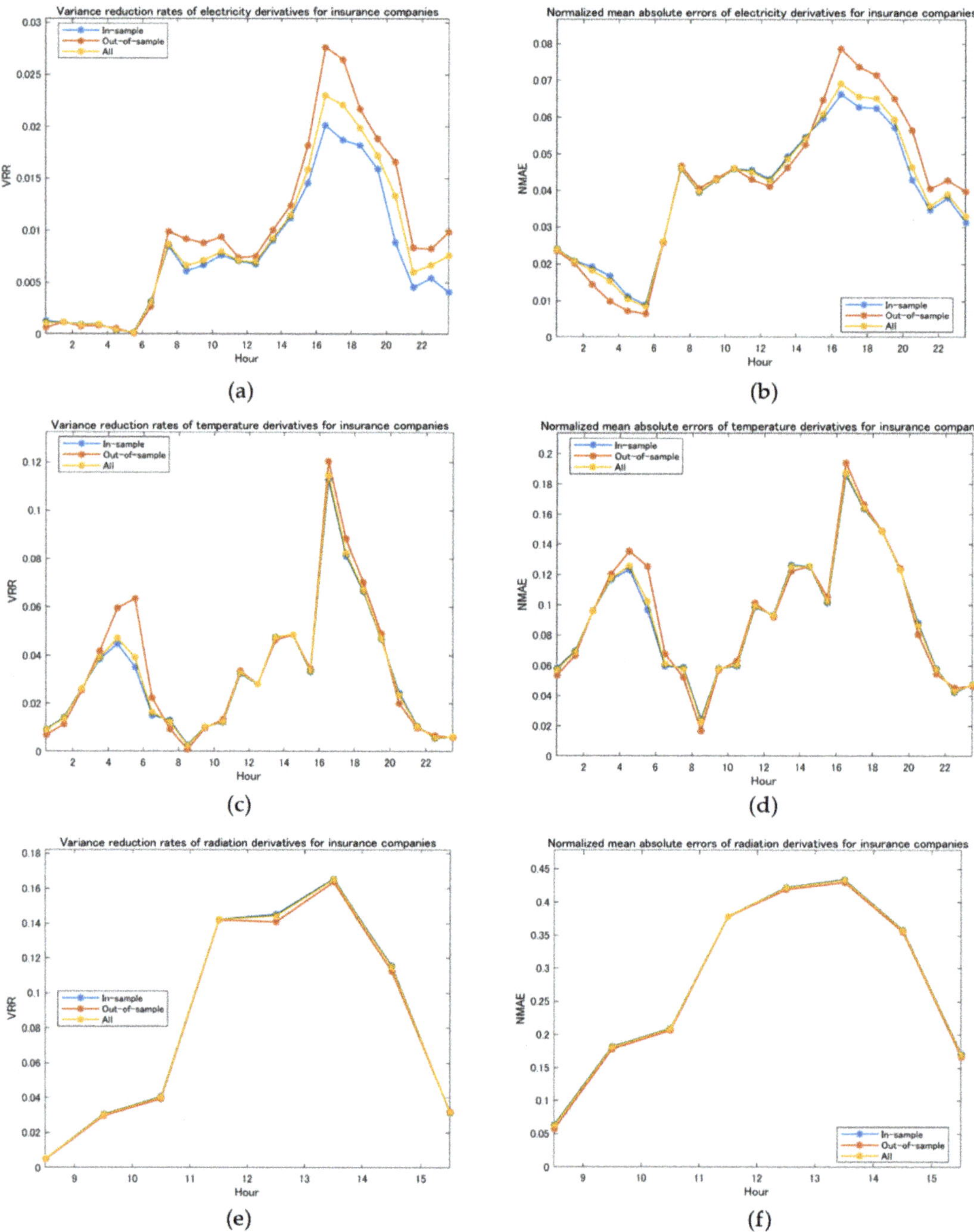

Figure 13. Variance reduction rates (VRRs) and normalized mean absolute errors (NMAEs) for insurance companies' cash flows (CFs): (**a**) VRRs for CFs of electricity derivatives' payoffs; (**b**) NMAEs for CFs of electricity derivatives' payoffs; (**c**) VRRs for CFs of temperature derivatives' payoffs; (**d**) NMAEs for CFs of temperature derivatives' payoffs; (**e**) VRRs for CFs of radiation derivatives' payoffs; (**f**) NMAEs for CFs of radiation derivatives' payoffs.

6. Discussion

In this study, we have systematically organized the theoretical aspects of our previous studies in [33,34] and developed a unified approach using derivatives and forwards on the spot electricity price and weather data. We aim not only to clarify the applicability of our proposed methods, but also to provide a new and useful perspective on hedging schemes involving various electricity utilities, such as power retailers, solar PV generators, and thermal generators. In our empirical analysis, we have measured the hedging effects on their cashflow management using electricity and weather derivatives as well as forward contracts. The key findings of our analysis are summarized below.

1. For the hedging problems using derivatives for the power retailers and the thermal generators, the payoff functions of the electricity derivatives increase monotonically with the underlying electricity price, but a nonlinear dependence is observed when the electricity price is low during the day. This seems to reflect the relationship between the PV generation and electricity prices. In general, the electricity price increases with demand, but in the daytime solar radiation tends to increase, resulting in pushing the electricity price in the lower direction;

2. The coefficients of the electricity forwards for power retailers' and thermal generators' hedges have two peaks in a year, which correspond to the demand increases in summer and winter. On the other hand, the coefficients of temperature forwards incorporate the correlation between the temperature and the demand. That is, the demand increases with a higher temperature and decreases with a lower temperature in summer, whereas in winter it increases with a lower temperature;

3. Both derivatives and forwards are generally effective for reducing the cashflow fluctuations, but in the cases of power retailers' and thermal generators' hedging problems the out-of-sample VRRs and NMAEs were better for hedging problems using forwards. This may be explained by the fact that both cashflows for the power retailers and the thermal generators are largely dependent on the electricity demand, which may be better explained using the cyclic trend for the forwards than the spline functions for derivatives;

4. On the other hand, in the case of the solar PV generators' hedging problem, the hedge errors were smaller for derivatives in terms of the VRRs and NMAEs. The reason for this difference is that it seems that the radiation derivatives are more effective for reducing the cashflow fluctuations for the solar PV generations. The same phenomenon was observed in the hedging problem for the thermal generators, where the radiation derivatives are more effective for reducing the risk of cashflow fluctuations based on out-of-sample VRRs and NMAEs.

In our analysis, we have assumed that there exist counter parties of derivative and forward transactions, such as insurance companies, and that the electricity utility players can execute electricity and weather derivative transactions with any payoff functions. Such insurance companies can profit if a commission is purchased for every transaction. Moreover, as explained in Sections 2 and 5, their risks may be averaged out by executing derivative contracts with power retailers and generators simultaneously. This is because their cash flow directions may be different or opposite for the electricity purchase and the payoffs of derivatives may be canceled out. We have illustrated insurance company risk reduction using empirical simulations and obtained the following:

5. The fluctuations in the aggregate cashflow of the electricity derivative's payoffs from the hedging problems for power retailers, solar PV generators, and thermal generators were reduced significantly compared to the sum of independent cashflow fluctuations. This indicates that the insurance company can take and cancel out the risk in electricity purchase by combining appropriate positions;

6. For temperature and radiation derivatives, the risk reduction effect for insurance companies is not as significant as in the case of electricity derivatives; however, their risks were reasonably reduced. Moreover, weather derivatives are useful products

for insurance companies compared to other financial instruments because weather indexes are not affected by human activities, at least in a short period. Therefore, fair prices may be set using their mean values, and the risk of cashflow fluctuation may be averaged out if the transaction period is sufficiently long.

Although we have incorporated the seasonal trend (i.e., the cyclic trend) in the coefficients of forwards in our analysis, we should be able to apply the result of [34] for derivatives with cyclic trends using tensor product spline functions. Then, the hedging effect could be further enhanced by designing derivatives with nonlinear payoffs that change gradually by date. However, it is necessary to consider the tradeoff of different advantages between derivatives and forwards in this regard. That is, while forwards have a payoff function that depends only on the underlying asset (i.e., the hedger optimizes the contract volume), derivatives have a payoff function that depends on the hedger's profit function (i.e., the hedger optimizes the payoff function itself). This means that the forward market may allow liquid transactions among multiple players, while derivatives are subject to bilateral contracts between the risk taker (insurance company) and the hedger. Thus, whether to use derivatives to improve the hedging effectiveness or to use forwards for the liquid transactions is an issue to be considered based on not only the results of the empirical analysis but also the actual market environment and practical needs.

In addition, as a first step to verify the effectiveness of an efficient market-wide hedging scheme, this study conducted an empirical analysis targeting the Tokyo area, where a certain percentage of solar power generation exists, and the necessary public data is sufficiently available. However, further improvements in the design of derivative products, such as increasing the number of observation points to be taken into account in the creation of the weather index, may be necessary when targeting areas with relatively low population density. Moreover, if the introduction of solar power continues to increase, the effectiveness of solar radiation derivatives for hedging solar volume risk will become increasingly effective, and there is a possibility that this method can be applied more widely. The expansion of such application areas and empirical analysis regarding the verification of the versatility of the method will be a future task.

Furthermore, as described at the end of Section 2, it would be interesting to introduce market makers (e.g., insurance companies or financial institutions) who provide fair bid and ask prices and accept sell and buy orders from power generators and retailers in the forward market. If the numbers of short and long positions are the same for the same product, the market makers do not have any risks. Therefore, the balance between the long and short positions is important for estimating the market maker's risk. Moreover, the inefficiencies of the market that can be assumed in practice, and the bid–ask spreads (or the premiums demanded by the insurers) that they bring about, may be an additional issue to be further investigated. These will be left for future study.

Author Contributions: Conceptualization, Y.Y. and T.M.; methodology, Y.Y. and T.M.; software, Y.Y.; validation, Y.Y.; formal analysis, Y.Y.; investigation, Y.Y. and T.M.; resources, Y.Y.; data curation, Y.Y.; writing—original draft preparation, Y.Y.; writing—review and editing, Y.Y. and T.M.; visualization, Y.Y.; supervision, Y.Y.; project administration, Y.Y.; funding acquisition, Y.Y. and T.M.; All authors have read and agreed to the published version of the manuscript.

Funding: This work was funded by a Grant-in-Aid for Scientific Research (A) 20H00285, Grant-in-Aid for Challenging Research (Exploratory) 19K22024, and Grant-in-Aid for Young Scientists 21K14374 from the Japan Society for the Promotion of Science (JSPS).

Institutional Review Board Statement: Not applicable.

Informed Consent Statement: Not applicable.

Data Availability Statement: Publicly available datasets were analyzed in this study. This data can be found here: http://www.jepx.org/market/index.html (accessed on 27 October 2021); https://www.data.jma.go.jp/gmd/risk/obsdl/ (accessed on 27 October 2021); https://www.tepco.co.jp/forecast/html/area_data-j.html (accessed on 27 October 2021).

Conflicts of Interest: The authors declare no conflict of interest.

References

1. Eydeland, A.; Wolyniec, K. *Energy and Power Risk Management: New Developments in Modeling, Pricing and Hedging*; John Wiley & Sons: New York, NY, USA, 2003.
2. Clewlow, L.; Strickland, C. *Energy Derivative*; Lacima Group: Sydney, Australia, 2000; ISBN 0953889602.
3. Di Masi, G.B.; Kabanov, Y.M.; Runggaldier, W.J. Mean-variance hedging of options on stocks with Markov volatility. *Theory Probab. Its Appl.* **1994**, *39*, 173–181. [CrossRef]
4. Guo, X. Information and option pricing. *Quant. Financ.* **2001**, *1*, 38–44. [CrossRef]
5. Duan, J.C.; Popova, I.; Ritchken, P. Option pricing under regime switching. *Quant. Financ.* **2002**, *2*, 116–132. [CrossRef]
6. Buffington, J.; Elliott, R.J. American options with regime switching. *Int. J. Theor. Appl. Financ.* **2002**, *5*, 497–514. [CrossRef]
7. Oum, Y.; Oren, S.; Deng, S. Hedging Quantity Risk with Standard Power Options in a Competitive Wholesale Electricity Market. *Nav. Res. Logist.* **2006**, *53*, 697–712. [CrossRef]
8. Hinderks, W.J.; Wagner, A. Pricing German Energiewende products: Intraday cap/floor futures. *Energy Econ.* **2019**, *81*, 287–296. [CrossRef]
9. Matsumoto, T.; Bunn, D.W.; Yamada, Y. Pricing electricity day-ahead cap futures with multifactor skew-t densities. *Quant. Financ.* **2021**, in press. [CrossRef]
10. Berhane, T.; Shibabaw, A.; Awgichew, G.; Walelgn, A. Pricing of weather derivatives based on temperature by obtaining market risk factor from historical data. *Model. Earth Syst. Environ.* **2020**, *7*, 871–884. [CrossRef]
11. Göncü, A. Pricing temperature-based weather derivatives in China. *J. Risk Financ.* **2011**, *13*, 32–44. [CrossRef]
12. Groll, A.; Cabrera, B.L.; Meyer-Brandis, T. A consistent two-factor model for pricing temperature derivatives. *Energy Econ.* **2016**, *55*, 112–126. [CrossRef]
13. Hell, P.; Meyer-Brandis, T.; Rheinländer, T. Consistent factor models for temperature markets. *Int. J. Theor. Appl. Financ.* **2012**, *15*, 1250027. [CrossRef]
14. Meissner, G.; Burke, J. Can we use the Black-Scholes-Merton model to value temperature options? *Int. J. Financ. Mark. Deriv.* **2011**, *2*, 298. [CrossRef]
15. Lee, Y.; Oren, S.S. An equilibrium pricing model for weather derivatives in a multi-commodity setting. *Energy Econ.* **2009**, *31*, 702–713. [CrossRef]
16. Lee, Y.; Oren, S.S. A multi-period equilibrium pricing model of weather derivatives. *Energy Syst.* **2010**, *1*, 3–30. [CrossRef]
17. Davis, M. Pricing weather derivatives by marginal value. *Quant. Financ.* **2001**, *1*, 305–308. [CrossRef]
18. Platen, E.; West, J. A Fair Pricing Approach to Weather Derivatives. *Asia-Pac. Financ. Mark.* **2004**, *11*, 23–53. [CrossRef]
19. Brockett, P.L.; Wang, M.; Yang, C.; Zou, H. Portfolio Effects and Valuation of Weather Derivatives. *Financ. Rev.* **2006**, *41*, 55–76. [CrossRef]
20. Kanamura, T.; Ohashi, K. Pricing summer day options by good-deal bounds. *Energy Econ.* **2009**, *31*, 289–297. [CrossRef]
21. Benth, F.E.; Di Persio, L.; Lavagnini, S. Stochastic Modeling of Wind Derivatives in Energy Markets. *Risks* **2018**, *6*, 56. [CrossRef]
22. Gersema, G.; Wozabal, D. An equilibrium pricing model for wind power futures. *Energy Econ.* **2017**, *65*, 64–74. [CrossRef]
23. Hess, M. A New Model for Pricing Wind Power Futures. *SSRN Electron. J.* **2019**. [CrossRef]
24. Rodríguez, Y.E.; Pérez-Uribe, M.A.; Contreras, J. Wind Put Barrier Options Pricing Based on the Nordix Index. *Energies* **2021**, *14*, 1177. [CrossRef]
25. Boyle, C.F.; Haas, J.; Kern, J.D. Development of an irradiance-based weather derivative to hedge cloud risk for solar energy systems. *Renew. Energy* **2020**, *164*, 1230–1243. [CrossRef]
26. Leobacher, G.; Ngare, P. On Modelling and Pricing Rainfall Derivatives with Seasonality. *Appl. Math. Financ.* **2011**, *18*, 71–91. [CrossRef]
27. Bhattacharya, S.; Gupta, A.; Kar, K.; Owusu, A. Risk management of renewable power producers from co-dependencies in cashflows. *Eur. J. Oper. Res.* **2020**, *283*, 1081–1093. [CrossRef]
28. Yamada, Y. Valuation and hedging of weather derivatives on monthly average temperature. *J. Risk* **2007**, *10*, 101–125. [CrossRef]
29. Yamada, Y. Optimal Hedging of Prediction Errors Using Prediction Errors. *Asia Pac. Financ. Mark.* **2008**, *15*, 67–95. [CrossRef]
30. Matsumoto, T.; Yamada, Y. Cross Hedging Using Prediction Error Weather Derivatives for Loss of Solar Output Prediction Errors in Electricity Market. *Asia Pac. Financ. Mark.* **2018**, *26*, 211–227. [CrossRef]
31. Yamada, Y. Simultaneous hedging of demand and price by temperature and power derivative portfolio: Consideration of periodic correlation using GAM with cross variable. In Proceedings of the 50th JAFEE Winter Conference, Tokyo, Japan, 22–23 February 2019.
32. Matsumoto, T.; Yamada, Y. Hedging strategies for solar power businesses in electricity market using weather derivatives. In Proceedings of the 2019 IEEE 2nd International Conference on Renewable Energy and Power Engineering (REPE), Toronto, ON, Canada, 2–4 November 2019; pp. 236–240.
33. Matsumoto, T.; Yamada, Y. Simultaneous hedging strategy for price and volume risks in electricity businesses using energy and weather derivatives. *Energy Econ.* **2021**, *95*, 105101. [CrossRef]
34. Matsumoto, T.; Yamada, Y. Customized yet Standardized Temperature Derivatives: A Non-Parametric Approach with Suitable Basis Selection for Ensuring Robustness. *Energies* **2021**, *14*, 3351. [CrossRef]

35. Hastie, T.; Tibshirani, R. *Generalized Additive Models*; Chapman & Hall: Boca Raton, FL, USA, 1990.
36. Wood, S.N. *Generalized Additive Models: An Introduction with R*; Chapman and Hall: New York, NY, USA, 2017.
37. Wood, S.N. Package 'mgcv' v. 1.8–38. Available online: https://cran.r-project.org/web/packages/mgcv/mgcv.pdf (accessed on 18 October 2021).

MDPI

Review

Comprehensive Review on Electricity Market Price and Load Forecasting Based on Wind Energy

Hakan Acaroğlu [1] and Fausto Pedro García Márquez [2,*]

1　Department of Economics, Faculty of Economics and Administrative Sciences, Eskisehir Osmangazi University, Eskisehir 26480, Turkey; hacaroglu@ogu.edu.tr
2　Ingenium Research Group, University of Castilla-La Mancha, 13004 Ciudad Real, Spain
*　Correspondence: FaustoPedro.Garcia@uclm.es

Abstract: Forecasting the electricity price and load has been a critical area of concern for researchers over the last two decades. There has been a significant economic impact on producers and consumers. Various techniques and methods of forecasting have been developed. The motivation of this paper is to present a comprehensive review on electricity market price and load forecasting, while observing the scientific approaches and techniques based on wind energy. As a methodology, this review follows the historical and structural development of electricity markets, price, and load forecasting methods, and recent trends in wind energy generation, transmission, and consumption. As wind power prediction depends on wind speed, precipitation, temperature, etc., this may have some inauspicious effects on the market operations. The improvements of the forecasting methods in this market are necessary and attract market participants as well as decision makers. To this end, this research shows the main variables of developing electricity markets through wind energy. Findings are discussed and compared with each other via quantitative and qualitative analysis. The results reveal that the complexity of forecasting electricity markets' price and load depends on the increasing number of employed variables as input for better accuracy, and the trend in methodologies varies between the economic and engineering approach. Findings are specifically gathered and summarized based on researches in the conclusions.

Keywords: electricity price; electricity load; electricity price forecasting; wind energy; day-ahead market; intra-day market; balancing power market

Citation: Acaroğlu, H.; García Márquez, F.P. Comprehensive Review on Electricity Market Price and Load Forecasting Based on Wind Energy. *Energies* **2021**, *14*, 7473. https://doi.org/10.3390/en14227473

Academic Editor: Yuji Yamada

Received: 7 October 2021
Accepted: 3 November 2021
Published: 9 November 2021

Publisher's Note: MDPI stays neutral with regard to jurisdictional claims in published maps and institutional affiliations.

1. Introduction

The government-controlled and monopolistic characteristics of the power sector has been changing since the beginning of the 1990s with the introduction of competitive market and deregulation processes [1]. The free-competitive market rules reshape electricity trade, as electricity is a non-storable commodity in economic terms, and its consumption and production require a balance dependent on power system stability [2,3]. In line with these changes, generating electricity from the renewable energy resources, mainly wind and solar powers, is rapidly increasing in the world [4,5]. This increase can be attributed to the environmentally friendly characteristics of renewable energy resources, that can be expressed by increasing energy demand triggering global warming in the world [6].

Energy demand can be supplied by electricity production through wind energy [7]. However, electricity production is affected by weather conditions (e.g., speed of wind, precipitation, and temperature) and industrial activities (e.g., business work hours, weekdays, holidays, weekends, etc.) [1,8]. These elements are particular to the electricity commodity, making it unique and different from other commodities in terms of forecasting related price dynamics. It leads to researchers developing new prediction methods. Besides, in both financial and academic institutions, electricity price forecasts (EPFs) have become a basic information for energy companies and energy researchers in their decision-making systems and agendas [1,9,10].

Various methods have been tried and developed for EPFs through renewable energy, and it will continue as the new techniques are studied [11]. A contribution of this paper to the literature is to analyze the relationship between EPF and wind energy. This paper presents, as scientific novelty, a review on recent trends of EPF techniques considering wind energy and updated references. The advances in EPF and load techniques are comparatively discussed, and it is concluded with the main future works to cover in:

- Short-term, middle-term, and long-term price and load forecasting approaches;
- Simulation, equilibrium, production cost and fundamental models for middle and long terms;
- Statistical, artificial intelligence, and hybrid models in the framework of time series for short terms;
- Moving trends of EPF and load techniques that are in the span of economics and engineering fields;
- Working principles of electricity markets through country-specific examples.

Forecasting methods in electricity market and renewable energy resources have gained a forward acceleration and attracted attention from market participants and decision makers [12]. To this end, the motivation of this paper is to present a comprehensive review for electricity markets considering price and load forecasting mechanisms through wind energy, which is one of the fastest growing renewable energy resources due to a growing wind power integration into the electrical grids [13]. For the determined hypothesis, it is observed that forecasting approaches vary between economic terms (i.e., demand [14], supply [15], profit [16], producer, and consumer surplus [17]) and engineering techniques (i.e., power systems [18,19], optimization [20], control [21], and meta-heuristics algorithms [22,23]). As a methodology, this review follows the historical and structural development of electricity markets (i.e., day-ahead markets, intra-day markets, balancing power markets), price and load forecasting methods, and recent trends in wind energy generation, transmission, and consumption, being a novel contribution to the literature. The difficulties of predicting wind power [24], i.e., wind power has a stochastic nature [25] and its prediction is contigent upon weather conditions, e.g., wind speed; precipitation; temperature, may have some adverse effects on the market operations such as fast fluctations of wind power and loads in the new designed power grid [18]. Nonetheless, wind energy resource applications require extremely rigarous and accurate data [26].

Findings are discussed and compared through the use of quantitative and qualitative analysis, and they reveal that the complexity of forecasting electricity markets price and load depends on the increasing number of employed variables as input for better accuracy, and the trend in methodologies varies between the economic and engineering approaches, and specifically includes mathematics, statistics, econometrics, and electrical engineering and computer science.

The content of the work is presented as follows: Section 2 presents a literature analysis on electricity market mechanism, components, and instruments, considering the day-ahead market (DAM), or spot market; the intra-day market (IDM), or future market; the balancing power market (BPM), or balance market; price of electricity; and electric load. Section 3 shows the electricity market price and load forecasting through wind energy generation. Section 4 analyzes the forecasting models of the electricity markets through wind energy, where several case studies are considered and discussed.

2. Electricity Market Mechanism, Components, and Instruments

2.1. Electricity Market: Structure and Components

The short-term electricity market structure includes day-ahead and intraday markets which are often known as "spot markets" [27]. However, these markets' designs show differences. While DAMs have been coupled for the last few years, IDMs have gained traction by going global from being national [28]. Moreover, DAMs are organized as auctions, whereas IDMs operate as trades and enable market participants to balance demand and supply variations in the short-term to decrease exposure to an imbalance penalty [28,29].

The reason being, DAMs are based on forecasts and forecasts include errors in their nature. Specifically, various and increasing number of parameters, intermittent production from wind power plants can be given as the factors. However, the closer to real-time, the more accurate the forecast is possible. The bilateral basis with continuous trading enables market participants to adjust their last updated positions [27]. In addition to these markets, the eventual balancing of the supply and demand is accomplished by the BPMs, which are regulated by the transmission system operator (TSO). The system stability is provided in the context of security in these markets [30] (see [31,32] for detailed information).

2.1.1. Day-Ahead Markets

DAMs are organized markets that are used for electricity trading and balancing activities just one day before the delivery date of electricity, operated by a transmission system operator. DAMs include auctions that are conducted simultaneously 24 h in a day. The market participants are able to adjust their own transaction schedule by selling or buying power with the short-term price forecasts thereby maximizing their profits [33]. The main reasons that DAMs are needed and their purposes are summarized as follows [34]:

- Determining the electrical energy reference price.
- To provide market participants with the opportunity to balance themselves by giving them selling and buying energy options for the next day in addition to their bilateral agreements.
- To provide the system operator with a balanced system the day before.
- To provide the system operator with the opportunity to manage the constraints in the day before, by creating bid zones for large-scale and continuous constraints.

DAMs are developing through institutions, regulations, software and web applications daily. For instance, currently, a DAM software and optimization model on the DAM for the Turkish electricity sector, which has a user-friendly interface design and is amenable to flexibility and improvements, since it is designed and written entirely by the domestic resources, has been completed [35]. Table 1 shows the various DAMs electricity markets over the world.

Table 1. Various DAMs in the world.

Country	Name (Year)
UK	England and Wales Electricity Pool (1990)
Norway	Nord Pool (1992)
Sweden	Nord Pool (1996)
Spain	Operadora del Mercado Español de Electricidad (OMEL) (1998)
Finland	Nord Pool (1998)
USA	California Power Exchange (CalPX) (1998)
Netherlands	Amsterdam Power Exchange (APX) (1999)
USA	New York ISO (NYISO) (1999)
Germany	Leipzig Power Exchange (LPX) (2000)
Germany	European Energy Exchange (EEX) (2000)
Denmark	Nord Pool (2000)
Poland	Towarowa Gielda Energii (Polish Power Exchange, PolPX) (2000)
USA	Pennsylvania-New Jersey-Maryland (PJIM) Interconnection (2000)

Table 1. *Cont.*

Country	Name (Year)
UK	UK Power Exchange (UKPX) (2001)
UK	Automated Power Exchange (APX UK) (2001)
Slovenia	Borzen (2001)
France	Powernext (2002)
Austria	Energy Exchange Austria (EXAA) (2002)
USA	ISO New England (2003)
Italy	Italian Power Exchange (IPEX) (2004)
Chez Republic	Operator Trhu s Electrinou (OTE) (2004)
USA	Midwest ISO (MISO) (2005)
Belgium	Belgian Power Exchange (Belpex) (2006)

Source: Adapted from [1].

The liberalization of the electricity markets in Europe began three decades ago [36]. Before the 1990s, the markets had a monopolistic characteristic and were dictated by governments. This transformation led to electricity generation, transmission and distribution along with the law of supply/demand, which enabled competition and price reductions [37]. It is noteworthy that the DAMs in the world have adapted to this transformation and quickly became larger markets, and some of their names that are mentioned in Table 1 changed due to integrations, where detailed information can be found in [1].

2.1.2. Intra-Day Markets

In addition to the currently operating DAM, Ancillary Services, and balancing power market, the intra-day market (IDM) enables near real-time trading and offers market participants the opportunity to balance their portfolios in the short term. The IDM works as a bridge between the DAM and the BPM, and it contributes greatly to sustainability of the whole system.

The functionality of the IDM changes the role of the factors that cause imbalances, such as power plant failures, changes in the production of renewable energy sources, and unpredictable changes in the amount of consumption, as they will be eliminated in a near real time, and the participants will be given the opportunity to balance or minimize the negative or positive imbalances that they may face. Additional trading space will be provided by giving the participants the chance to evaluate their capacities, which they cannot use in the DAM, in the IDM after the closing time of the DAM. It will contribute to the increase of liquidity in the markets. It will also be of significant assistance to the TSO in providing a balanced system prior to real-time balancing.

IDMs are developing daily in terms of institutions, regulations, software, and web applications. The market designs in IDM might strongly deviate between countries [38]. For instance, a new software, named "Intraday Market Software", on IDM for the Turkish electricity sector was developed and has been in use by Energy Exchange Istanbul (EPIAS) since 2016 [39]. More information can be found in [40] for the German IDM, in [41] for the European IDM, and in [42] for the Swedish IDM.

2.1.3. Balancing Power Markets (Balance Markets)

Real-time balancing consists of balancing power market (BPM) and ancillary services. The system operator is provided the spare capacity that can be activated in a couple of minutes (i.e., around 15 min) by the BPM for real-time balancing. Ancillary services provide demand and frequency control services. The balancing market prices are determined hourly based on upward and downward regulating power offers evaluated by the TSO in real-time balancing [43].

Although a market with balanced production and consumption amounts is given to the TSO with the DAM and IDM, there are deviations in real time. For example, if a power plant is out of service, or when a large amount of consumption causes the plant to stop (start), the balance is disrupted [44]. For instance, on BPM for Turkish electricity sector [45]:

- All market participants participating in the BPM must present their available capacities.
- Balancing units that can receive or load independently in a couple of minutes (around 15 min) are obliged to engage in the BPM.

More information can be found in [46] for the European BPMs.

2.2. Electricity Market Instruments through Country-Specific Researches

2.2.1. Electricity Price

Electricity prices, or market clearing price (MCP), are determined by the law of supply/demand curves. The place for this is the DAM, which is managed by the system operators of the countries. The system operators gather hourly offers for the following day from sellers and buyers, and the supply/demand curves are analytically built in this way. The intersection of the supply and demand curves gives the MCP. While the buying and selling amounts are named as equilibrium quantities of electricity, the electricity trade volume is determined by multiplication of the equilibrium quantity and MCP. However, forecasting electricity prices is not easy because price series show characteristics such as variance, nonconstant mean, significant outliers, and volatility [47]. The common characteristics of electricity prices can be summarized as follows [1,48,49]:

- Seasonal effects for prices;
- Mean reversion;
- Spikes and volatilities due to changes in fuel price, load uncertainty, outages, market power, and market participant's behavior;
- Correlation between electricity load and price.

More detailed information can be found in [1] for various countries, and in [50] for the Turkish electricity markets, in [51] for the England and Wales electricity markets, in [52,53] for the Nordic electricity (Nord Pool) markets, in [54] for the New Zealand electricity markets, in [55] for Danish electricity markets, and in [56] for the US electricity markets.

2.2.2. Electricity Load

Forecasting the electricity load has been a key role in the operation of power systems, and it includes forecasts on various time scales (i.e., minutely, hourly, and yearly) [57]. Several decisions are based on load forecasts, for instance, reliability analysis, dispatch planning of generating capacity, and operation and maintenance plans for power systems. With the free competition and deregulation of the electric power industry, load forecasting increased its viability and importance all around the world. An accurately predicted load is vital data for the EPF, since market shares, profits, and shareholder value can easily be influenced by forecast errors. Nevertheless, due to the nonstationary and variability of the load series, forecasting procedures of the electric load is increasingly difficult. Time-varying prices, price-dependent loads, and the dynamic bidding strategies of market participants make this complexity [58]. Therefore, more accurate results are needed by more sophisticated forecasting instruments for the electrical power systems and the motivation behind more accurate forecast methods is hidden in the economic effect of the forecast errors [59]. However, a substantial amount of research has been done (see [60,61] for reviews and [58,62,63] for methods and techniques of short-term load forecasting and modeling, respectively).

Moreover, electric power should be stored or consumed very close-after from its generation. The cost of storing electric power is expensive, therefore, electricity markets, through system operators, exist for allocating the transactions between market participants. This mechanism provides a possible distribution of loads, freeing networks will be avoided from excessive loads. This review is focused on renewable energy through wind energy. Weather conditions, e.g., wind speed, precipitation, and temperature, have an important influence on electricity production from wind energy. The countries that supply a considerable share of electricity demand from wind energy (e.g., Spain, Denmark, Germany [4]) and have wind energy potential (e.g., Turkey) should consider this energy source, mitigating global warming. More details can be found in [1] for various countries, and in [50] for the Turkish electricity markets.

3. Electricity Market Price and Load Forecasting through Wind Energy Production

The EPF studies can be categorized in the following two main groups: Long/middle terms and short terms. While long/middle models can be gathered into: simulation, equilibrium, production cost, and fundamental models. Short term models, or time series models, can be gathered into: statistical, artificial intelligence, and hybrid models [64], see Figure 1. This review paper follows the approach presented in [64]. Tables 2 and 3 presents a literature review through statistical models. However, it differs from the mentioned approach by merging the artificial intelligence and hybrid models into one category, as shown in Table 4. Table 5 presents a literature review through middle/long term models on electricity market price and load forecasting through wind energy.

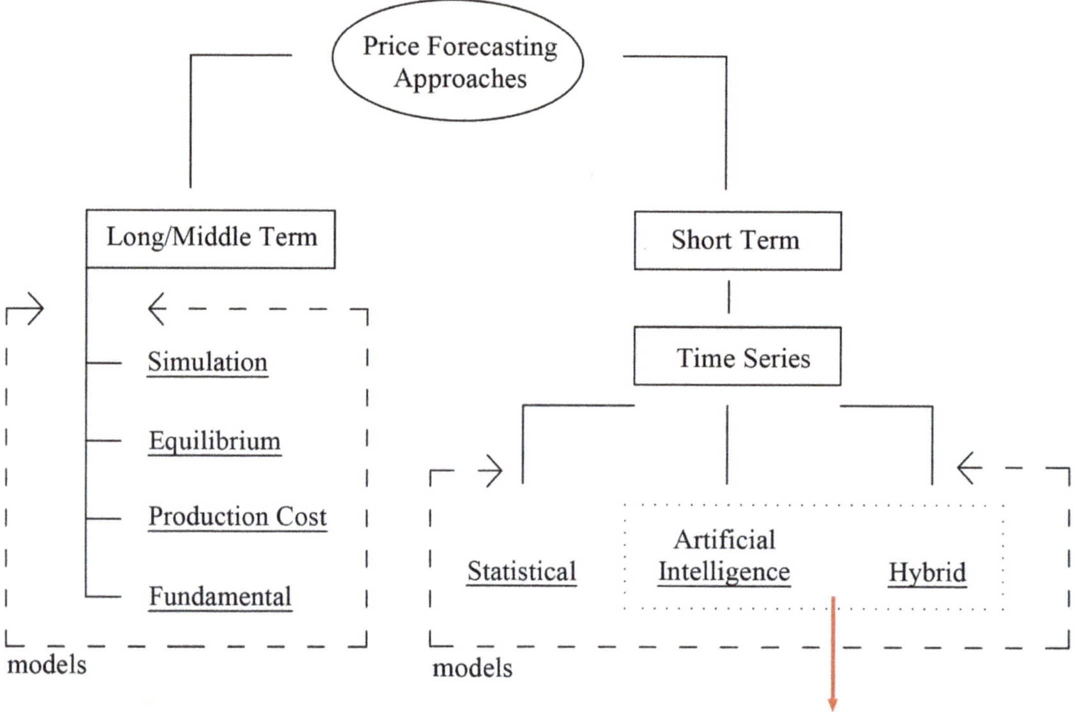

Figure 1. A classification for EPF approaches. Source: Adapted from [64].

Various statistical model examples are shown in Tables 2 and 3 (Table 2 contains more simple models, represents the first part of the statistical models and Table 3 contains more advanced models, represents the second part of the statistical models). These models can be gathered in a main title named as time series analysis. Specifically, ordinary least squares (OLS) regressions, autoregressive distributed lag (ARDL) regressions, panel data analysis, vector autoregressive (VAR) analysis, generalized autoregressive conditional heteroskedasticity (GARCH) analysis, multiple linear regressions, auto-regressive with eXternal model input (ARX) analysis, logit-probit regressions, quantile regressions, autoregression (AR) models, exponential generalized autoregressive conditional heteroskedasticity (eGARCH) analysis, autoregressive moving average model with exogenous regressors (ARMAX) analysis, least absolute shrinkage and selection operator (LASSO) analysis, seasonal component autoregressive (SCAR) analysis, and univariate and multivariate regressions.

The studies concentrating on merit-order effect for wind power on electricity market price are viable among researchers. Positive merit order effects were found with OLS analysis and time series regressions for Italy [31,65] and for US (California) [66], with time series analysis for Australia [67], and Germany [68], and with ARDL model and demand/supply framework for Australia [69,70], and with quantile regression model for Germany [71] and for US (California) [72]. A different type of time series analysis with panel data analysis through fixed effect regression was applied in [31] for Germany, and a dampening effect of wind power with reduced forecasting errors, which led to decreased price volatility. The VAR model was applied in [42] for Sweden with Granger causality analysis (i.e., unit root tests and impulse-response functions), and it was shown that the prices in the IDMs responded to wind power forecast errors. The same model was applied in [73] for Denmark, Sweden, and Finland. It was found that wind forecast errors did not affect price spreads in locations with large amounts of wind power generation. Studies for Germany [74,75] and Australia [76] with GARCH and eGARCH models showed that an increase in wind generation decreased the prices and increased the price volatility. A multiple linear regression model was applied for Germany's electricity markets [32,77], which showed that 15 min scale helped significantly to reduce imbalances in intraday trading, and a considerable share of spot price variance was explained by fundamental modelling. The ARX models, which are linear models, were applied for Germany [30,78], Poland [78], European countries, and the US [79], and the findings supported more accurate EPPs in the mentioned electricity markets. The ARMAX model was applied for Germany, where it showed that wind energy generation decreased market spot prices [80]. The AR models were applied for Denmark, Finland, Norway, and Sweden, and the used models were better performed compared to commonly-used EPF models [81,82]. The LASSO models were applied for Denmark, Finland, Norway, and Sweden, Germany, and the European Countries, and they demonstrated that LASSO models lead to better performance compared to the typically considered EPF models [83–85]. The SCAR models were applied for Denmark, Finland, Norway, and Sweden, where the SCAR models significantly outperformed the autoregressive benchmark [86]. The multivariate and univariate models were applied for the European countries and some guidelines were provided to structuring better performing models [87].

Table 2. Statistical models (first-part) on electricity market price and load forecasting through wind energy.

Author (s)	Data/Period	Country	Method (s)	Findings
Clo et al. (2015), [65].	GME/2005–2013	Italy	Time series (OLS) analysis	The merit-order effect for wind power was found.
Cludius et al. (2014a), [67].	AEMO/ 2011–2013	Australia	Time series regression analysis	The merit-order effect for wind power was found.
Cludius et al. (2014b), [68].	EEX/2008–2016	Germany	Time series regression analysis	The merit-order effect for wind power was found.
Csereklyei et al. (2019), [69].	NEM/2010–2018	Australia	ARDL model	The merit-order effect for wind power was found.
Forrest and MacGill (2013), [70].	AEMO and NEM /2009–2011	Australia	Econometric analysis techniques (a supply/demand analysis for electricity markets)	The merit-order effect for wind power was found and wind generation had an impact on the MCPs.

Table 2. *Cont.*

Author (s)	Data/Period	Country	Method (s)	Findings
Gianfreda et al. (2016), [31].	ENTSO-E/ 2012–2014	Italy	Time series regression analysis	It was found that wind generation power induced high imbalance values.
Gürtler et al. (2018), [88].	ENTSO-E/2010–2016	Germany	Panel data analysis (fixed effect regression)	It was found that there were dampening effects of wind power on MCPs, however this effect started to decrease after 2013.
Hu et al. (2018), [42].	Nord Pool FTP server and ENTSO-E/2015–2018	Sweden	VAR framework (Granger causality tests and impulse response functions)	It was found that intraday prices responded to wind power forecast errors.
Koch and Hirth, (2019), [32].	ENTSO-E and TSO/2012–2017	Germany	A multiple linear regression model	It was shown that the 15 min scale became common in intraday trading and helped significantly to reduce imbalances.
Maciejowska (2020), [71].	EPEX and ENTSO-E/ 2015–2018	Germany	Quantile regression model	It was found that wind energy generations had a negative effect on the MCPs.
Pape et al. (2016), [77].	ENTSO-E, EEX, EPEX/2012–2013	Germany	Multiple linear regression models (Fundamental price modeling)	It was shown that the used models well explained the spot price variance.
Serafin et al. (2019), [89].	Nord Pool, PJM/2013–2018	Denmark, Finland, Norway, and Sweden	Quantile Regression Averaging and Quantile Regression Machine	It was shown that QRM was both more efficient and had more accurate distributional predictions.
Spodniak et al. (2021), [73].	ENTSO-E, Nord Pool/2015–2017	Denmark, Sweden, and Finland	VAR model	It was found that wind forecast errors had no impact on price spreads in locations with a big amount of wind power generation.
Westgaard et al. (2021), [72].	LCG Consulting, OASIS/ 2013–2016	US (California)	Quantile regression	Wind generation had a negative effect on electricity prices.
Woo et al. (2016), [66].	CAISO/ 2012–2015	US (California)	OLS Regression	It was found that trading efficiency could be enhanced by DAM forecasts.
Ziel and Steinert, (2018a), [90].	EPEX/2012–2015	Germany and Austria	Time series models (supply/demand curves)	It was found that using the law of supply/demand curve yields realistic patterns for electricity prices and leads to promising results.
Ziel and Weron, (2018b), [87].	EPEX, Nord Pool, BELPEX/ 2011–2013	European Countries	Multivariate and univariate models.	More powerful variables identified and guidelines were provided for better performing models.
AEMO: Australia Energy Market Operator ARDL: Autoregressive distributed lag models BELPEX: EPEX Spot Belgium DAM: Day-ahead market EEX: The European Energy Exchange ENTSO-E: European Network of Transmission System Operators for Electricity			EPEX: The European Power Exchange GME: Gestore dei Mercati Energetici MCPs: Market clearing prices NEM: The Australian National Electricity Market's	PJM: The Pennsylvania–New Jersey–Maryland Interconnection OLS: Ordinary least squares QRM: Quantile regression machine VAR: The vector autoregressive

Table 3. A literature review through statistical models (second-part) on electricity market price and load forecasting through wind energy.

Author (s)	Data/Period	Country	Method (s)	Findings
Ketterer (2014), [74].	EEX and ENTSO-E/2006–2012	Germany	GARCH model	Wind power generation had a positive effect on decreasing the wholesale electricity price; however, increased its volatility.
Kyritsis et al. (2017), [75].	Phelix Day Base, EEX, and ENTSO-E/ 2010–2015	Germany	GARCH-in-Mean model	It was found that wind power Granger cause of MCPs and the volatility of electricity prices were increased by wind power generation
Maciejowska et al. (2019), [78].	TGE, PSE, EPEX SPOT and ENTSO-E/2016–2017	Germany and Poland	Econometric models (i.e., ARX and probit)	It was shown that the price spread could be forecasted by ARX and probit models.
Maciejowska et al. (2021), [30].	EPEX and ENTSO-E/2015–2019	Germany	Econometric models (ARX)	It was shown that variables that were forecasted gave biased results; however, they could be corrected with regression models.
Marcjasz et al. (2018), [81].	Nord Pool, PJM Interconnection and EPEX/2013–2018	Denmark, Finland, Norway, and Sweden	Autoregression Models	It was the extended model of Hubicka et al. (2019), [91] analysis with much longer datasets.
Mwampashi et al. (2021), [76].	NEM/2011–2020	Australia	eGARCH model	It was found that wind generation increase decreased daily prices and increased price volatility
Nowotarski et al. (2014), [79].	Nord Pool, EEX, and PJM/1998–2012	European Countries and US	ARX model (Constrained least squares regression)	The findings supported more accurate results and the used models were well performed for EPFs in the electricity markets.
Paraschiv et al. (2014), [80].	EEE, TSO, Bloomberg/ 2010–2013	Germany	ARMAX model	It was found that wind energy generation decreased market spot prices.
Uniejewski et al. (2016), [82].	GEFCom, Nord Pool/2011–2013	Denmark, Finland, Norway, and Sweden	Autoregression (ridge regression; stepwise regression, LASSO; elastic net) models	The used models performed well in comparison to previous preferred EPF models.
Uniejewski and Weron (2018), [83].	Nord Pool, PJM/2013–2017	Denmark, Finland, Norway, and Sweden	LASSO models	It was shown that LASSO models performed well in comparison to previous preferred EPF models.
Uniejewski et al. (2019a), [86].	GEFCom, Nord Pool/2013–2015	Denmark, Finland, Norway, and Sweden	SCAR models	SCAR models significantly outperformed the autoregressive benchmark.
Uniejewski et al. (2019b), [84].	EPEX/2015–2018	Germany	LASSO models	Some recommendations were provided for very short-term EPF with LASSO models.
Ziel, (2016), [85].	EPEX/2009–2014	European Countries	Time series model -Linear regression (LASSO)	It was shown that the LASSO forecasting technique performed well.

Table 3. *Cont.*

Author (s)	Data/Period	Country	Method (s)	Findings
ARMAX: Autoregressive moving average model with exogenous regressors ARX: Auto-regressive with eXternal model input EEX: The European Energy Exchange GARCH: A generalized autoregressive conditional heteroskedasticity model eGARCH: An exponential generalized autoregressive conditional heteroskedasticity) model			ENTSO-E: European Network of Transmission System Operators for Electricity EPEX: The European Power Exchange EPF: Electricity price forecasting LASSO: The least absolute shrinkage and selection operator	NEM: The Australian National Electricity Market's PJM: The Pennsylvania–New Jersey–Maryland Interconnection SCAR: The Seasonal Component AutoRegressive

The first part of the statistical models that are shown in Table 2 are closer to the research perspective of the fields of economics, and the traditionally used regression models by OLS (i.e., the difference between actual and predicted values are squared), VAR (i.e., the causality relationships), quantile regressions (i.e., the nonlinear relationships between electricity prices and variables are possible), and univariate and multivariate models (i.e., multivariate models are accepted as more accurate than the univariate ones but each approaches have its own advantages or disadvantages). However, when the number of regressors become large, these models were insufficient and, thereby, linear models via LASSO [92], ARX [93], SCAR (introduced by [94] and built on the ARX framework), GARCH [95–98] and eGARCH (i.e., proposed by [99]), and ARMAX [100] models were preferred, as it is shown in the second part of the statistical models with Table 3. Therefore, to obtain more accurate findings, statistical models should be more advanced and, since the complexity increases, artificial intelligence and hybrid models are required for more accurate and sensitive forecasts that are shown in Table 4. However, this time the subject becomes closer to the research perspective of the engineering field.

Various artificial intelligence and hybrid/ensemble models on electricity market price and load forecasting through wind energy examples are shown in Table 4. These models can be gathered in a main title named as time series analysis. Specifically, ensemble learning methods for Austria [101], deep neural networks analysis for Germany [102] and US (New York) [103], sensitivity analysis for Mexico [104], and deep learning models for US (New York) [105] can be given as country-specific examples. General findings for the studies showed that the proposed method could provide an effective forecast.

Table 4. A literature review through artificial intelligence and hybrid/ensemble models on electricity market price and load forecasting through wind energy.

Author (s)	Data/Period	Country	Method (s)	Findings
Bhatia et al. (2021), [101].	ENTSO-E/2015–2016	Austria	A real-time hourly resolution model (ensemble learning model)	The developed forecasting model showed more consistency, accuracy, and validity.
Bublits et al. (2017), [106].	EPEX, ENTSO-E/2011–2015	Germany	Agent based modelling and multiple regression analysis	The effect of renewable energy prices has been as half low as the coal and carbon prices on electricity prices in Germany in the duration of analysis.
Li and Becker (2021), [102].	Nord Pool, ENTSO-E, Thomson Reuters Eikon/2015–2019	Germany	LSTM deep neural networks	It was shown that feature selection is useful for more accurate forecasts.

Table 4. *Cont.*

Author (s)	Data/Period	Country	Method (s)	Findings
May et al. (2022), [104].	CONAGUA, CENACE, AND CRE/2017–2018	Mexico	Artificial Intelligence Techniques (Sensitivity Analysis)	It was found that the effects of the variables fluctuated due to consumption market conditions.
Nowotarski and Weron, (2018), [107].	GEFCom/2011–2013	-	Neural network and autoregression	The study was an update of EPF techniques of Weron (2014), [108].
Osorio et al. (2015), [109].	Portuguese TSO (REN)/2007–2008	Portugal	Hybrid evolutionary-adaptive method	A new hybrid method was tested and reduce the uncertainty of wind power predictions.
Yang and Schell, (2021), [103].	NYISO/ historical data	US (New York)	Deep neural networks	It was displayed that TL improved accuracy across all network representations.
Yang and Schell, (2022), [105].	NYISO/ historical data	US (New York)	Deep learning model	The deep learning model was developed and it was shown that it performed well on time series for EPF.
Zhang et al. (2012), [110].	NSW/2006	Australia	WT, ARIMA and LSDVM	It was shown that the preferred method performed well on EPF.
ARIMA: Autoregressive integrated moving average CENACE: Natural Center for Energy Control CONAGUA: Natural Water Commission CRE: Energy Regulatory Commission ENTSO-E: European Network of Transmission System Operators for Electricity		EPEX: The European Power Exchange LSSVM: Shrinkage and selection operator least squares support vector machine		NYISO: The New York Independent System Operator GEFCom: The Global Energy Forecasting Competition NSW: New South Wales TSO: Transmission system operator WT: Wavelet transform

The need for artificial intelligence models comes from the non-linear characteristics of electricity price. Since the large number of time series models have linear predictors, the time series techniques lack the ability to capture the behavior of the price signal [64]. Neural [47] and fuzzy neural networks [111] are proposed due to solving this problem. Nonetheless, due to functional relationship of electricity price with time and the nature (characteristics) of electricity price, it is a time variant signal; therefore, neural and fuzzy neural network solutions may not be sufficient for precise forecasting results [64], and it needs hybrid models, which are the combination of non-linear and linear modelling capabilities occurs.

Hybrid models have a very complex forecasting structure, including several algorithms for decomposing or cluster data, feature selection, combined forecasting models, and heuristic optimization [112]. The most commonly preferred decomposition method is the wavelet transform [113–122]. Other decomposition studies that used empirical mode are given in [123–129]. The most widely preferred feature selection methods are the correlation analysis are presented in [118,123,130–132], and the mutual information method in [121,123,130,133–135]. The algorithms for the clustering data are based on: (1) k-means [136,137]; (2) enhanced game [136]; (3) self-organizing maps [114,136,138]; and (4) fuzzy [121,139]. Combined forecasting models for hybrid models that build on more than one method are very common. Some examples can be found in [114,116,124,135,140,141]. The heuristic optimization studies can be found in [126,131,133,139]. The major problems in employing hybrid model are [112]: (1) The proposed methods avoid to be compared with well-build models; (2) the used data sets are small; (3) lack of analysis of the effect of selecting different components.

Various middle/long term models on electricity market price and load forecasting through wind energy examples are shown in Table 5. These models can be gathered by time series analysis. Specifically, a case study for US (Texas) [142], the sensitivity analysis through scenarios for Australia [143], balancing the cost of electricity demand with large amount of wind energy for Australia [144], data analysis techniques through electricity demand models for Australia [145], WILMAR model through scenarios for Ireland and Great Britain [146]. Monte Carlo simulations for Mykonos (Greece) and La Ventosa (Mexico) [147], and for Denmark [148]. Simulations with stochastic and robust optimization for China [149], a market equilibrium model for China [150]. A modelling demand response utility function for Iran [151], and a dispatch model for Colombia [152] can be given as country specific examples.

Table 5. A literature review through middle/long term models on electricity market price and load forecasting through wind energy.

Author (s)	Data/Period	Country	Method (s)	Findings
Baldick (2012), [142].	ERCOT empirical data	US	Case study for Texas	Cost predictions are developed for using wind energy to mitigate CO_2 emissions.
Banaei et al. (2021), [4].	Game theory data	-	The supply function model (pricing models)	Results showed that the applied method reduced the market players profit that depended on uncertainties.
Bell et al. (2017), [143].	WRF data/2015	Australia	The sensitivity analysis through scenarios	The average wholesale spot price in the NEM decreased due to the increase in wind power generation.
Blakers et al. (2021), [144].	NEM/2006–2010	Australia	Balancing the cost of electricity demand with high levels of wind energy	It is found that wind energy generation led deployment on the MCP, but it was modest.
Cutler et al. (2011), [145].	AEOM/2008–2010	Australia	Various data analysis techniques through electricity demand models	Wind power generation became a significant secondary influence (the relationship is inverse with spot prices) after electricity demand on spot prices.
Denny et al. (2010), [146].	AIGS	Ireland and Great Britain	WILMAR model through scenarios	It was found that the increased interconnection reduced both average prices and the volatility of those prices in countries.
Elfarra and Kaya (2021), [147].	Akdağ et al. (2010), [153]/2008–2009	Mykonos (Greece) and La Ventosa (Mexico)	Annual energy production through Monte Carlo simulations	The PDFs (i.e., spline based) produced minimum fitting error
Ji et al. (2021), [149].	Simulation forecast data	China	Simulations with stochastic and robust optimization	The validity and superiority of the recommended models were shown in case studies.
Khosravi et al. (2022), [148].	WF power generation and West Denmark electricity markets	Denmark	Stochastic scheduling, simulations with Monte-Carlo method	Increase in the profit was observed from the wind power management method.
Liu and Xu (2021), [150].	CMDC/2013	China	A market equilibrium model	The impact of wind power development on the spot market price results were explored for both long and short terms.
Niromandfam et al. (2020), [151].	Ordoudis et al. (2016), [154].	Iran	Modelling demand response utility function	It was shown that the proposed demand response utility function improved the wind generation profit in the DAM.

Table 5. *Cont.*

Author (s)	Data/Period	Country	Method (s)	Findings
Perez and Garcia-Rendon, (2021), [152].	Provided by the authors through the XM data/2018–2019	Colombia	Dispatch model	New bid prices in the market were determined by the firms through the structural model.
AEMO: Australia Energy Market Operator AIGS: All Island Grid Study CMDC: The China Meteorological Data Service Center DAM: Day-ahead market ERCOT: The electric reliability council of Texas			MCP: Market clearing price NEM: The Australian National Electricity Market PDFs: Probability density functions	WILMAR: A stochastic unit commitment model WRF: Mesoscale numerical weather prediction system

The long/middle term models include simulations (i.e., Monte Carlo simulations), market equilibrium models, production cost models, and fundamental models such as game theoretical approaches. The duration is longer or at least the considered period is middle-term in these models. They have remarkable theoretical contributions to the development of the EPF models by using economics terminology and approaches. Table 6 gives the main pros and cons of the reviewed methods and techniques based on the references that are given with Tables 2–5. Additionally, the last row of Table 6 shows the error comparison of the models that are selected among Tables 2–5.

Table 6. Main pros and cons of the reviewed methods based on the references in Tables 2–5.

Pros and Cons of the Reviewed Methods	Statistical Models (First-Part)	Statistical Models (Second-Part)	Artificial Intelligence and Hybrid/Ensemble Models	Middle/Long Term Models
Prons-1	Models allows the use of data by converting them from hourly to daily, which reduce unwanted and excessive noise. Their implementtion are easy.	Conditional heteroscedasticity models truly explain the volatilities in prices (i.e., seasonality, mean reversion, and jumps). Dynamic effects can be considered.	These models display improved forcasting performance in terms of consistency, accuracy, and statistical tests). High-frequency electricity price data forecasts are possible.	More realistic modes can be possible to visualize the market players' behaviours (i.e., risk management preferences).
Prons-2	Model allows omitting variables which their inclusion in regressions may generate an endogeneity problem. They are wide-spread preferred models.	The negative electricity prices can be included into the models, which helps to conduct analysis without shifting or cutting off the series.	Private information and imperfect market structure (i.e., oligopolies) can be included and represented with these models.	Theorethical economic models (i.e., Nash Equilibrium conditions) can be implemented with simulations.
Prons-3	Models allows to control the seasonal effects by introducing time dummies.	The causality tests can be implemented in the context of multivariate during off-peak hours, peak hours, and all hours.	These methods are capable of learning lon-term dependencies. They cen control how information is abandoned or memorized throughout time.	Strategical behaviours of the market participants can be modeled and simulated.
Prons-4	Binary variables for the weekend can be included in models.	More accurate estimations of load and wind with these models might improve EPF.	These models are reliable and robust for the system's complexity. Specifically, the ensemble methods have better results than their individual equivalents.	Parametric and nonparamatric methods can be simultaneously implemented.
Prons-5	Yearly, monthly, daily, and hourly dummies can be used to control for systematic demand changes.	These models (i.e., ARX) can utilize both the information on system forecasts and actual past realizations of these variables.	Decision-making strategies can be done with these model and these models can be implemented for other regions to improve EPF efficiency.	Seasonal effects can be simulated effectively.

Table 6. *Cont.*

Pros and Cons of the Reviewed Methods	Statistical Models (First-Part)	Statistical Models (Second-Part)	Artificial Intelligence and Hybrid/Ensemble Models	Middle/Long Term Models
Cons-1	There can be a lack of certainity on estimations of net effects for individual consumers. Estimated prices can be different (i.e., higher) than observed spot market prices.	The stochastic nature of weather conditions causes the volatilities of wind power. This effects electiricity prices electricity price spikes occur.	The decion-making rules are difficult to validate. The implementations might be time-consuming.	The models might be case dependent and different findings can be obtained for other situtaitions.
Cons-2	The differences in wind load profiles can affect the hours of the day and electricity prices can be dependent on these changes.	Mean absolute errors might not work properly when the models with more variables are considered.	These methods have a significantly increased computational burden.	Prediction of wind power effect on prices is difficult due to the wide range of factors (i.e., uncertain demand, several contingencies depend on long-term forecasting intervals).
Cons-3	Many of the variables tend to show near-unit root, or autoregrsssive properties; therefore, lags of the variables should be included into the models.	The system of equations need many parameters and the estimation of the coefficients are reletively difficult or complex.	Irrelevent asssumtions might block or decrease the performance of the estimator.	If the computation time increases with problem size, this might weaken the solution capabilitiy of the concentrated problem.
Cons-4	Possible endogeneity problems cause from either omitted variables or reverse causalities (i.e., the aggregate or average electricity demand).	ARMA type models are bounded by the assumption of constant variance that yields inconsistancy through volatility.	Various open-source software platforms might be needed, so that any researchers can implement the codes as benchmarks in their individual studies.	
Error comparison of the models	–	Lasso (Ziel, 2016) [85], MAAPE (%): 6.604, RMSE: 2.715, MAE: 1.819	Ensemble learning model (Bhatia, 2021) [101], MAAPE (%): 5.132, RMSE: 2.156, MAE: 1.385	–

Note: The last row of Table 6 shows the comparison of the Lasso and Ensemble learning models in terms of mean arctangent absolute percentage error (MAAPE), mean absolute error (MAE), and root mean squared error (RMSE).

4. Discussion of Forecasting Models on Electricity Markets

Electricity price and load are determined by day-ahead, intra-day, and balancing markets all around the world; however, research shows that, although its data are usually publicly available, market clearing price forecasting is more complex (i.e., fuel prices; equipment outages; and the nature of the market clearing price depends on the hourly loads creates this complexity [155]) than the load price forecasting.

Forecasting the electricity market's prices is needed as a result of the dynamic features of markets, moving from deregulated to regulated form, that cause price volatility. Thereby, well performed MCP estimation and its confidence interval prediction may help power producers and its utilities when submitting bids in cases that are more risk-free (i.e., they can adjust their producers' supply and profits) [155]. Moreover, with reliable daily price forecasting, energy service companies or producers are able to lay out better financial contracts or bilateral ones. The complexity of forecasting electricity markets price and load is also dependent on the increasing number of employed variables as input for better accuracy [64,112]. Thereby, the trend in methodologies moves to more sophisticated instruments, such as hybrid models, as shown and discussed in this review.

In addition to the explanation of operating principles of the electricity market, it is understood from the papers examined in this review that renewable energy resources should be preferred, transforming the structure of electricity markets for better environment conditions with low-carbon levels. Incentives and supply security can be the instruments for all countries [156].

Many methods and models have been developed for the EPF of markets for the last two decades. As a result of the stochastic and nonlinear nature of statistical models and price series, autoregression, moving average, exponential smoothing, and their variants [33,157] have shown to be insufficient [49]. The artificial intelligence models are able to capture non-linearity and complexities and flexible [47,158–160].

Artificial neural networks are outstanding for short-term forecasting, and they are efficiently applicable for electricity markets [161], being more accurate and robust than autoregressive (AR) models. The research [48] uses artificial neural network models to display the strong impact of electricity price on the trend load and MCP. Singhal and Swarup [48] apply artificial neural network models to study the dependency of electricity price in MCP and electricity load. Wang et al. [159] implement a deep neural network model to forecast the price in US electricity markets, differently from conventional models of neural networks. This model supports vector regression. On the other hand, since the price series are volatile, the neural network models have potential to lose the properties of the value of prices [64]. Moreover, neural networks are not convenient for too short-term predictions, since they need high training time. As a result of the aforementioned issues, artificial intelligence models have handicaps in perfect price forecasting [108].

Relying on a sole forecasting electricity price model may fail in the treatment of network features in the short term. In those circumstances, hybrid models can be a better alternative for price forecasting. An example of a hybrid model which is a composition of a stochastic approach with a neural network model is given in [135]. Ghayekhloo et al. [136] show hybrid models that include game theoretic approaches. Signal decomposition methods are also used in hybrid models such as empirical mode decomposition and wavelet transform; the examples are given in [115,162,163]. Although the performance is significantly improved by those models, the computational cost can be disadvantageous [101].

5. Conclusions

The power industry is rapidly growing all over the world, and renewable energy resources are one of the most vital components in electricity production. Besides, renewable energy has environmentally friendly features (i.e., a considerable reduction of emission helps to mitigate global warming). To this end, increasing wind energy utilization is a challenge to provide electricity power for electricity markets. For the last two decades, the electricity market mechanisms have been faced with regulation procedures designed by decision and policy-making processes. The competition is the key factor to decreasing the cost of electricity and reliably meeting-demand solutions. However, the price spikes and price volatilities, due to various environmental and business factors, are the handicaps of this commodity. These handicaps encourage researchers to produce more effective instruments, techniques, and solutions.

This review paper gathers the latest electricity price and load forecasting techniques and discusses their strengths and weaknesses. Nevertheless, electricity trading markets are becoming more sophisticated, with novel types of contracts in the bilateral transactions or organized markets due to an existing free market competition rule. The independent transmission system operators for each specific market have the responsibility of controlling the entire transmission networks. The price mechanism operates with market clearing price, which is obtained by the law of supply and demand curves that are determined in the day-ahead markets. The price deviations caused by supply and demand forces are corrected in balancing power markets by transmission system operators. Moreover, the intra-day markets are functioning as a bridge between the day-ahead markets and balancing markets. Market participants, who do not sell their entire power or do not take their positions in the day-ahead markets, have the alternative to sell or buy the needed power in the intra-day markets.

As a methodology, this review paper follows the historical and structural development of electricity markets, price and load forecasting methods, and recent trends in wind energy generation, transmission, and consumption. The findings that are based on the considered studies in this review reveal that:

The merit order effect is found for wind power generation, which means that wind power decreases wholesale price of electricity, however, it increases its volatility.

The volatility of wind power is induced by the stochastic character of weather conditions; therefore, both the parametric and non-parametric techniques might be needed in the calculations. Moreover, this indirectly effects the market clearing prices; however, the volatility of electricity prices is driven by the market design.

Technically, the models can be calibrated by transforming data, known as variance stabilizing transformation, which yields more accurate predictions along with less spikes and lower variation features of data.

As the EPF and load methods tend to be explained more dimensionally (i.e., hybrid methods including deep learning and artificial intelligence), the performance of the methods increase in terms of accuracy, stability, and consistency. Besides, both the linear and the non-linear nature of electricity price data can be observed in this way.

The regulatory interventions due to Covid-19 pandemic and the carbon pricing mechanism might have an adverse effect on electricity price dynamics. However, inventions of new vaccines and pills and prevalent use of renewable energy sources (i.e., wind and solar energy) will lessen the unpredicted effects of Covid-19 and carbon emissions.

Nevertheless, extreme weather events that are related with climate change seem a barrier for electricity market participants through wind energy production in the near future. Therefore, future studies may consider those facts and propose new forecasting techniques and improvements for better market operations. As a practical solution proposal, a cooperation between government, energy producers, manufacturers, and researchers in developing countries might lead to the start of arrangements whereby produced power can be directly delivered to energy-intensive factories, such as fertilizer factories (i.e., fertilizer industry require significant electricity in the world). Therefore, energy transfer losses can be prevented and, with special agreements, the manufacturers can benefit from these arrangements as a means of production cost reduction and wind farm owners can benefit from the utilization of produced electricity without any restriction. As a theoretical solution proposal, research has demonstrated that a large installed capacity of wind energy might reduce wind power variability. Thereby, smooth wind generation could be possible by utilizing storage optimization systems and flexible electricity interconnections (i.e., high voltage direct current systems with voltage source converters operating for wind farms).

Author Contributions: Study conception and design: H.A. and F.P.G.M.; acquisition, analysis, drawing figures, and interpretation of data: H.A. and F.P.G.M.; drafting of manuscript: H.A. and F.P.G.M.; critical revision: H.A. and F.P.G.M. All authors have read and agreed to the published version of the manuscript.

Funding: This research received no external funding.

Institutional Review Board Statement: Not applicable.

Informed Consent Statement: Not applicable.

Data Availability Statement: Not applicable.

Acknowledgments: The work reported herewith has been financially by the Dirección General de Universidades, Investigación e Innovación of Castilla-La Mancha, under Research Grant ProSeaWind project (Ref.: SBPLY/19/180501/000102). We are grateful to three anonymous reviewers that helped us to improve the paper.

Conflicts of Interest: The authors declare no conflict of interest.

Nomenclature

AEMO	Australia Energy Market Operator
AIGS	All Island Grid Study
APX	Amsterdam Power Exchange
AR	Autoregression
ARDL	Autoregressive distributed lag
ARMAX	Autoregressive moving average model with exogenous regressors
ARX	Auto-regressive with eXternal model input
Belpex	Belgian Power Exchange
BPM	Balancing power market
CalPX	California Power Exchange
CMDC	The China Meteorological Data Service Center
CRE	Energy Regulatory Commission
DAM	Day-ahead market
EEX	European Energy Exchange
eGARCH	Exponential generalized autoregressive conditional heteroskedasticity
ENTSO-E	European Network of Transmission System Operators for Electricity
EPFs	Electricity price forecasts
EEX	European Energy Exchange
EPEX	European Power Exchange
EPIAS	Energy Exchange Istanbul
ERCOT	The electric reliability council of Texas
EXAA	Energy Exchange Austria
GARCH	Generalized autoregressive conditional heteroskedasticity
GEFCom	Global Energy Forecasting Competition
GME	Gestore dei Mercati Energetici
IDM	Intra-day market
IPEX	Italian Power Exchange
LASSO	Least absolute shrinkage and selection operator
LPX	Leipzig Power Exchange
LSSVM	Shrinkage and selection operator least squares support vector machine
MCP	Market clearing price
MISO	Midwest ISO
NYISO	New York ISO
NEM	Australian National Electricity Market
NSW	New South Wales
NYISO	New York Independent System Operator
OLS	Ordinary least squares
OMEL	Operadora del mercado espanol de electricidad
PJM	Pennsylvania-New Jersey Maryland Interconnection
PolPX	Polish Power Exchange
QRM	Quantile Regression Machine
SCAR	Seasonal component autoregressive
TSO	Transmission system operator
UKPX	UK Power Exchange
VAR	Vector autoregressive
WILMAR	A stochastic unit commitment model
WRF	Mesoscale numerical weather prediction system

References

1. Weron, R. *Modeling and Forecasting Electricity Loads and Prices: A Statistical Approach*; Wiley Finance Series; John Wiley & Sons: Chichester, UK, 2006.
2. Kaminski, V. *Energy Markets/Vincent Kaminski*; Risk Books: London, UK, 2012.
3. Mohammad, S.; Hatim, Y.; Zuyi, L. *Market Operations in Electric Power Systems: Forecasting, Scheduling, and Risk Management*; John Wiley & Sons: Hoboken, NJ, USA, 2002.
4. Banaei, M.; Raouf-Sheybani, H.; Oloomi-Buygi, M.; Boudjadar, J. Impacts of large-scale penetration of wind power on day-ahead electricity markets and forward contracts. *Int. J. Electr. Power Energy Syst.* **2021**, *125*, 106450. [CrossRef]

5. Márquez, F.P.G.; Karyotakis, A.; Papaelias, M. *Renewable Energies: Business Outlook 2050*; Springer: Berlin, Germany, 2018.
6. Salam, R.A.; Amber, K.P.; Ratyal, N.I.; Alam, M.; Akram, N.; Muñoz, C.Q.G.; Márquez, F.P.G. An Overview on Energy and Development of Energy Integration in Major South Asian Countries: The Building Sector. *Energies* **2020**, *13*, 5776. [CrossRef]
7. Dey, B.; Márquez, F.P.G.; Basak, S.K. Smart Energy Management of Residential Microgrid System by a Novel Hybrid MG-WOSCACSA Algorithm. *Energies* **2020**, *13*, 3500. [CrossRef]
8. Dey, B.; Raj, S.; Mahapatra, S.; Márquez, F.P.G. Optimal scheduling of distributed energy resources in microgrid systems based on electricity market pricing strategies by a novel hybrid optimization technique. *Int. J. Electr. Power Energy Syst.* **2022**, *134*, 107419. [CrossRef]
9. Bunn, D.W. *Modelling Prices in Competitive Electricity Markets*; Wiley Finance Series; John Wiley & Sons: London, UK, 2004.
10. Eydeland, A.; Wolyniec, K. *Energy and Power Risk Management: New Developments in Modeling, Pricing, and Hedging*; John Wiley & Sons: Hoboken, NJ, USA, 2003.
11. Singh, S.; Fozdar, M.; Malik, H.; Fernández Moreno, M.D.V.; García Márquez, F.P. Influence of Wind Power on Modeling of Bidding Strategy in a Promising Power Market with a Modified Gravitational Search Algo-rithm. *Appl. Sci.* **2021**, *11*, 4438. [CrossRef]
12. Golmohamadi, H.; Asadi, A. A multi-stage stochastic energy management of responsive irrigation pumps in dynamic electricity markets. *Appl. Energy* **2020**, *265*, 114804. [CrossRef]
13. Qian, Z.; Pei, Y.; Zareipour, H.; Chen, N. A review and discussion of decomposition-based hybrid models for wind energy forecasting applications. *Appl. Energy* **2019**, *235*, 939–953. [CrossRef]
14. Albadi, M.; El-Saadany, E. A summary of demand response in electricity markets. *Electr. Power Syst. Res.* **2008**, *78*, 1989–1996. [CrossRef]
15. Soloviova, M.; Vargiolu, T. Efficient representation of supply and demand curves on day-ahead electricity markets. *J. Energy Mark.* **2021**, *14*. [CrossRef]
16. Oskouei, M.Z.; Mirzaei, M.A.; Mohammadi-Ivatloo, B.; Shafiee, M.; Marzband, M.; Anvari-Moghaddam, A. A hybrid robust-stochastic approach to evaluate the profit of a multi-energy retailer in tri-layer energy markets. *Energy* **2021**, *214*, 118948. [CrossRef]
17. Grimm, V.; Rückel, B.; Sölch, C.; Zöttl, G. The impact of market design on transmission and generation investment in electricity markets. *Energy Econ.* **2021**, *93*, 104934. [CrossRef]
18. Elsisi, M.; Bazmohammadi, N.; Guerrero, J.M.; Ebrahim, M.A. Energy management of controllable loads in multi-area power systems with wind power penetration based on new supervisor fuzzy nonlinear sliding mode control. *Energy* **2021**, *221*, 119867. [CrossRef]
19. Basit, A.; Hansen, A.D.; Sørensen, P.E.; Giannopoulos, G. Real-time impact of power balancing on power system operation with large scale integration of wind power. *J. Mod. Power Syst. Clean Energy* **2015**, *5*, 202–210. [CrossRef]
20. Elsisi, M.; Soliman, M. Optimal design of robust resilient automatic voltage regulators. *ISA Trans.* **2021**, *108*, 257–268. [CrossRef]
21. Elsisi, M.; Soliman, M.; Aboelela, M.; Mansour, W. Improving the grid frequency by optimal design of model predictive control with energy storage devices. *Optim. Control. Appl. Methods* **2018**, *39*, 263–280. [CrossRef]
22. Elsisi, M. New variable structure control based on different meta-heuristics algorithms for frequency regulation considering nonlinearities effects. *Int. Trans. Electr. Energy Syst.* **2020**, *30*, 12428. [CrossRef]
23. Elsisi, M. New design of robust PID controller based on meta-heuristic algorithms for wind energy conversion system. *Wind. Energy* **2019**, *23*, 391–403. [CrossRef]
24. Lei, M.; Shiyan, L.; Chuanwen, J.; Hongling, L.; Yan, Z. A review on the forecasting of wind speed and generated power. *Renew. Sustain. Energy Rev.* **2009**, *13*, 915–920. [CrossRef]
25. Foley, A.M.; Leahy, P.G.; Marvuglia, A.; McKeogh, E.J. Current methods and advances in forecasting of wind power generation. *Renew. Energy* **2012**, *37*, 1–8. [CrossRef]
26. Al-Yahyai, S.; Charabi, Y.; Gastli, A. Review of the use of Numerical Weather Prediction (NWP) Models for wind energy assessment. *Renew. Sustain. Energy Rev.* **2010**, *14*, 3192–3198. [CrossRef]
27. Peter, Z.; Aaron, P.; Georg, E. *Energy Economics: Theory and Applications (Springer Texts in Business and Economics)*; Springer: Berlin/Heidelberg, Germany, 2017. (In English)
28. Ocker, F.; Jaenisch, V. The way towards European electricity intraday auctions—Status quo and future developments. *Energy Policy* **2020**, *145*, 111731. [CrossRef]
29. Chaves-Ávila, J.P.; Fernandes, C. The Spanish intraday market design: A successful solution to balance renewable generation? *Renew. Energy* **2015**, *74*, 422–432. [CrossRef]
30. Maciejowska, K.; Nitka, W.; Weron, T. Enhancing load, wind and solar generation for day-ahead forecasting of electricity prices. *Energy Econ.* **2021**, *99*, 105273. [CrossRef]
31. Gianfreda, A.; Parisio, L.; Pelagatti, M.; Gianfreda, A.; Parisio, L.; Pelagatti, M. The Impact of RES in the Ital-ian Day-Ahead and Balancing Markets. *Energy J.* **2016**, *37*, 161–184. Available online: https://stanford.idm.oclc.org/login?url=https://search.ebscohost.com/login.aspx?direct=true&site=eds-live&db=edsjsr&AN=edsjsr.26606234 (accessed on 1 October 2021).
32. Koch, C.; Hirth, L. Short-term electricity trading for system balancing: An empirical analysis of the role of intraday trading in balancing Germany's electricity system. *Renew. Sustain. Energy Rev.* **2019**, *113*, 109275. [CrossRef]

33. Girish, G. Spot electricity price forecasting in Indian electricity market using autoregressive-GARCH models. *Energy Strat. Rev.* **2016**, *11–12*, 52–57. [CrossRef]
34. Day ahead Market Web Application, Used Guide. 2016. Available online: https://www.epias.com.tr/wp-content/uploads/2017/09/ENG-DAM-User-Guide_vol_5.pdf (accessed on 27 September 2021).
35. EPİAŞ. Day-ahead Market. EPİAŞ. Available online: https://www.epias.com.tr/en/day-ahead-market/introduction/ (accessed on 27 September 2021).
36. Green, R. Electricity liberalisation in Europe—How competitive will it be? *Energy Policy* **2006**, *34*, 2532–2541. [CrossRef]
37. Moreno, B.; López, A.J.; García-Álvarez, M.T. The electricity prices in the European Union. The role of renewable energies and regulatory electric market reforms. *Energy* **2012**, *48*, 307–313. [CrossRef]
38. Weber, C. Adequate intraday market design to enable the integration of wind energy into the European power systems. *Energy Policy* **2010**, *38*, 3155–3163. [CrossRef]
39. EPİAŞ. Intra-Day Market. EPİAŞ. Available online: https://www.epias.com.tr/en/intra-day-market/introduction/ (accessed on 27 September 2021).
40. Hagemann, S.; Weber, C. *An Empirical Analysis of Liquidity and Its Determinants in the German Intraday Market for Electricity*; EWL Working Paper No. 17/2013; University of Duisburg-Essen: Duisburg, Germany, 2013. [CrossRef]
41. Le, H.L.; Ilea, V.; Bovo, C. Integrated European intra-day electricity market: Rules, modeling and analysis. *Appl. Energy* **2019**, *238*, 258–273. [CrossRef]
42. Hu, X.; Jaraitė, J.; Kažukauskas, A. The effects of wind power on electricity markets: A case study of the Swedish intraday market. *Energy Econ.* **2021**, *96*, 105159. [CrossRef]
43. Dinler, A. Reducing balancing cost of a wind power plant by deep learning in market data: A case study for Turkey. *Appl. Energy* **2021**, *289*, 116728. [CrossRef]
44. Dey, B.; Bhattacharyya, B.; Márquez, F.P.G. A hybrid optimization-based approach to solve environment constrained economic dispatch problem on microgrid system. *J. Clean. Prod.* **2021**, *307*, 127196. [CrossRef]
45. EPİAŞ. Balancing Market. Available online: https://www.epias.com.tr/genel-esaslar/ (accessed on 27 September 2021).
46. Vandezande, L.; Meeus, L.; Belmans, R.; Saguan, M.; Glachant, J.-M. Well-functioning balancing markets: A prerequisite for wind power integration. *Energy Policy* **2010**, *38*, 3146–3154. [CrossRef]
47. Anbazhagan, S.; Kumarappan, N. Day-Ahead Deregulated Electricity Market Price Forecasting Using Recurrent Neural Network. *IEEE Syst. J.* **2012**, *7*, 866–872. [CrossRef]
48. Singhal, D.; Swarup, S. Electricity price forecasting using artificial neural networks. *Int. J. Electr. Power Energy Syst.* **2011**, *33*, 550–555. [CrossRef]
49. Chan, S.-C.; Tsui, K.M.; Wu, H.C.; Hou, Y.; Wu, Y.C.; Wu, F.F. Load/Price Forecasting and Managing Demand Response for Smart Grids: Methodologies and Challenges. *IEEE Signal Process. Mag.* **2012**, *29*, 68–85. [CrossRef]
50. Kalay, O. Electricity Load and Price Forecasting of Turkish Electricity Markets. Master's Thesis, Middle East Technical University, Ankara, Turkey, 2018.
51. Tashpulatov, S. Estimating the volatility of electricity prices: The case of the England and Wales wholesale electricity market. *Energy Policy* **2013**, *60*, 81–90. [CrossRef]
52. Hellström, J.; Lundgren, J.; Yu, H. Why do electricity prices jump? Empirical evidence from the Nordic electricity market. *Energy Econ.* **2012**, *34*, 1774–1781. [CrossRef]
53. Weron, R.; Zator, M. Revisiting the relationship between spot and futures prices in the Nord Pool electricity market. *Energy Econ.* **2014**, *44*, 178–190. [CrossRef]
54. Philpott, A.; Read, G.; Batstone, S.; Miller, A. The New Zealand Electricity Market: Challenges of a Renewable Energy System. *IEEE Power Energy Mag.* **2019**, *17*, 43–52. [CrossRef]
55. Karabiber, O.A.; Xydis, G. Electricity Price Forecasting in the Danish Day-Ahead Market Using the TBATS, ANN and ARIMA Methods. *Energies* **2019**, *12*, 928. [CrossRef]
56. Su, W. The Role of Customers in the U.S. Electricity Market: Past, Present and Future. *Electr. J.* **2014**, *27*, 112–125. [CrossRef]
57. Olsson, M.; Perninge, M.; Söder, L. Modeling real-time balancing power demands in wind power systems using stochastic differential equations. *Electr. Power Syst. Res.* **2010**, *80*, 966–974. [CrossRef]
58. Fan, S.; Chen, L. Short-Term Load Forecasting Based on an Adaptive Hybrid Method. *IEEE Trans. Power Syst.* **2006**, *21*, 392–401. [CrossRef]
59. Ranaweera, D.K.; Karady, G.G.; Farmer, R.G. Economic impact analysis of load forecasting. *IEEE Trans. Power Syst.* **1997**, *12*, 1388–1392. [CrossRef]
60. Abu-El-Magd, M.A.; Sinha, N.K. Short-Term Load Demand Modeling and Forecasting: A Review. *IEEE Trans. Syst. Man Cybern.* **1982**, *12*, 370–382. [CrossRef]
61. Moghram, I.; Rahman, S. Analysis and evaluation of five short-term load forecasting techniques. *IEEE Trans. Power Syst.* **1989**, *4*, 1484–1491. [CrossRef]
62. Amjady, N. Short-Term Bus Load Forecasting of Power Systems by a New Hybrid Method. *IEEE Trans. Power Syst.* **2007**, *22*, 333–341. [CrossRef]
63. Yun, Z.; Quan, Z.; Caixin, S.; Shaolan, L.; Yuming, L.; Yang, S. RBF Neural Network and ANFIS-Based Short-Term Load Forecasting Approach in Real-Time Price Environment. *IEEE Trans. Power Syst.* **2008**, *23*, 853–858. [CrossRef]

64. Cerjan, M.; Krzelj, I.; Vidak, M.; Delimar, M. A literature review with statistical analysis of electricity price forecasting methods. In Proceedings of the Eurocon 2013, Zagreb, Croatia, 1–4 July 2013; pp. 756–763.
65. Clò, S.; Cataldi, A.; Zoppoli, P. The merit-order effect in the Italian power market: The impact of solar and wind generation on national wholesale electricity prices. *Energy Policy* **2015**, *77*, 79–88. [CrossRef]
66. Woo, C.; Moore, J.; Schneiderman, B.; Ho, T.; Olson, A.; Alagappan, L.; Chawla, K.; Toyama, N.; Zarnikau, J. Merit-order effects of renewable energy and price divergence in California's day-ahead and real-time electricity markets. *Energy Policy* **2016**, *92*, 299–312. [CrossRef]
67. Cludius, J.; Forrest, S.; MacGill, I. Distributional effects of the Australian Renewable Energy Target (RET) through wholesale and retail electricity price impacts. *Energy Policy* **2014**, *71*, 40–51. [CrossRef]
68. Cludius, J.; Hermann, H.; Matthes, F.C.; Graichen, V. The merit order effect of wind and photovoltaic electricity generation in Germany 2008–2016: Estimation and distributional implications. *Energy Econ.* **2014**, *44*, 302–313. [CrossRef]
69. Csereklyei, Z.; Qu, S.; Ancev, T. The effect of wind and solar power generation on wholesale electricity prices in Australia. *Energy Policy* **2019**, *131*, 358–369. [CrossRef]
70. Forrest, S.; MacGill, I. Assessing the impact of wind generation on wholesale prices and generator dispatch in the Australian National Electricity Market. *Energy Policy* **2013**, *59*, 120–132. [CrossRef]
71. Maciejowska, K. Assessing the impact of renewable energy sources on the electricity price level and variability—A quantile regression approach. *Energy Econ.* **2020**, *85*, 104532. [CrossRef]
72. Westgaard, S.; Fleten, S.-E.; Negash, A.; Botterud, A.; Bogaard, K.; Verling, T.H. Performing price scenario analysis and stress testing using quantile regression: A case study of the Californian electricity market. *Energy* **2021**, *214*, 118796. [CrossRef]
73. Spodniak, P.; Ollikka, K.; Honkapuro, S. The impact of wind power and electricity demand on the relevance of different short-term electricity markets: The Nordic case. *Appl. Energy* **2021**, *283*, 116063. [CrossRef]
74. Ketterer, J.C. The impact of wind power generation on the electricity price in Germany. *Energy Econ.* **2014**, *44*, 270–280. [CrossRef]
75. Kyritsis, E.; Andersson, J.; Serletis, A. Electricity prices, large-scale renewable integration, and policy implications. *Energy Policy* **2017**, *101*, 550–560. [CrossRef]
76. Mwampashi, M.M.; Nikitopoulos, C.S.; Konstandatos, O.; Rai, A. Wind generation and the dynamics of electricity prices in Australia. *Energy Econ.* **2021**, *103*, 105547. [CrossRef]
77. Pape, C.; Hagemann, S.; Weber, C. Are fundamentals enough? Explaining price variations in the German day-ahead and intraday power market. *Energy Econ.* **2016**, *54*, 376–387. [CrossRef]
78. Maciejowska, K.; Nitka, W.; Weron, T. Day-Ahead vs. Intraday—Forecasting the Price Spread to Maximize Economic Benefits. *Energies* **2019**, *12*, 631. [CrossRef]
79. Nowotarski, J.; Raviv, E.; Trueck, S.; Weron, R. An empirical comparison of alternative schemes for combining electricity spot price forecasts. *Energy Econ.* **2014**, *46*, 395–412. [CrossRef]
80. Paraschiv, F.; Erni, D.; Pietsch, R. The impact of renewable energies on EEX day-ahead electricity prices. *Energy Policy* **2014**, *73*, 196–210. [CrossRef]
81. Marcjasz, G.; Serafin, T.; Weron, R. Selection of Calibration Windows for Day-Ahead Electricity Price Forecasting. *Energies* **2018**, *11*, 2364. [CrossRef]
82. Uniejewski, B.; Nowotarski, J.; Weron, R. Automated Variable Selection and Shrinkage for Day-Ahead Electricity Price Forecasting. *Energies* **2016**, *9*, 621. [CrossRef]
83. Uniejewski, B.; Weron, R. Efficient Forecasting of Electricity Spot Prices with Expert and LASSO Models. *Energies* **2018**, *11*, 2039. [CrossRef]
84. Uniejewski, B.; Marcjasz, G.; Weron, R. Understanding intraday electricity markets: Variable selection and very short-term price forecasting using LASSO. *Int. J. Forecast.* **2019**, *35*, 1533–1547. [CrossRef]
85. Ziel, F. Forecasting Electricity Spot Prices Using Lasso: On Capturing the Autoregressive Intraday Structure. *IEEE Trans. Power Syst.* **2016**, *31*, 4977–4987. [CrossRef]
86. Uniejewski, B.; Marcjasz, G.; Weron, R. On the importance of the long-term seasonal component in day-ahead electricity price forecasting. *Energy Econ.* **2019**, *79*, 171–182. [CrossRef]
87. Ziel, F.; Weron, R. Day-ahead electricity price forecasting with high-dimensional structures: Univariate vs. multivariate modeling frameworks. *Energy Econ.* **2018**, *70*, 396–420. [CrossRef]
88. Gürtler, M.; Paulsen, T. The effect of wind and solar power forecasts on day-ahead and intraday electricity prices in Germany. *Energy Econ.* **2018**, *75*, 150–162. [CrossRef]
89. Serafin, T.; Uniejewski, B.; Weron, R. Averaging Predictive Distributions Across Calibration Windows for Day-Ahead Electricity Price Forecasting. *Energies* **2019**, *12*, 2561. [CrossRef]
90. Ziel, F.; Steinert, R. Probabilistic mid- and long-term electricity price forecasting. *Renew. Sustain. Energy Rev.* **2018**, *94*, 251–266. [CrossRef]
91. Hubicka, K.; Marcjasz, G.; Weron, R. A Note on Averaging Day-Ahead Electricity Price Forecasts Across Calibration Windows. *IEEE Trans. Sustain. Energy* **2019**, *10*, 321–323. [CrossRef]
92. Tibshirani, R. Regression Shrinkage and Selection Via the Lasso. *J. R. Stat. Soc. Ser. B* **1996**, *58*, 267–288. [CrossRef]
93. Raviv, E.; Bouwman, K.E.; van Dijk, D. Forecasting day-ahead electricity prices: Utilizing hourly prices. *Energy Econ.* **2015**, *50*, 227–239. [CrossRef]

94. Nowotarski, J.; Weron, R. On the importance of the long-term seasonal component in day-ahead electricity price forecasting. *Energy Econ.* **2016**, *57*, 228–235. [CrossRef]
95. Swider, D.J.; Weber, C. Extended ARMA models for estimating price developments on day-ahead electricity markets. *Electr. Power Syst. Res.* **2007**, *77*, 583–593. [CrossRef]
96. Worthington, A.; Kay-Spratley, A.; Higgs, H. Transmission of prices and price volatility in Australian electricity spot markets: A multivariate GARCH analysis. *Energy Econ.* **2005**, *27*, 337–350. [CrossRef]
97. Higgs, H.; Lien, G.; Worthington, A.C. Australian evidence on the role of interregional flows, production capacity, and generation mix in wholesale electricity prices and price volatility. *Econ. Anal. Policy* **2015**, *48*, 172–181. [CrossRef]
98. Han, L.; Kordzakhia, N.; Trück, S. Volatility spillovers in Australian electricity markets. *Energy Econ.* **2020**, *90*, 104782. [CrossRef]
99. Nelson, D.B. Conditional Heteroskedasticity in Asset Returns: A New Approach. *Econom. J. Econom. Soc.* **1991**, *59*, 347. [CrossRef]
100. Bordignon, S.; Bunn, D.W.; Lisi, F.; Nan, F. Combining day-ahead forecasts for British electricity prices. *Energy Econ.* **2013**, *35*, 88–103. [CrossRef]
101. Bhatia, K.; Mittal, R.; Varanasi, J.; Tripathi, M. An ensemble approach for electricity price forecasting in markets with renewable energy resources. *Util. Policy* **2021**, *70*, 101185. [CrossRef]
102. Li, W.; Becker, D.M. Day-ahead electricity price prediction applying hybrid models of LSTM-based deep learning methods and feature selection algorithms under consideration of market coupling. *Energy* **2021**, *237*, 121543. [CrossRef]
103. Yang, H.; Schell, K.R. Real-time electricity price forecasting of wind farms with deep neural network transfer learning and hybrid datasets. *Appl. Energy* **2021**, *299*, 117242. [CrossRef]
104. May, E.C.; Bassam, A.; Ricalde, L.J.; Soberanis, M.E.; Oubram, O.; Tzuc, O.M.; Alanis, A.Y.; Livas-García, A. Global sensitivity analysis for a real-time electricity market forecast by a machine learning approach: A case study of Mexico. *Int. J. Electr. Power Energy Syst.* **2022**, *135*, 107505. [CrossRef]
105. Yang, H.; Schell, K.R. GHTnet: Tri-Branch deep learning network for real-time electricity price forecasting. *Energy* **2022**, *238*, 122052. [CrossRef]
106. Bublitz, A.; Keles, D.; Fichtner, W. An analysis of the decline of electricity spot prices in Europe: Who is to blame? *Energy Policy* **2017**, *107*, 323–336. [CrossRef]
107. Nowotarski, J.; Weron, R. Recent advances in electricity price forecasting: A review of probabilistic forecasting. *Renew. Sustain. Energy Rev.* **2018**, *81*, 1548–1568. [CrossRef]
108. Weron, R. Electricity price forecasting: A review of the state-of-the-art with a look into the future. *Int. J. Forecast.* **2014**, *30*, 1030–1081. [CrossRef]
109. Osório, G.; Matias, J.; Catalão, J.P.S. Short-term wind power forecasting using adaptive neuro-fuzzy inference system combined with evolutionary particle swarm optimization, wavelet transform and mutual information. *Renew. Energy* **2015**, *75*, 301–307. [CrossRef]
110. Zhang, J.; Tan, Z.; Yang, S. Day-ahead electricity price forecasting by a new hybrid method. *Comput. Ind. Eng.* **2012**, *63*, 695–701. [CrossRef]
111. Amjady, N. Day-Ahead Price Forecasting of Electricity Markets by a New Fuzzy Neural Network. *IEEE Trans. Power Syst.* **2006**, *21*, 887–896. [CrossRef]
112. Lago, J.; Marcjasz, G.; De Schutter, B.; Weron, R. Forecasting day-ahead electricity prices: A review of state-of-the-art algorithms, best practices and an open-access benchmark. *Appl. Energy* **2021**, *293*, 116983. [CrossRef]
113. Chang, Z.; Zhang, Y.; Chen, W. Electricity price prediction based on hybrid model of adam optimized LSTM neural network and wavelet transform. *Energy* **2019**, *187*, 115804. [CrossRef]
114. Nazar, M.S.; Fard, A.E.; Heidari, A.; Shafie-Khah, M.; Catalão, J.P. Hybrid model using three-stage algorithm for simultaneous load and price forecasting. *Electr. Power Syst. Res.* **2018**, *165*, 214–228. [CrossRef]
115. Yang, Z.; Ce, L.; Lian, L. Electricity price forecasting by a hybrid model, combining wavelet transform, ARMA and kernel-based extreme learning machine methods. *Appl. Energy* **2017**, *190*, 291–305. [CrossRef]
116. Olamaee, J.; Mohammadi, M.; Noruzi, A.; Hosseini, S.M.H. Day-ahead price forecasting based on hybrid prediction model. *Complex.* **2016**, *21*, 156–164. [CrossRef]
117. Singh, N.; Mohanty, S.R.; Shukla, R.D. Short term electricity price forecast based on environmentally adapted generalized neuron. *Energy* **2017**, *125*, 127–139. [CrossRef]
118. Bento, P.; Pombo, J.; Calado, M.D.R.; Mariano, S.J.P.S. A bat optimized neural network and wavelet transform approach for short-term price forecasting. *Appl. Energy* **2018**, *210*, 88–97. [CrossRef]
119. Peter, S.E.; Raglend, I.J. Sequential wavelet-ANN with embedded ANN-PSO hybrid electricity price forecasting model for Indian energy exchange. *Neural Comput. Appl.* **2016**, *28*, 2277–2292. [CrossRef]
120. Anamika; Peesapati, R.; Kumar, N. Electricity Price Forecasting and Classification Through Wavelet–Dynamic Weighted PSO–FFNN Approach. *IEEE Syst. J.* **2018**, *12*, 3075–3084. [CrossRef]
121. Gao, W.; Sarlak, V.; Parsaei, M.R.; Ferdosi, M. Combination of fuzzy based on a meta-heuristic algorithm to predict electricity price in an electricity markets. *Chem. Eng. Res. Des.* **2018**, *131*, 333–345. [CrossRef]
122. Zhang, J.; Tan, Z.; Li, C. A Novel Hybrid Forecasting Method Using GRNN Combined With Wavelet Transform and a GARCH Model. *Energy Sources Part B Econ. Plan. Policy* **2015**, *10*, 418–426. [CrossRef]

123. Hong, Y.-Y.; Liu, C.-Y.; Chen, S.-J.; Huang, W.-C.; Yu, T.-H. Short-term LMP forecasting using an artificial neural network incorporating empirical mode decomposition. *Int. Trans. Electr. Energy Syst.* **2015**, *25*, 1952–1964. [CrossRef]

124. Zhang, J.-L.; Zhang, Y.-J.; Li, D.-Z.; Tan, Z.-F.; Ji, J.-F. Forecasting day-ahead electricity prices using a new integrated model. *Int. J. Electr. Power Energy Syst.* **2019**, *105*, 541–548. [CrossRef]

125. Kurbatsky, V.G.; Sidorov, D.; Spiryaev, V.; Tomin, N. Forecasting nonstationary time series based on Hilbert-Huang transform and machine learning. *Autom. Remote Control.* **2014**, *75*, 922–934. [CrossRef]

126. Gobu, B.; Jaikumar, S.; Arulmozhi, N.; Kanimozhi, P. Two-Stage Machine Learning Framework for Simultaneous Forecasting of Price-Load in the Smart Grid. In Proceedings of the 2018 17th IEEE International Conference on Machine Learning and Applications (ICMLA), Orlando, FL, USA, 17–20 December 2018; pp. 1081–1086.

127. Lahmiri, S. Comparing Variational and Empirical Mode Decomposition in Forecasting Day-Ahead Energy Prices. *IEEE Syst. J.* **2017**, *11*, 1907–1910. [CrossRef]

128. Varshney, H.; Sharma, A.; Kumar, R. A hybrid approach to price forecasting incorporating exogenous variables for a day ahead electricity Market. In Proceedings of the 2016 IEEE 1st International Conference on Power Electronics, Intelligent Control and Energy Systems (ICPEICES), Delhi, India, 4–6 July 2016; pp. 1–6.

129. Xiao, L.; Shao, W.; Yu, M.; Ma, J.; Jin, C. Research and application of a hybrid wavelet neural network model with the improved cuckoo search algorithm for electrical power system forecasting. *Appl. Energy* **2017**, *198*, 203–222. [CrossRef]

130. Khajeh, M.G.; Maleki, A.; Rosen, M.A.; Ahmadi, M.H. Electricity price forecasting using neural networks with an improved iterative training algorithm. *Int. J. Ambient. Energy* **2018**, *39*, 147–158. [CrossRef]

131. Bisoi, R.; Dash, P.K.; Das, P.P. Short-term electricity price forecasting and classification in smart grids using optimized multikernel extreme learning machine. *Neural Comput. Appl.* **2018**, *32*, 1457–1480. [CrossRef]

132. Kim, M.K. Short-term price forecasting of Nordic power market by combination Levenberg–Marquardt and Cuckoo search algorithms. *IET Gener. Transm. Distrib.* **2015**, *9*, 1553–1563. [CrossRef]

133. Pourdaryaei, A.; Mokhlis, H.; Illias, H.A.; Kaboli, S.H.A.; Ahmad, S. Short-Term Electricity Price Forecasting via Hybrid Backtracking Search Algorithm and ANFIS Approach. *IEEE Access* **2019**, *7*, 77674–77691. [CrossRef]

134. Ebrahimian, H.; Barmayoon, S.; Mohammadi, M.; Ghadimi, N. The price prediction for the energy market based on a new method. *Econ. Res.-Ekon. Istraživanja* **2018**, *31*, 313–337. [CrossRef]

135. Abedinia, O.; Amjady, N.; Shafie-Khah, M.; Catalão, J. Electricity price forecast using Combinatorial Neural Network trained by a new stochastic search method. *Energy Convers. Manag.* **2015**, *105*, 642–654. [CrossRef]

136. Ghayekhloo, M.; Azimi, R.; Ghofrani, M.; Menhaj, M.; Shekari, E. A combination approach based on a novel data clustering method and Bayesian recurrent neural network for day-ahead price forecasting of electricity markets. *Electr. Power Syst. Res.* **2019**, *168*, 184–199. [CrossRef]

137. Itaba, S.; Mori, H. An Electricity Price Forecasting Model with Fuzzy Clustering Preconditioned ANN. *Electr. Eng. Jpn.* **2018**, *204*, 10–20. [CrossRef]

138. Ghofrani, M.; Azimi, R.; Najafabadi, F.M.; Myers, N. A new day-ahead hourly electricity price forecasting framework. In Proceedings of the 2017 North American Power Symposium (NAPS), Morgantown, WV, USA, 17–19 September 2017; pp. 1–6. [CrossRef]

139. Itaba, S.; Mori, H. A Fuzzy-Preconditioned GRBFN Model for Electricity Price Forecasting. *Procedia Comput. Sci.* **2017**, *114*, 441–448. [CrossRef]

140. Zhou, L.; Wang, B.; Wang, Z.; Wang, F.; Yang, M. Seasonal classification and RBF adaptive weight based parallel combined method for day-ahead electricity price forecasting. In Proceedings of the 2018 IEEE Power & Energy Society Innovative Smart Grid Technologies Conference (ISGT), Washington, DC, USA, 19–22 February 2018; pp. 1–5.

141. Naz, A.; Javed, M.U.; Javaid, N.; Saba, T.; Alhussein, M.; Aurangzeb, K. Short-Term Electric Load and Price Forecasting Using Enhanced Extreme Learning Machine Optimization in Smart Grids. *Energies* **2019**, *12*, 866. [CrossRef]

142. Baldick, R. Wind and Energy Markets: A Case Study of Texas. *IEEE Syst. J.* **2011**, *6*, 27–34. [CrossRef]

143. Bell, W.P.; Wild, P.; Foster, J.; Hewson, M. Revitalising the wind power induced merit order effect to reduce wholesale and retail electricity prices in Australia. *Energy Econ.* **2017**, *67*, 224–241. [CrossRef]

144. Blakers, A.; Stocks, M.; Lu, B.; Cheng, C. The observed cost of high penetration solar and wind electricity. *Energy* **2021**, *233*, 121150. [CrossRef]

145. Cutler, N.J.; Boerema, N.D.; MacGill, I.F.; Outhred, H.R. High penetration wind generation impacts on spot prices in the Australian national electricity market. *Energy Policy* **2011**, *39*, 5939–5949. [CrossRef]

146. Denny, E.; Tuohy, A.; Meibom, P.; Keane, A.; Flynn, D.; Mullane, A.; O'Malley, M. The impact of increased interconnection on electricity systems with large penetrations of wind generation: A case study of Ireland and Great Britain. *Energy Policy* **2010**, *38*, 6946–6954. [CrossRef]

147. Elfarra, M.A.; Kaya, M. Estimation of electricity cost of wind energy using Monte Carlo simulations based on nonparametric and parametric probability density functions. *Alex. Eng. J.* **2021**, *60*, 3631–3640. [CrossRef]

148. Khosravi, M.; Afsharnia, S.; Farhangi, S. Stochastic power management strategy for optimal day-ahead scheduling of wind-HESS considering wind power generation and market price uncertainties. *Int. J. Electr. Power Energy Syst.* **2022**, *134*, 107429. [CrossRef]

149. Ji, Y.; Xu, Q.; Zhao, J.; Yang, Y.; Sun, L. Day-ahead and intra-day optimization for energy and reserve scheduling under wind uncertainty and generation outages. *Electr. Power Syst. Res.* **2021**, *195*, 107133. [CrossRef]

150. Liu, T.; Xu, J. Equilibrium strategy based policy shifts towards the integration of wind power in spot electricity markets: A perspective from China. *Energy Policy* **2021**, *157*, 112482. [CrossRef]
151. Niromandfam, A.; Yazdankhah, A.S.; Kazemzadeh, R. Modeling demand response based on utility function considering wind profit maximization in the day-ahead market. *J. Clean. Prod.* **2020**, *251*, 119317. [CrossRef]
152. Perez, A.; Garcia-Rendon, J.J. Integration of non-conventional renewable energy and spot price of electricity: A counterfactual analysis for Colombia. *Renew. Energy* **2021**, *167*, 146–161. [CrossRef]
153. Akdag, S.A.; Bagiorgas, H.; Mihalakakou, G. Use of two-component Weibull mixtures in the analysis of wind speed in the Eastern Mediterranean. *Appl. Energy* **2010**, *87*, 2566–2573. [CrossRef]
154. Ordoudis, C.; Pinson, P.; Morales, J.M.; Zugno, M. *An Updated Version of the IEEE RTS 24-Bus System for Electricity Market and Power System Operation Studies*; Technical University of Denmark: Kongens Lyngby, Denmark, 2016.
155. Amjady, N.; Hemmati, M. Energy price forecasting—Problems and proposals for such predictions. *IEEE Power Energy Mag.* **2006**, *4*, 20–29. [CrossRef]
156. International Energy Agency. Electricity Market Report. Available online: www.iea.org (accessed on 1 September 2021).
157. Zhao, Z.; Wang, C.; Nokleby, M.; Miller, C.J. Improving short-term electricity price forecasting using day-ahead LMP with ARIMA models. In Proceedings of the 2017 IEEE Power & Energy Society General Meeting, Chicago, IL, USA, 16–20 July 2017; pp. 1–5.
158. Lin, W.-M.; Gow, H.-J.; Tsai, M.-T. An enhanced radial basis function network for short-term electricity price forecasting. *Appl. Energy* **2010**, *87*, 3226–3234. [CrossRef]
159. Wang, L.; Zhang, Z.; Chen, J. Short-Term Electricity Price Forecasting With Stacked Denoising Autoencoders. *IEEE Trans. Power Syst.* **2016**, *32*, 2673–2681. [CrossRef]
160. Yan, X.; Chowdhury, N.A. Mid-term electricity market clearing price forecasting: A multiple SVM approach. *Int. J. Electr. Power Energy Syst.* **2014**, *58*, 206–214. [CrossRef]
161. Sahay, K.B.; Bhushan, S.K. One hour ahead price forecast of Ontario electricity market by using ANN. In Proceedings of the 2015 International Conference on Energy Economics and Environment (ICEEE), Greater Noida, India, 27–28 March 2015; pp. 1–6.
162. He, K.; Yu, L.; Tang, L. Electricity price forecasting with a BED (Bivariate EMD Denoising) methodology. *Energy* **2015**, *91*, 601–609. [CrossRef]
163. Qiu, X.; Suganthan, P.; Amaratunga, G.A. Short-term Electricity Price Forecasting with Empirical Mode Decomposition based Ensemble Kernel Machines. *Procedia Comput. Sci.* **2017**, *108*, 1308–1317. [CrossRef]

energies

MDPI

Article

Short-Term Electricity Prices Forecasting Using Functional Time Series Analysis

Faheem Jan [†], Ismail Shah [*,†] and Sajid Ali [†]

Department of Statistics, Quaid-i-Azam University, Islamabad 45320, Pakistan;
faheemjan93@yahoo.com (F.J.); sajidali@qau.edu.pk (S.A.)
* Correspondence: ishah@qau.edu.pk
† These authors contributed equally to this work.

Abstract: In recent years, efficient modeling and forecasting of electricity prices became highly important for all the market participants for developing bidding strategies and making investment decisions. However, as electricity prices exhibit specific features, such as periods of high volatility, seasonal patterns, calendar effects, nonlinearity, etc., their accurate forecasting is challenging. This study proposes a functional forecasting method for the accurate forecasting of electricity prices. A functional autoregressive model of order $\mathbb{P}$ is suggested for short-term price forecasting in the electricity markets. The applicability of the model is improved with the help of functional final prediction error (FFPE), through which the model dimensionality and lag structure were selected automatically. An application of the suggested algorithm was evaluated on the Italian electricity market (IPEX). The out-of-sample forecasted results indicate that the proposed method performs relatively better than the nonfunctional forecasting techniques such as autoregressive (AR) and naïve models.

Keywords: functional autoregressive model; functional principle component analysis; vector autoregressive model; functional final prediction error (FFPE); naive method

Citation: Jan, F.; Shah, I.; Ali, S. Short-Term Electricity Prices Forecasting Using Functional Time Series Analysis. *Energies* **2022**, *15*, 3423. https://doi.org/10.3390/en15093423

Academic Editors: Yuji Yamada and Ricardo J. Bessa

Received: 7 April 2022
Accepted: 6 May 2022
Published: 7 May 2022

Publisher's Note: MDPI stays neutral with regard to jurisdictional claims in published maps and institutional affiliations.

1. Introduction

In the late 1980s, the worldwide electricity industry had undergone numerous fundamental changes when the state-owned monopolistic structure was restructured into the deregulated and competitive electricity market. The main driving force behind the restructuring of the electricity market was to promote competition among producers, retailers, and consumers by boosting private investments in production, supply, and retail sectors. Liberalization of this sector brought many benefits to the stakeholders in terms of reliable, secure, and economical electricity trading. However, due to electricity's inherent physical characteristic of non-storability in large volumes, the uncertainty related to electricity prices and demand forecasting increased. In addition, electricity prices and demand series generally exhibit specific features, such as multiple periodicities, long-trend, bank holiday effect, spikes, jumps, etc. In the presence of these features, the forecasting problem is challenging in all three forecasting horizons, i.e., short term, medium term, and long term [1].

In electricity markets, short-term forecasting refers to forecasting electricity prices from a few minutes to a week ahead. Apart from the power scheduling, management, and risk assessment, a short-term forecast is essential for market participants to optimize their bidding strategies. Medium-term forecast generally refers to the forecast made for a few weeks to a few months ahead. It is usually vital for expanding generation plants, scheduling maintenance, developing investment, fuel contracting, bilateral contracting, and hedging strategies. Forecasts ranging from a few months ahead to a few years ahead are commonly referred to as long-term-ahead forecasts. They are used for planning and investment profitability analysis, i.e., making decisions for future investments in power plants, inducing sites, and fuel sources [2,3]. In the literature, short-term forecasting has received greater research attention as the maximum electricity trade takes place in this market.

The literature concerning electricity price forecasting reported several statistical, machine learning, econometric, and hybrid models used to forecast short-term electricity prices [4–7]. Different linear time series models, including AR, ARMA, ARIMA, SARIMA, and ARIMAX [8–12], and nonlinear time series models, such as NPAR, ARCH, GARCH, and their extensions [13–15], are extensively used for forecasting electricity prices. Parametric and nonparametric regression-type models considering multiple, local polynomial, kernel, smoothing spline, and quantile regression are easy to implement and are widely studied in the case of electricity price forecasting [2,16–21]. In addition, models based on exponential smoothing including simple, double, and triple Holt's winters models that account for various periodicities [22–27] are often used for forecasting purposes. Artificial intelligence models have also been used to predict day-ahead electricity prices [28–32], as well as state-space models [33,34]. Various researchers combined the characteristics of two or more models to build a new model generally referred to as a hybrid model [3,35–39]. Generally, the above-stated models have their own functional and structural form, and the forecasting performance varies from market to market [40].

In the last three decades, technological developments simplified and decreased the cost of data collection and storage processes. Such advancements helped us to examine and record practical life activities in great detail. Examples include curves, images, surfaces, or anything else varying over a continuum. Consequently, classical statistical analysis techniques are inadequate and inefficient due to the large dimensions of data. To analyze such datasets, some suitable statistical methods are required, and functional data analysis (FDA) is one of the prominent methods to tackle such data in an efficient way. The FDA presents the essential statistical background for the analysis of functional variables, where every observation is a continuous function. The application of the FDA exists in almost every field of science, including economics, environment, engineering, energy, etc. [41,42]. In this research work, the application of the FDA is proposed for the electricity market, which is of primary interest for many researchers working in this field, especially after the liberalization of this market.

Given the temporal dependence, the FAR models have been suggested for the time series of trajectories. The autoregressive Hilbertian (ARH) process proposed by [43], also called the FAR model under Hilbert space, is likely the most popular pioneering work that plays an important role in the FDA context. The FAR is an extension of the AR process to infinite-dimensional space and is also used in electricity price forecasting. For example, using functional analysis of variance (FANOVA) and FAR model, Ref. [44] studied the seasonal patterns and improved prediction accuracy for electricity demand time series used from the Nord Pool electricity market. The application of a local linear method with functional explanatory variables was studied by [45]. They compared their proposed approach with the functional Nadaraya–Watson (NW) method and other finite-dimensional nonparametric techniques. For empirical analysis, monthly electricity consumption data of the United States of America (USA) were used, and the results suggest the superior performance of their proposed methods. The forecasting performances of different parametric and nonparametric functional models for electricity demand were studied by [46]. The authors used data from the Italian and British electricity markets and concluded that the nonparametric functional models give superior performance to their parametric counterparts. In another study, Ref. [47] used different functional models and compared their results with the finite univariate dimension (univariate and multivariate) models. Data from four different electricity markets, namely, the Nord Pool electricity market (NP), Pennsylvania–New Jersey–Maryland electricity market (PJM), the Italian electricity market (IPEX), and the British electricity market (APX Power UK), were used, and the results were summarized using different descriptive measures. The results suggested that the functional approach produces better results than the rest. Ref. [48] used the electricity demand curves data from Southern Australia. The author sliced the univariate time series into curves and reduced their dimensionality by applying the functional principal components technique. Finally, the author used univariate time series models to predict short-term electricity demand.

The main aim of this research work is to propose a functional model that can efficiently predict electricity prices. To this end, a method based on a two-components estimation procedure is proposed. The first component, known as the deterministic component, is computed using the additive modeling technique. The stochastic component, on the other hand, is modeled using an FAR($\mathbb{P}$) model where the selection of the dimension and lags is automatic. Finally, the model is tested for a whole year to see its forecasting performance. The rest of this paper is organized as follows. Section 2 provides an overview of the preliminaries. Section 3 describes a comprehensive review of the FAR($\mathbb{P}$) and functional final prediction error (FFPE). Section 4 provides the application of the proposed method, while Section 5 concludes the study.

2. Functional Modeling

2.1. Preliminaries

Let $\{\mathbb{Z}_i(t) : i \in \mathbb{N}, t \in \mathcal{J}\}$ be an arbitrary stationary $\mathbb{N}$-dimensional time series where $\mathcal{J}$ represents a continuum bounded within a finite interval. For each i, the functional observation $\mathbb{Z}_i$ belongs to a Hilbert space $\mathbb{H} = \mathcal{L}^2([0,1], \|\cdot\|)$ of square integrable functions which is equipped with a norm $\|\cdot\|$ induced by the inner product $< g, h > = \int g(t)h(t)dt$. The object $\{\mathbb{Z}_i(t)\}$ is referred to as FTS with i as the time index [49,50]. Furthermore, all stochastic functions are defined on a common probability space$(\Omega, \mathcal{A}, P)$. The notation $\mathbb{Z} \in \mathcal{L}_{\mathbb{H}}^p(\Omega, \mathcal{A}, P)$ is used to indicate $\mathbb{E}(\|\mathbb{Z}\|^p) < \infty$ for some $p > 0$. When $p = 1$, $\mathbb{Z}(t)$ has the mean curve $\mu(t)$; when $p = 2$, the covariance operators $C(t,s)$ are defined as in Equations (1) and (2) as under

$$\mu(t) = \mathbb{E}[\mathbb{Z}(t)] \tag{1}$$

$$C(t,s) = \mathbb{E}[(\mathbb{Z}(t) - \mu(t))(\mathbb{Z}(s) - \mu(s))] \tag{2}$$

Mercer's theorem [51] provides the following convenient spectral decomposition of Equation (2):

$$C(t,s) = \sum_{j=1}^{\infty} \kappa_j \varphi_j(t) \varphi_j(s) \tag{3}$$

where φ_j denotes the jth orthonormal principal component, and κ_j denotes the jth eigenvalue. The principal component scores (PCSs) $\gamma_{i,j}$ are given by the projection of $[\mathbb{Z}_i(t) - \mu(t)]$ in the direction of the jth eigenfunction φ_j, i.e., $\gamma_{i,j} = \langle \mathbb{Z}_i - \mu, \varphi_j \rangle$. Based on the separability of the Hilbert space, the Karhunen–Loève (KL) expansion [52,53] of the random function $\mathbb{Z}(t)$ can be expressed as

$$\mathbb{Z}(t) = \mu(t) + \sum_{j=1}^{\infty} \gamma_{i,j} \varphi_j(t) \tag{4}$$

The KL expansion provides the theoretical background for FPCA; see [54,55] for more details about FPCA and its practical demonstration.

Expansion (4) facilitates dimension reduction as the first $\mathbb{D}$ terms often provide a good approximation to the infinite sums, and, thus, the information contained in $\mathbb{Z}(t)$ can be adequately summarized by the jth-dimensional vector $(\gamma_1, \ldots, \gamma_j)$. The approximated processes can be defined as

$$\mathbb{Z}(t) = \mu(t) + \sum_{j=1}^{\mathbb{D}} \gamma_j \varphi_j(t) + \epsilon(t) \tag{5}$$

where $\epsilon(t)$ denotes the zero-mean white noise function that captures the variation excluded from the first $\mathbb{D}$ leading functional principal components (FPCs). There are different methods available in the literature for choosing the value of $\mathbb{D}$: (i) scree plots or the fraction of variation explained by first few PCSs [56], (ii) using the Akaike information and Bayesian information criteria [57], (iii) cross-validation with one-curve-leave-out or k-fold method [58], or (iv) bootstrap techniques [59].

Once the sample functional data are available, the sample mean can be obtained as

$$\hat{\mu}(t) = \frac{1}{N}\sum_{i=1}^{N}\mathbb{Z}_i(t), \quad t \in [0,1], \tag{6}$$

and the sample covariance function is defined as

$$\hat{C}(t,s) = \frac{1}{N-1}\sum_{i=1}^{N}(\mathbb{Z}_i(t) - \hat{\mu}(t))(\mathbb{Z}_i(s) - \hat{\mu}(s)) \tag{7}$$

Ref. [60] proved that the estimators are consistent for weakly dependent process.

2.2. Functional Autoregressive Model

Autoregressive (AR) models are one of the most popular forecasting models used in time series analysis. In the AR modeling framework, the response variable is linearly dependent on it past p lags with an error term. The theory of AR and more general linear processes in Hilbert spaces is developed in the monograph of [50], containing sufficient technical details. In addition, more relevant information can also be found in [49,61].

Recall a sequence of stationary random curves $(\mathbb{Z}_i(t), i \in \mathcal{N})$ in $\mathcal{L}^2([0,1])$ defined in Section 2.1. The functional AR model of order $\mathbb{P}$ (FAR($\mathbb{P}$)) can be written as [50]:

$$\mathbb{Z}_i(t) - \mu(t) = \sum_{k=1}^{\mathbb{P}}\Psi_k(\mathbb{Z}_{i-k}(t) - \mu(t)) + \xi_i(t) \tag{8}$$

where $\Psi_k(k = 1,\ldots,\mathbb{P})$ are the FAR operators (functional parameters), $\mu(t)$ is the mean function of $\mathbb{Z}_i(t)$, $\mathbb{Z}_{i-k}(t)$ denotes kth lag of curve $\mathbb{Z}_i$, and $\xi_i(t)$ is a strong $\mathbb{H}$-white noise with zero mean and finite second moment ($\mathbb{E}\|\xi_i(t)\|^2 < \infty$). For the prediction and forecasting of the model given in Equation (8), the following forecasting algorithm is used, which is based on Equations (5)–(7) [62].

1. First, the dimension which is denoted by $\mathbb{D}$ is fixed by using the method described in Section 2.3, and the estimated FPC scores are obtained as $\hat{\gamma}_{i,j} = \int \hat{\mathbb{Z}}_i(t)\hat{\phi}_j(t)dt$ for each observation $\hat{\mathbb{Z}}_i(t), i = 1,\ldots,N, j = 1,\ldots,\mathbb{D}$, and the estimated j-variate FPC scores vectors $\hat{\boldsymbol{\gamma}}_i = (\hat{\gamma}_{1,i},\ldots,\hat{\gamma}_{\mathbb{D},i})^t, i = 1,\ldots,N$.
2. Next, the order $\mathbb{P}$ is fixed using the technique described in Section 2.3 and we fit the vector AR model, VAR($\mathbb{P}$), as $\boldsymbol{\gamma}_i = \sum_{k=1}^{\mathbb{P}}\Psi_k\boldsymbol{\gamma}_{i-k} + \epsilon_i$ for eigenscores vectors to produce forecasting $\hat{\boldsymbol{\gamma}}_{N+1} = (\hat{\gamma}_{N+1,1},\ldots,\hat{\gamma}_{N+1,\mathbb{D}})^t$. Durbin–Levinson and innovations algorithm can be readily applied here, given the vectors $\hat{\boldsymbol{\gamma}}_1,\ldots,\hat{\boldsymbol{\gamma}}_N$.
3. In the last step, the multivariate time series are converted back to functional version using the KL theorem $\hat{\mathbb{Z}}_{N+1}(t) = \hat{\mu}(t) + \hat{\gamma}_{N+1,1}\hat{\phi}_1(t) + \cdots + \hat{\gamma}_{N+1,\mathbb{D}}\hat{\phi}_{\mathbb{D}}(t)$. The FPC scores and sample eigenfunctions result in $\hat{\mathbb{Z}}_{N+1}(t)$, which is then used as a one-step-ahead forecast of $\mathbb{Z}_{N+1}(t)$.

As can be seen, the selection of the dimension $\mathbb{D}$ and lags $\mathbb{P}$ is an important step in the above algorithm. The following section illustrates how to select the optimal values for these variables.

2.3. Selection of Order and Dimension of FAR($\mathbb{P}$)

The main goal of the current article is the accurate forecasting through FAR($\mathbb{P}$), which requires the appropriate order $\mathbb{P}$ selection as well as the dimension $\mathbb{D}$, in such a way that the mean square error (MSE) is minimized.

As the eigenfunctions φ_j and the PCS's $\gamma_{\mathbb{N},j}$ are uncorrelated, the MSE can be decomposed as

$$\mathbb{E}\left\{\left\|\mathbb{Z}_{\mathbb{N}+1}-\hat{\mathbb{Z}}_{\mathbb{N}+1}\right\|^2\right\} = \mathbb{E}\left\{\left\|\sum_{j=1}^{\infty}\gamma_{\mathbb{N}+1,j}\varphi_j-\sum_{j=1}^{\mathbb{D}}\hat{\gamma}_{\mathbb{N}+1,j}\varphi_j\right\|^2\right\}$$

$$= \mathbb{E}\left\{\left\|\mathbf{Z}_{\mathbb{N}+\mathbb{K}}-\hat{\mathbf{Z}}_{\mathbb{N}+1}\right\|^2\right\}+\sum_{j=\mathbb{D}+1}^{\infty}\kappa_j$$

where $\|.\|^2$ denotes the usual l-2 Euclidean norm of vectors. We suppose that the vector $\mathbf{Z}_{\mathbb{N}}$ is stationary and follows a $\mathbb{D}$-variables vector AR of order $\mathbb{P}$, VAR($\mathbb{P}$), that can be written as

$$\mathbf{Z}_{\mathbb{N}+1} = \mathbf{\Phi}_1\mathbf{Z}_{\mathbb{N}}+\mathbf{\Phi}_2\mathbf{Z}_{\mathbb{N}-1}+,\ldots,+\mathbf{\Phi}_{\mathbb{P}}\mathbf{Z}_{\mathbb{N}-\mathbb{P}+1}+\mathbf{Y}_{\mathbb{N}+1}. \tag{9}$$

Ref. [63] showed that $(\mathbf{Y}_{\mathbb{N}})$ is a white noise process such that

$$\sqrt{\mathbb{N}}(\hat{\rho}-\rho)\xrightarrow{D}\mathbf{N}(\mathbf{0},\Sigma_{\mathbf{Y}}\otimes\Delta_{\mathbb{P}}^{-1}) \tag{10}$$

where $\rho=\text{vec }[\mathbf{\Phi}_1,\ldots,\mathbf{\Phi}_{\mathbb{P}}]^t$ and $\hat{\rho}=\text{vec }[\hat{\mathbf{\Phi}}_1,\ldots,\hat{\mathbf{\Phi}}_{\mathbb{P}}]^t$ is the least squares estimator in vector form, and $\Delta_{\mathbb{P}}=\text{var}[\text{vec}(\mathbf{Z}_{\mathbb{P}},\ldots,\mathbf{Z}_1)]$ and $\Sigma_{\mathbf{Y}}=\mathbb{E}[\mathbf{Y}_1,\mathbf{Y}_1^t]$. Assume that the $\hat{\rho}$ are estimated from independent training sample $(\mathbf{X}_1,\ldots,\mathbf{X}_{\mathbb{N}})\overset{D}{=}(\mathbf{Z}_1,\ldots,\mathbf{Z}_{\mathbb{N}})$. It follows then that

$$\mathbb{E}\left\{\left\|\mathbf{Z}_{\mathbb{N}+1}-\hat{\mathbf{Z}}_{\mathbb{N}+1}\right\|^2\right\} = \mathbb{E}\left\{\left\|\mathbf{Z}_{\mathbb{N}+1}-(\hat{\mathbf{\Phi}}_1\mathbf{Z}_{\mathbb{N}}+\cdots+\hat{\mathbf{\Phi}}_{\mathbb{P}}\mathbf{Z}_{\mathbb{N}-\mathbb{P}+1})\right\|^2\right\}$$

$$= \mathbb{E}\left\{\|\mathbf{Y}_{\mathbb{N}+1}\|^2\right\}+\mathbb{E}\left\{\left\|(\mathbf{\Phi}_1-\hat{\mathbf{\Phi}}_1)\mathbf{Z}_{\mathbb{N}}+\cdots+(\mathbf{\Phi}_{\mathbb{P}}-\hat{\mathbf{\Phi}}_{\mathbb{P}})\mathbf{Z}_{\mathbb{N}-\mathbb{P}+1}\right\|^2\right\}$$

$$= \text{trace}\{\Sigma_{\mathbf{Y}}\}+\mathbb{E}\left\{\left\|I_{\mathbb{P}}\otimes(\mathbf{Z}_{\mathbb{N}}^t,\ldots,\mathbf{Z}_{\mathbb{N}-\mathbb{P}+1}^t)(\rho-\hat{\rho})\right\|^2\right\} \tag{11}$$

For some further derivation by using Equation (10), Ref. [64] showed that Equation (11) can be approximated as

$$\mathbb{E}\left\|\mathbf{Z}_{\mathbb{N}+1}-\hat{\mathbf{Z}}_{\mathbb{N}+1}\right\|^2\approx\frac{\mathbb{N}+\mathbb{P}*\mathbb{D}}{\mathbb{N}-\mathbb{P}*\mathbb{D}}\text{ trace}(\hat{\Sigma}_{\mathbf{Y}})+\sum_{j>\mathbb{D}}\kappa_j.$$

The suggested functional final prediction error selects order $\mathbb{P}$ and dimension $\mathbb{D}$ simultaneously by minimizing error term.

$$fFPE(\mathbb{P},\mathbb{D})=\frac{\mathbb{N}+\mathbb{P}*\mathbb{D}}{\mathbb{N}-\mathbb{P}*\mathbb{D}}\text{ trace}(\hat{\Sigma}_{\mathbf{Y}})+\sum_{j>\mathbb{D}}\kappa_j \tag{12}$$

Using the fFPE method, the suggested forecasting procedure works in a completely data-driven-based way and does not require any subjective specification of parameters. It is specifically important that the choice of $\mathbb{D}$ depends upon the sample size $\mathbb{N}$. For more technical details, the interested readers are referred to [64] and the references cited therein.

3. Modeling Framework

This section provides the general modeling framework used to model and forecast electricity prices. As described in Section 1, electricity prices exhibit specific features, e.g., extreme values (outliers), multiple periodicities, bank holidays effect, etc. Incorporating these specific features in the model greatly improves the forecasting accuracy [47]. To this end, the price time series is first filtered using the moving window filter on prices discussed in the following section.

3.1. Moving Window Filter on Prices

The identification of outliers, also known as the extreme values, in the data is one of the growing research areas. Various methods and ideas have been used in the literature to detect and impute outliers in the data. The significant developments in terms of outliers detection techniques in time series are suggested by [65–68]. Generally, the presence of outliers in the original electricity price data can substantially influence most forecasting models, which can result in poor forecasting performance. Therefore, identifying and analyzing outliers in the data is an essential step in constructing a forecasting model.

The moving window filter on price (MFP) [69] is an extension of the standard deviation filter on prices (SFP) technique. The SFP technique is based on the idea that the prices whose absolute deviation is taken from the mean $\hat{\mu}$ and are greater than some multiple of the sample standard deviation $\hat{\sigma}$ are referred to as outliers. However, the MFP technique differs from the SFP in the sense that it works out with the rolling window having fixed width of intervals. Using the MFP technique, the original price series is divided into $\mathbb{N} = \mathbb{T}/\mathbb{M}$ parts, where $\mathbb{M}$ is the width of the windows. Then, the SFP technique is applied to the first window of the given time series. Next, the window is shifted into the next fixed interval of $\mathbb{M}$ width, and the SFP is applied. Finally, the process is repeated until the last window is treated. Our work considers the same predictive interval used in [69], with the width of the window being equal to ten weeks. Thus, the subset of outliers $\mathbb{Z}^*$, obtained by the MFP with a moving window of width $\mathbb{M}$, is obtained as

$$\mathbb{Z}_i^o = \bigcup_{i=1,\dots,\mathbb{N}} \{ \mathbb{Z}_{\tau_i} : |\mathbb{Z}_{\tau_i} - \hat{\mu}_i| \geqslant 1.64 \cdot \hat{\sigma}_i$$
$$\tau_i \in ((i-1) \cdot \mathbb{M} + 1, i \cdot \mathbb{M}) \} \tag{13}$$

Once the outliers are identified, they are replaced by normal values [70]. In this work, they are replaced by the median value price of the specific window period.

3.2. The Model

Once the filtered price series is obtained, it is modeled using the following model.

$$\mathbb{Y}_i = D_i + \mathbb{Z}_i \quad i = 1, \cdots, N \tag{14}$$

where $\mathbb{Y}_i$ is the filtered time series and $\mathbb{Z}_i$ is a stochastic term. The deterministic component captures the long trend, the yearly and weekly periodicity, and the bank holidays effect. Mathematically, it is defined as

$$D_i = l_i + y_i + w_i + b_i$$

where the terms $l_i, y_i, w_i,$ and b_i represent the long-term trend, yearly periodicity, weekly periodicity, and bank holidays effect, respectively. In this work, the estimation procedure for the deterministic component described in [71] is used.

Once the deterministic component is estimated, the stochastic component $\mathbb{Z}_i$ is obtained as

$$\mathbb{Z}_i = \mathbb{Y}_i - D_i \tag{15}$$

which is modeled using the aforementioned FAR($\mathbb{P}$) and two alternate competing models. The alternate competing models used in this work are the univariate AR(P) model and a naïve benchmark model. The details of the competing models are as below.

3.2.1. Autoregressive (AR) Model

The univariate AR is one of the popular forecasting models used in time series analysis. It is similar to a regression model where the response variable is regressed over its lagged values. More specifically, in the AR modeling, a response variable is linearly dependent on its P lagged (past) values and an error term. Denoted by AR(P), mathematically, it can be written as

$$Z_i = \beta + \sum_{k=1}^{P} \alpha_k Z_{i-k} + \varepsilon_i \tag{16}$$

where Z_i is a univariate stationary time series, β is a constant, $\alpha_k (k = 1, \ldots, P)$ are the autoregressive parameters, and ε_i is a white noise process having zero mean and a constant variance. The choice of appropriate lag order selection is one of the most important steps in AR modeling. Different methods, including the Akaike information criterion (AIC) or Bayesian information criterion (BIC), or residual plots, e.g., autocorrelation function (ACF) and partial autocorrelation function (PACF), can be used to determine the lag order to be used in the model. In our work, the ACF and PACF are used, which indicate to use a restricted AR(7) model with $\alpha_k = 0$ for $k = 3, 4, 5, 6$. The maximum likelihood estimation (MLE) method is used to estimate the parameters of the above model.

Once both the deterministic and stochastic components are modeled and forecasted, the final forecast is obtained as

$$\hat{\mathbb{Y}}_{i+1} = \hat{D}_{i+1} + \hat{\mathbb{Z}}_{i+1} \quad i = 1, \cdots, N. \tag{17}$$

The flowchart of the proposed general modeling framework is given in Figure 1.

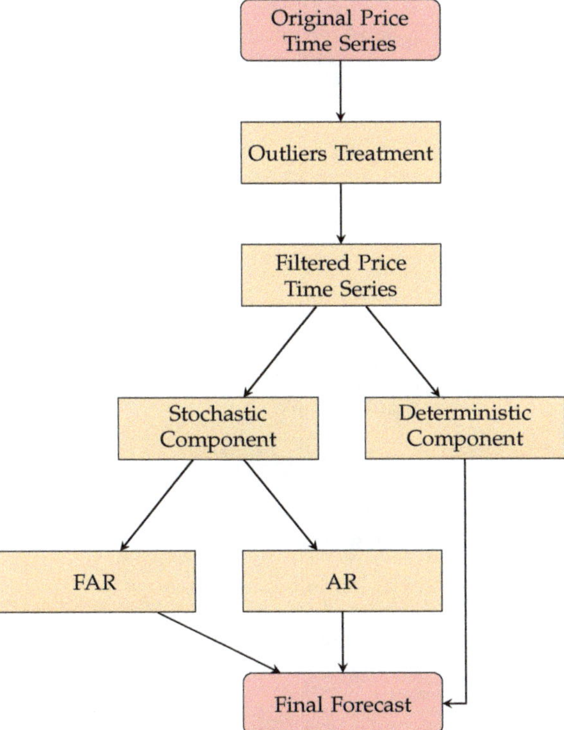

Figure 1. Flowchart of the proposed modeling framework.

3.2.2. The Naïve Benchmark

This section provides details about a naïve forecasting method that belongs to a similar day technique and has reported greater accuracy than other naïve methods [2]. This method works as follows.

1. To forecast a given day, for example, Thursday, select the day before Thursday, which is Wednesday, and denote it by x^*.
2. From the validation dataset, select all the Wednesdays (except x^*) and compare them with x^* using the mean absolute error (MAE).
3. Obtain a value of the MAE for each comparison that will result in a vector of the MAE values.
4. Find and locate the smallest value of the MAE in the vector. Once the Wednesday having the lowest MAE is located, use its next day, i.e., Thursday, as the forecast for the concerned Thursday. This process is repeated for all the remaining days of the week.

4. Out-of-Sample Forecast

The dataset used in this empirical study includes electricity prices data called "Prezzo Unico Nazionale (PUN)" from the Italian Electricity Market (IPEX), collected from 1 January 2012 until 31 December 2017. Each day consists of 24 observations, where each observation corresponds to a load period. For modeling and forecasting purposes, we split the data into two periods. The period from 1 January 2012 to 31 December 2016 (1827 days) is used for model estimation. This period is used to optimize the parameters of the models. The out-of-sample period ranges from 1 January 2017 to 31 December 2017 (365 days). This period is used for forecasting the performance of the models. The one-day-ahead out-of-sample forecast is obtained through the window expending technique. In Figure 2, the spot electricity prices series is depicted for six years with a sample of functional (smoothed) curves for a week plotted on the right-hand side. The weekly periodicity is evident in the price time series as the prices profile for working days is relatively different from the non-working days.

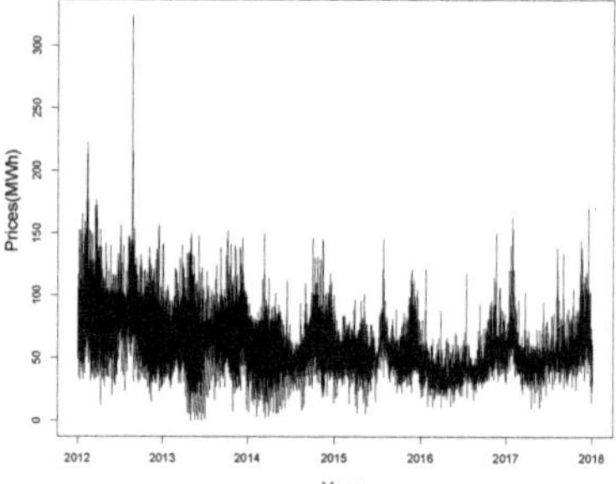
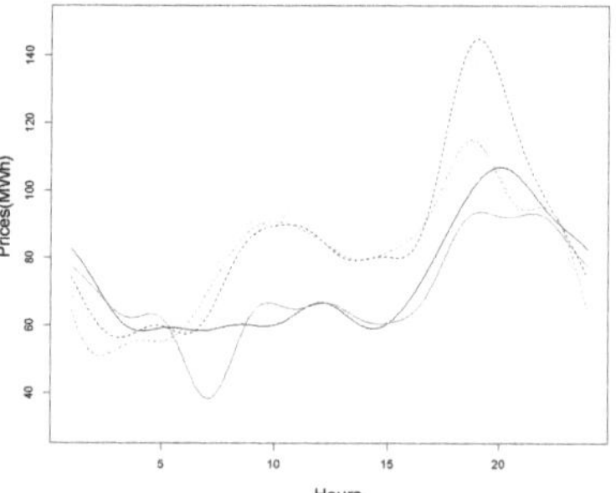

Figure 2. Electricity prices: (**left**) the original time series of 52,608 hourly electricity spot prices and (**right**) electricity prices smoothed curves for one week.

The forecasting performance of the proposed and alternative models is compared using three standard descriptive forecast error measures. The point forecast accuracy is evaluated using three standard accuracy measures, namely, mean absolute percentage error (MAPE), MAE, and root mean square error (RMSE). Mathematically, the MAPE, MAE, and RMSE are given as

$$\mathrm{MAPE} \;=\; \frac{1}{N}\sum_{i=1}^{N}\frac{\left|\mathbb{Z}_{i,j}-\widehat{\mathbb{Z}}_{i,j}\right|}{\mathbb{Z}_{i,j}}\times 100$$

$$\mathrm{MAE} \;=\; \frac{1}{N}\sum_{i=1}^{N}\left|\mathbb{Z}_{i,j}-\widehat{\mathbb{Z}}_{i,j}\right|$$

$$\mathrm{RMSE} \;=\; \sqrt{\frac{1}{N}\sum_{i=1}^{N}[\mathbb{Z}_{i,j}-\widehat{\mathbb{Z}}_{i,j}]^{2}}$$

where N represents the number of observations in the out-of-sample forecasting period, $\mathbb{Z}_{i,j}$ denotes the original observed prices of the ith day and jth hour, and $\widehat{\mathbb{Z}}_{i,j}$ denotes the forecasted price of the aforementioned day and hour with $j = 1, 2, \ldots, 24$.

In addition, directional forecast statistics can be very beneficial for traders in the electricity market in making investment decisions. These direction moments or turning points can be measured using directional statistic defined as [72]

$$D_{stat} = \frac{1}{N}\sum_{i=1}^{N}\alpha_i * 100$$

where

$$\alpha_i = \begin{cases} 1, & \text{if} \quad (\mathbb{Z}_{i+1,j}-\mathbb{Z}_{i,j})(\widehat{\mathbb{Z}}_{i+1,j}-\mathbb{Z}_{i,j}) \geq 1 \\ 0, & \text{otherwise} \end{cases}$$

The electricity prices forecast through the FAR($\mathbb{P}$) model have the following steps. In the first step, the moving window filter method was used for the identification and accommodation of outliers. In the second step, a logarithm (log) transformation was performed to stabilize the variance of the series. In the third step, model (17) is applied to the data and the series $\mathbb{Z}_i$ is obtained using Equation (15). In the fourth step, the Fourier basis functions are used to transform the discrete data into functional data to obtain 2192 daily functional trajectories, say, $\mathbb{Z}_1(t), \ldots, \mathbb{Z}_{2192}(t)$, $t \in J$. Once the functional data are obtained, the FAR($\mathbb{P}$) model described in Section 2.1 is applied, and one-day-ahead forecasts are obtained for the whole year. In the case of the competing models, the univariate AR model and the naïve benchmark are applied directly to $\mathbb{Z}_i$, and the one-day-ahead forecasts are obtained for the whole out-of-sample period.

Figure 3 highlights the population mean function $\mu(t)$ and the functions obtained by adding and subtracting a suitable multiple of the eigenfunctions to the mean. Such plots are helpful to understand the variability in the direction of certain eigenfunctions. The first eigenfunction is positive, indicating that subjects with positive scores on this component will contribute to obtaining a consistently larger proportion (77.1%) of the total variation of the data. The second eigenfunction displays an oscillatory behavior, suggesting that subjects with positive scores will have lower electricity prices from midnight till early morning and then slightly more between hours 7 a.m. and 10 a.m., and explain 10.5% of the total variation of the data. Similarly, the third and fourth eigenfunctions explain 4.8% and 2.7% of the total variation of the data, respectively. The first four eigenfunctions collectively explain more than 95% of the total variability in the electricity prices data.

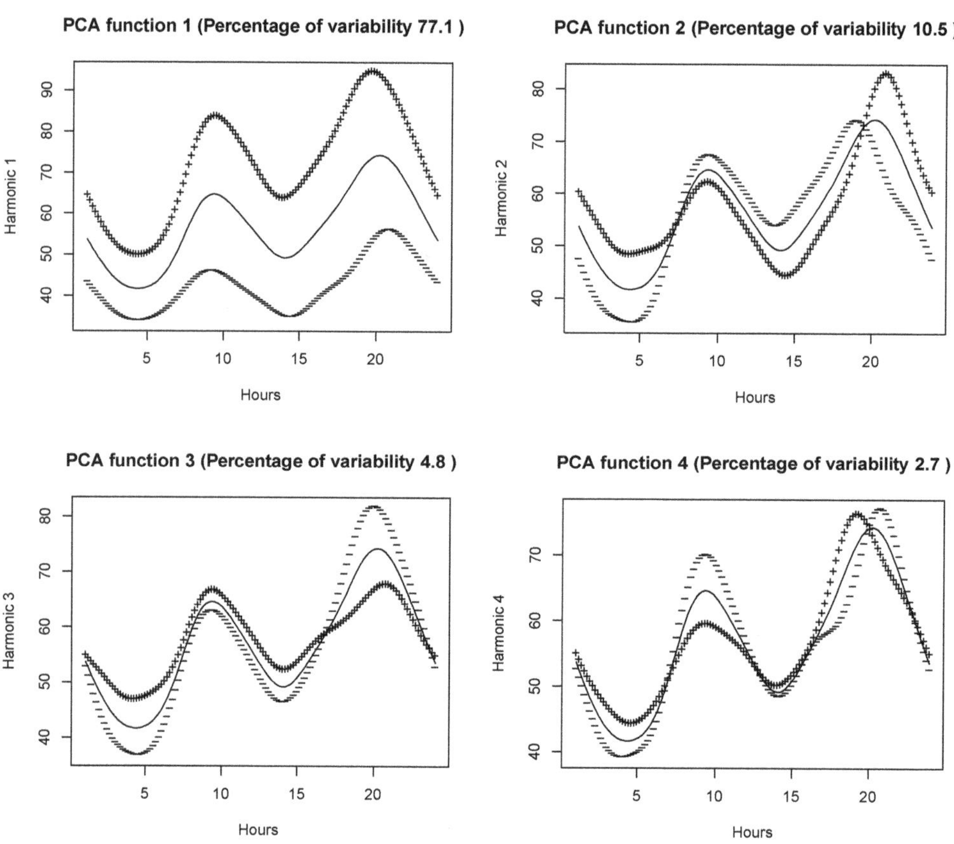

Figure 3. The effect of 1st FPC (**upper left** panel), the effect of 2nd FPC (**upper right** panel), the effect of 3rd FPC (**lower left** panel), and the effect of 4th FPC (**lower right** panel).

Concerning the forecasting results for the proposed and alternative models, Table 1 compares the overall forecasting ability of the FAR($\mathbb{P}$), AR(7), and naïve models through out-of-sample forecasting errors computed by MAE, MAPE, and RMSE. The table also provides the directional forecasting performance for these models. From the results, it is evident that our proposed functional model performs significantly better than the other competing models. The proposed FAR($\mathbb{P}$) models produce MAE, MAPE, and RMSE of 5.16, 8.99, and 8.65, respectively. Although the univariate AR model produces better results than the naïve model, it produces considerably higher forecasting errors compared to the proposed functional model. Looking at the directional forecasting results, note that the value of D_{stat} for FAR($\mathbb{P}$) is 88.34%, whereas values of 82.96% and 53.64% are obtained in the case of AR and naïve models, respectively. Hence, our proposed functional model performs relatively well compared to the competing models. From the number of forecast direction moments, it can be seen that the FAR($\mathbb{P}$) forecast 1525 out of the total 8760 load periods accurately (the "SAME" in Table 1 refers to the absolute difference of the forecasted value minus the actual value to be less than EUR 1), whereas this value for the AR and naïve models is 1272 and 385, respectively. The number of over-forecasted values for FAR($\mathbb{P}$) and AR(7) are 3743 and 4084, respectively. Again, the poor performance of the naïve model is evident from the results of the directional forecast.

Table 1. IPEX electricity prices: out-of-sample forecasting errors MAE, MAPE, and RMSE for FAR($\mathbb{P}$), AR(7), naïve models, and the directional statistics D_{stat} with number of forecasting directions (same, up, down).

Model	MAE	MAPE	RMSE	D_{stat} (%)	SAME	UP	DOWN
FAR($\mathbb{P}$)	5.16485	8.99009	8.65032	88.34342	1525	3743	3492
AR(7)	5.65833	10.09469	9.20305	82.95525	1272	4084	3404
Naive	6.86278	12.63467	10.09929	53.63626	385	4137	4238

Table 2 reports the daily forecast accuracy for the electricity prices using different models. From the table, one can see that the FAR($\mathbb{P}$) model produces lower forecasting errors compared to the univariate AR(7) and naïve models. Although the forecast errors vary from day to day, they are lower on Thursday and Friday when considering MAPE. The poor performance of the naïve model is evident from this table. The hourly forecast errors for different models are listed in Table 3, which shows that the forecast errors vary throughout the day. Although the FAR($\mathbb{P}$) model produces better results on most hours, the AR(7) has better results on two hours when considering the MAPE. It is worth mentioning that the proposed FAR($\mathbb{P}$) model performs significantly well during peak hours compared to the competing models. Again, the poor performance of the naïve model is evident from the results.

Table 2. IPEX electricity prices: daily forecast errors for FAR($\mathbb{P}$), AR(7), and naïve models.

Model	Error	Days of a Week						
		Monday	Tuesday	Wednesday	Thursday	Friday	Saturday	Sunday
FAR($\mathbb{P}$)		5.63112	5.87795	6.53440	5.00897	4.88412	4.06196	4.17448
AR(7)	MAE	6.16619	6.34358	6.95023	5.86286	5.45145	4.54586	4.31402
Naive		7.58172	7.38245	6.39990	6.93643	6.40897	6.86697	6.47056
FAR($\mathbb{P}$)		9.87044	9.23202	9.30545	7.72602	7.85736	8.58661	10.32703
AR(7)	MAPE	11.14407	10.06261	10.35827	9.02434	8.86721	9.91458	11.26919
Naive		13.82605	12.47558	11.01666	10.93947	11.04299	13.52881	18.26894
FAR($\mathbb{P}$)		8.72362	9.56184	11.93833	9.59608	7.97935	5.43820	5.36185
AR(7)	RMSE	9.34377	10.08267	11.97317	10.87680	8.79962	5.86412	5.60787
Naive		11.60842	10.63521	10.37988	10.76548	8.88560	9.54221	8.54465

Finally, the results obtained by our proposed functional model in this study are compared with the results listed in the literature. Here, it is worth mentioning that such a comparison is only to evaluate the performance of our model, as different authors considered different forecasting horizons, different periods, and different error summary measures. Using the Italian electricity market and considering a one-day-ahead forecast, Ref. [2] obtained an MAPE value of 9.74 using the NPAR model, which is significantly higher than our proposed model MAPE value of 8.99. The research work of [73] used the Italian electricity market data, and their proposed model produced an MAE of 8.58, whereas our proposal reported an MAE value of 5.16, 60% lower. For a one-day-ahead forecast, Ref. [70] reported an MAPE value of 9.05, which is slightly higher than our obtained MAPE value. Using an ARX-EGARCH model for the Italian electricity prices time series, Ref. [74] obtained an RMSE of 11.58, whereas our proposed model produced an RMSE value of 8.65. The work of [75] reported RMSE values of 16.72 and 15.79 using ARMA and GARCH models, respectively, significantly higher than our value of 8.65.

Table 3. IPEX electricity prices: hourly forecast errors for FAR($\mathbb{P}$), AR(7), and naïve models.

Model	Hour	MAE	MAPE	RMSE	Hour	MAE	MAPE	RMSE
FAR($\mathbb{P}$)		3.17448	10.32703	5.36185		5.03299	9.77343	7.50008
AR(7)	1	4.09499	8.20630	5.48352	13	5.50527	10.81530	8.32688
Naive		2.99810	5.71016	4.35740		5.89142	10.72254	8.54265
FAR($\mathbb{P}$)		5.63112	9.87044	8.72362		5.13495	10.97364	7.45864
AR(7)	2	3.90162	8.62680	5.04521	14	5.69565	12.52584	8.52711
Naive		3.99566	8.14583	5.04110		5.15640	10.87232	7.39582
FAR($\mathbb{P}$)		5.87795	9.23202	6.46184		4.91392	10.19534	9.29372
AR(7)	3	3.82670	9.05766	4.96036	15	6.52555	13.92227	10.37498
Naive		3.66855	8.12195	4.42199		5.40823	11.21776	7.62226
FAR($\mathbb{P}$)		3.39714	8.41575	4.46110		6.20807	11.99534	9.73345
AR(7)	4	3.97490	10.11668	5.28032	16	6.91072	13.74287	10.63356
Naive		3.52617	8.48418	4.64024		6.07377	11.75996	9.43945
FAR($\mathbb{P}$)		3.39652	8.35320	4.46989		6.63385	10.73122	10.96537
AR(7)	5	3.89939	9.77731	5.18900	17	7.02079	11.61225	11.34363
Naive		6.33318	12.83398	9.91261		6.47302	10.62784	10.48912
FAR($\mathbb{P}$)		4.88413	7.85736	7.97935		6.95788	9.73444	12.11111
AR(7)	6	3.73582	8.64474	5.02575	18	7.32959	10.47554	12.35075
Naive		7.59001	8.59831	4.74848		7.04453	13.45730	10.9536
FAR($\mathbb{P}$)		4.06196	8.58661	5.43820		7.83811	9.91660	14.00040
AR(7)	7	4.27194	8.50246	5.79461	19	7.99731	10.31619	13.91400
Naive		4.98340	11.28051	6.28747		7.35881	11.86122	11.68502
FAR($\mathbb{P}$)		4.94668	8.16684	7.92432		7.35007	9.7326	11.80230
AR(7)	8	5.60254	9.18891	9.23695	20	7.40072	9.98525	11.66830
Naive		6.25091	11.78682	8.76700		8.46851	12.49491	13.42239
FAR($\mathbb{P}$)		7.28609	10.34856	12.47580		6.28777	9.03979	9.63504
AR(7)	9	8.17881	11.92595	13.42361	21	6.26023	9.14635	9.46692
Naive		7.14544	11.18423	11.62178		8.05637	12.35241	11.55807
FAR($\mathbb{P}$)		6.73647	9.88567	11.63280		5.11291	7.86378	8.41742
AR(7)	10	7.67803	11.57003	12.60697	22	5.18257	8.05526	8.48756
Naive		6.95444	10.27270	11.25744		7.08351	10.66474	10.11036
FAR($\mathbb{P}$)		6.09630	9.80792	9.88950		3.86946	6.64648	6.09351
AR(7)	11	6.80914	11.15632	10.81531	23	3.93445	9.78030	6.12057
Naive		6.40122	10.11695	9.80316		6.14502	9.13875	9.47691
FAR($\mathbb{P}$)		5.79828	10.13717	8.84841		3.55292	6.70792	5.27639
AR(7)	12	6.45438	11.43950	9.81589	24	3.55885	6.69256	5.32193
Naive		5.94363	10.01007	8.73133		4.49332	7.98897	6.5533

5. Conclusions and Future Direction

In today's competitive electricity market, modeling and forecasting electricity prices are critical for market participants to optimize their strategies. However, electricity prices exhibit specific features, including long-trend, periodicities, spikes or jumps, bank holidays, etc. In the presence of these features, the forecasting problem is a great challenge for researchers. This paper proposes a functional model for modeling and forecasting electricity prices. To this end, the price time series is first treated for the extreme values. The filtered series is then divided into deterministic and stochastic parts. The deterministic part modeled the effects of long-trend, annual, and weekly periodicities, and bank holidays. For the stochastic component, a functional AR model (FAR) is proposed that is capable of automatic selection of lags and dimensions. To evaluate the performance of our proposed model, two alternate models, namely, the univariate AR and a naïve benchmark, are also used in this study. For empirical comparison, data from the Italian electricity market are used and the out-of-sample one-day-ahead forecast errors measured through MAPE, MAE, and RMSE are calculated for a complete year.

The empirical results suggest that the proposed FAR($\mathbb{P}$) model is significantly better than the competing model, as it produced considerably lower forecasting errors. Furthermore, the component estimation procedure is highly effective in forecasting electricity prices. Moreover, the directional forecast results suggest that this approach can significantly increase the number of accurate forecasts. Accurate forecasting can be very helpful for the traders (buyers and suppliers) to optimize their bidding strategies to maximize their gains and to use the resources required for electricity generation more effectively. Consequently, this will also benefit the end-user in terms of reliable and economical electricity facilities.

As the current study does not consider any exogenous variable effect in the model, this effect can be investigated in the future. Furthermore, as the current study only considers linear models, nonlinear models can also be compared with the proposed functional model.

Author Contributions: Conceptualization, I.S. and F.J.; methodology, F.J.; software, S.A.; validation, I.S., S.A. and F.J.; formal analysis, F.J.; investigation, F.J.; resources, S.A.; data curation, I.S.; writing-original draft preparation, F.J.; writing-review and editing, I.S. and S.A.; supervision, I.S.; project administration, I.S. and S.A. All authors have read and agreed to the published version of the manuscript.

Funding: This research received no external funding.

Institutional Review Board Statement: Not applicable.

Informed Consent Statement: Not applicable.

Data Availability Statement: The data is freely available from https://www.mercatoelettrico.org, accessed on 27 April 2022.

Conflicts of Interest: The authors declare no conflict of interest.

References

1. Bunn, D.W. *Modelling Prices in Competitive Electricity Markets*; Wiley: Hoboken, NJ, USA, 2004.
2. Shah, I.; Bibi, H.; Ali, S.; Wang, L.; Yue, Z. Forecasting one-day-ahead electricity prices for italian electricity market using parametric and nonparametric approaches. *IEEE Access* **2020**, *8*, 123104–123113. [CrossRef]
3. Lisi, F.; Nan, F. Component estimation for electricity prices: Procedures and comparisons. *Energy Econ.* **2014**, *44*, 143–159. [CrossRef]
4. Misiorek, A.; Trueck, S.; Weron, R. Point and interval forecasting of spot electricity prices: Linear vs. non-linear time series models. *Stud. Nonlinear Dyn. Econom.* **2006**, *10*, 1–36. [CrossRef]
5. Debnath, K.B.; Mourshed, M. Forecasting methods in energy planning models. *Renew. Sustain. Energy Rev.* **2018**, *88*, 297–325. [CrossRef]
6. Shah, I. Modeling and Forecasting Electricity Market Variables. Ph.D. Thesis, University of Padova, Padua, Italy, 2016.
7. Li, W.; Yang, X.; Li, H.; Su, L. Hybrid forecasting approach based on GRNN neural network and SVR machine for electricity demand forecasting. *Energies* **2017**, *10*, 44. [CrossRef]
8. Contreras, J.; Espinola, R.; Nogales, F.J.; Conejo, A.J. ARIMA models to predict next-day electricity prices. *IEEE Trans. Power Syst.* **2003**, *18*, 1014–1020. [CrossRef]
9. Conejo, A.J.; Plazas, M.A.; Espinola, R.; Molina, A.B. Day-ahead electricity price forecasting using the wavelet transform and ARIMA models. *IEEE Trans. Power Syst.* **2005**, *20*, 1035–1042. [CrossRef]
10. Babu, C.N.; Reddy, B.E. A moving-average filter based hybrid ARIMA–ANN model for forecasting time series data. *Appl. Soft Comput.* **2014**, *23*, 27–38. [CrossRef]
11. González, J.P.; San Roque, A.M.; Perez, E.A. Forecasting functional time series with a new Hilbertian ARMAX model: Application to electricity price forecasting. *IEEE Trans. Power Syst.* **2017**, *33*, 545–556. [CrossRef]
12. Wang, Q.; Li, S.; Li, R. Forecasting energy demand in China and India: Using single-linear, hybrid-linear, and non-linear time series forecast techniques. *Energy* **2018**, *161*, 821–831. [CrossRef]
13. Diongue, A.K.; Guégan, D. *The k-Factor Gegenbauer Asymmetric Power GARCH Approach for Modelling Electricity Spot Price Dynamics*; Universite Pantheon-Sorbonne: Paris, France, 2008.
14. Garcia, R.C.; Contreras, J.; Van Akkeren, M.; Garcia, J.B.C. A GARCH forecasting model to predict day-ahead electricity prices. *IEEE Trans. Power Syst.* **2005**, *20*, 867–874. [CrossRef]

15. Qu, H.; Duan, Q.; Niu, M. Modeling the volatility of realized volatility to improve volatility forecasts in electricity markets. *Energy Econ.* **2018**, *74*, 767–776. [CrossRef]

16. Bianco, V.; Manca, O.; Nardini, S. Electricity consumption forecasting in Italy using linear regression models. *Energy* **2009**, *34*, 1413–1421. [CrossRef]

17. Su, M.; Zhang, Z.; Zhu, Y.; Zha, D. Data-driven natural gas spot price forecasting with least squares regression boosting algorithm. *Energies* **2019**, *12*, 1094. [CrossRef]

18. Karakatsani, N.V.; Bunn, D.W. Forecasting electricity prices: The impact of fundamentals and time-varying coefficients. *Int. J. Forecast.* **2008**, *24*, 764–785. [CrossRef]

19. He, Y.; Liu, R.; Li, H.; Wang, S.; Lu, X. Short-term power load probability density forecasting method using kernel-based support vector quantile regression and Copula theory. *Appl. Energy* **2017**, *185*, 254–266. [CrossRef]

20. Yang, Y.; Li, S.; Li, W.; Qu, M. Power load probability density forecasting using Gaussian process quantile regression. *Appl. Energy* **2018**, *213*, 499–509. [CrossRef]

21. Lebotsa, M.E.; Sigauke, C.; Bere, A.; Fildes, R.; Boylan, J.E. Short term electricity demand forecasting using partially linear additive quantile regression with an application to the unit commitment problem. *Appl. Energy* **2018**, *222*, 104–118. [CrossRef]

22. Taylor, J.W.; McSharry, P.E. Short-term load forecasting methods: An evaluation based on european data. *IEEE Trans. Power Syst.* **2007**, *22*, 2213–2219. [CrossRef]

23. De Livera, A.M.; Hyndman, R.J.; Snyder, R.D. Forecasting time series with complex seasonal patterns using exponential smoothing. *J. Am. Stat. Assoc.* **2011**, *106*, 1513–1527. [CrossRef]

24. Taylor, J.W. Triple seasonal methods for short-term electricity demand forecasting. *Eur. J. Oper. Res.* **2010**, *204*, 139–152. [CrossRef]

25. Huang, J.; Srinivasan, D.; Zhang, D. Electricity Demand Forecasting Using HWT Model with Fourfold Seasonality. In Proceedings of the 2017 International Conference on Control, Artificial Intelligence, Robotics & Optimization (ICCAIRO), Prague, Czech Republic, 20–22 May 2017; IEEE: Piscataway, NJ, USA, 2017; pp. 254–258.

26. Jiang, W.; Wu, X.; Gong, Y.; Yu, W.; Zhong, X. Holt–Winters smoothing enhanced by fruit fly optimization algorithm to forecast monthly electricity consumption. *Energy* **2020**, *193*, 116779. [CrossRef]

27. Trull, O.; García-Díaz, J.C.; Troncoso, A. Initialization Methods for Multiple Seasonal Holt–Winters Forecasting Models. *Mathematics* **2020**, *8*, 268. [CrossRef]

28. Amor, S.B.; Boubaker, H.; Belkacem, L. Forecasting Electricity Spot Price with Generalized Long Memory Modeling: Wavelet and Neural Network. *Int. J. Econ. Manag. Eng.* **2018**, *11*, 2307–2323.

29. Gürbüz, F.; Öztürk, C.; Pardalos, P. Prediction of electricity energy consumption of Turkey via artificial bee colony: A case study. *Energy Syst.* **2013**, *4*, 289–300. [CrossRef]

30. Ugurlu, U.; Oksuz, I.; Tas, O. Electricity price forecasting using recurrent neural networks. *Energies* **2018**, *11*, 1255. [CrossRef]

31. Urolagin, S.; Sharma, N.; Datta, T.K. A combined architecture of multivariate LSTM with Mahalanobis and Z-Score transformations for oil price forecasting. *Energy* **2021**, *231*, 120963. [CrossRef]

32. Gholipour Khajeh, M.; Maleki, A.; Rosen, M.A.; Ahmadi, M.H. Electricity price forecasting using neural networks with an improved iterative training algorithm. *Int. J. Ambient. Energy* **2018**, *39*, 147–158. [CrossRef]

33. Dordonnat, V.; Koopman, S.J.; Ooms, M.; Dessertaine, A.; Collet, J. An hourly periodic state space model for modelling French national electricity load. *Int. J. Forecast.* **2008**, *24*, 566–587. [CrossRef]

34. Rigatos, G.G. *State-Space Approaches for Modelling and Control in Financial Engineering*; Springer: Berlin/Heidelberg, Germany, 2017.

35. Li, T.; Qian, Z.; Deng, W.; Zhang, D.; Lu, H.; Wang, S. Forecasting crude oil prices based on variational mode decomposition and random sparse Bayesian learning. *Appl. Soft Comput.* **2021**, *113*, 108032. [CrossRef]

36. Laouafi, A.; Mordjaoui, M.; Laouafi, F.; Boukelia, T.E. Daily peak electricity demand forecasting based on an adaptive hybrid two-stage methodology. *Int. J. Electr. Power Energy Syst.* **2016**, *77*, 136–144. [CrossRef]

37. Yang, Z.; Ce, L.; Lian, L. Electricity price forecasting by a hybrid model, combining wavelet transform, ARMA and kernel-based extreme learning machine methods. *Appl. Energy* **2017**, *190*, 291–305. [CrossRef]

38. Chang, Z.; Zhang, Y.; Chen, W. Electricity price prediction based on hybrid model of adam optimized LSTM neural network and wavelet transform. *Energy* **2019**, *187*, 115804. [CrossRef]

39. Bibi, N.; Shah, I.; Alsubie, A.; Ali, S.; Lone, S.A. Electricity Spot Prices Forecasting Based on Ensemble Learning. *IEEE Access* **2021**, *9*, 150984–150992. [CrossRef]

40. Hahn, H.; Meyer-Nieberg, S.; Pickl, S. Electric load forecasting methods: Tools for decision making. *Eur. J. Oper. Res.* **2009**, *199*, 902–907. [CrossRef]

41. Ramsay, J.O.; Silverman, B.W. *Applied Functional Data Analysis: Methods and Case Studies*; Springer: Berlin/Heidelberg, Germany, 2007.

42. Shah, I.; Lisi, F. Forecasting of electricity price through a functional prediction of sale and purchase curves. *J. Forecast.* **2020**, *39*, 242–259. [CrossRef]

43. Bosq, D. Modelization, nonparametric estimation and prediction for continuous time processes. In *Nonparametric Functional Estimation and Related Topics*; Springer: Berlin/Heidelberg, Germany, 1991; pp. 509–529.

44. Andersson, J.; Lillestøl, J. Modeling and forecasting electricity consumption by functional data analysis. *J. Energy Mark.* **2010**, *3*, 3. [CrossRef]

45. Aneiros-Pérez, G.; Cao, R.; Vilar-Fernández, J.M. Functional methods for time series prediction: a nonparametric approach. *J. Forecast.* **2011**, *30*, 377–392. [CrossRef]

46. Shah, I.; Lisi, F. Day-ahead electricity demand forecasting with nonparametric functional models. In Proceedings of the 12th International Conference on the European Energy Market (EEM), Lisbon, Portugal, 19–22 May 2015; IEEE: Piscataway, NJ, USA, 2015; pp. 1–5.

47. Lisi, F.; Shah, I. Forecasting next-day electricity demand and prices based on functional models. *Energy Syst.* **2020**, *11*, 947–979. [CrossRef]

48. Shang, H.L. Functional time series approach for forecasting very short-term electricity demand. *J. Appl. Stat.* **2013**, *40*, 152–168. [CrossRef]

49. Horváth, L.; Kokoszka, P. *Inference for Functional Data with Applications*; Springer Science & Business Media: Berlin/Heidelberg, Germany, 2012; Volume 200.

50. Bosq, D. *Linear Processes in Function Spaces: Theory and Applications*; Springer Science & Business Media: Berlin/Heidelberg, Germany, 2000; Volume 149.

51. Indritz, J. *Methods in Analysis*; Macmillan: New York, NY, USA, 1963.

52. Karhunen, K. *Under Lineare Methoden in der Wahr Scheinlichkeitsrechnung*; Annales Academiae Scientiarun Fennicae Series A1: Mathematia Physica; Universitat Helsinki: Helsinki, Finland, 1947; Volume 47.

53. Loeve, M. Functions aleatoires du second ordre. *Process. Stoch. Mouv. Brownien* **1948**, 366–420.

54. Ramsay, J.; Silverman, B. *Functional Data Analysis-Methods and Case Studies*; Springer: New York, NY, USA, 2002.

55. Shang, H.L. A survey of functional principal component analysis. *AStA Adv. Stat. Anal.* **2014**, *98*, 121–142. [CrossRef]

56. Chiou, J.M. Dynamical functional prediction and classification, with application to traffic flow prediction. *Ann. Appl. Stat.* **2012**, *6*, 1588–1614. [CrossRef]

57. Yao, F.; Müller, H.G.; Wang, J.L. Functional data analysis for sparse longitudinal data. *J. Am. Stat. Assoc.* **2005**, *100*, 577–590. [CrossRef]

58. Rice, J.A.; Silverman, B.W. Estimating the mean and covariance structure nonparametrically when the data are curves. *J. R. Stat. Soc. Ser. B (Methodol.)* **1991**, *53*, 233–243. [CrossRef]

59. Hall, P.; Vial, C. Assessing the finite dimensionality of functional data. *J. R. Stat. Soc. Ser. B (Stat. Methodol.)* **2006**, *68*, 689–705. [CrossRef]

60. Hörmann, S.; Kokoszka, P. Weakly dependent functional data. *Ann. Stat.* **2010**, *38*, 1845–1884. [CrossRef]

61. Chen, Y.; Koch, T.; Lim, K.G.; Xu, X.; Zakiyeva, N. A review study of functional autoregressive models with application to energy forecasting. *Wiley Interdiscip. Rev. Comput. Stat.* **2021**, *13*, e1525. [CrossRef]

62. Jiao, S.; Aue, A.; Ombao, H. Functional time series prediction under partial observation of the future curve. *J. Am. Stat. Assoc.* **2021**, 1–12. [CrossRef]

63. Lutkepohl, H. *New Introduction to Multiple Time Series Analysis*; Springer: Berlin/Heidelberg, Germany, 2006.

64. Aue, A.; Norinho, D.D.; Hörmann, S. On the prediction of stationary functional time series. *J. Am. Stat. Assoc.* **2015**, *110*, 378–392. [CrossRef]

65. Jau, Y.M.; Su, K.L.; Wu, C.J.; Jeng, J.T. Modified quantum-behaved particle swarm optimization for parameters estimation of generalized nonlinear multi-regressions model based on Choquet integral with outliers. *Appl. Math. Comput.* **2013**, *221*, 282–295. [CrossRef]

66. Bardwell, L.; Fearnhead, P. Bayesian detection of abnormal segments in multiple time series. *Bayesian Anal.* **2017**, *12*, 193–218. [CrossRef]

67. Blázquez-García, A.; Conde, A.; Mori, U.; Lozano, J.A. A review on outlier/anomaly detection in time series data. *arXiv* **2020**, arXiv:2002.04236.

68. Lai, K.H.; Zha, D.; Wang, G.; Xu, J.; Zhao, Y.; Kumar, D.; Chen, Y.; Zumkhawaka, P.; Wan, M.; Martinez, D.; et al. TODS: An Automated Time Series Outlier Detection System. *arXiv* **2020**, arXiv:2009.09822.

69. Borovkova, S.; Permana, F.J. Modelling electricity prices by the potential jump-diffusion. In *Stochastic Finance*; Springer: Berlin/Heidelberg, Germany, 2006; pp. 239–263.

70. Shah, I.; Akbar, S.; Saba, T.; Ali, S.; Rehman, A. Short-term forecasting for the electricity spot prices with extreme values treatment. *IEEE Access* **2021**, *9*, 105451–105462. [CrossRef]

71. Shah, I.; Iftikhar, H.; Ali, S.; Wang, D. Short-term electricity demand forecasting using components estimation technique. *Energies* **2019**, *12*, 2532. [CrossRef]

72. Yu, L.; Wang, S.; Lai, K.K. Forecasting crude oil price with an EMD-based neural network ensemble learning paradigm. *Energy Econ.* **2008**, *30*, 2623–2635. [CrossRef]

73. Petrella, A.; Sapio, S. *No PUN Intended: A Time Series Analysis of the Italian Day-Ahead Electricity Prices*; Florence School of Regulation: Florence, Italy, 2010.

74. Petrella, A.; Sapio, S. A time series analysis of day-ahead prices on the Italian power exchange. In Proceedings of the 6th International Conference on the European Energy Market, Porto, Portugal, 6–9 June 2016 ; IEEE: Piscataway, NJ, USA, 2009; pp. 1–6.

75. Cervone, A.; Santini, E.; Teodori, S.; Romito, D.Z. Electricity price forecast: A comparison of different models to evaluate the single national price in the Italian energy exchange market. *Int. J. Energy Econ. Policy* **2014**, *4*, 744–758.

 *energies*

Article

Designing a User-Centric P2P Energy Trading Platform: A Case Study—Higashi-Fuji Demonstration

Yasuhiro Takeda [1,2,*], **Yoichi Nakai** [2], **Tadatoshi Senoo** [2] and **Kenji Tanaka** [1]

1   Graduate School of Engineering, The University of Tokyo, 7-3-1 Hongo, Bunkyō, Tokyo 113-8654, Japan; tanaka@tmi.t.u-tokyo.ac.jp
2   TRENDE Inc., 1-16-7 Higashikanda, Chiyoda City, Tokyo 101-0031, Japan; yoichi.n.trende@gmail.com (Y.N.); senoo@trende.jp (T.S.)
*   Correspondence: yasu@g.ecc.u-tokyo.ac.jp or yasu@trende.jp

Abstract: Peer-to-peer (P2P) energy trading is gaining attention as a technology to effectively handle already existing distributed energy resources (DER). In order to manage a large number of DER, it is necessary to increase the number of P2P energy trading participants. For that, designing incentives for participants to engage in P2P energy trading is important. This paper describes a user-centric cooperative mechanism that enhances user participation in P2P energy trading. The key components of this incentive for participants to engage in P2P energy trading are described and evaluated in this study. The goal of the proposal is to make it possible to conduct economic transactions while reflecting the preferences of the traders in the ordering process, making it possible to conduct transactions with minimal effort. As a case study, the Higashi-Fuji demonstration experiment conducted in Japan verified the proposed mechanism. In this experiment, 19 households and 9 plugin hybrid vehicles (PHV) were evaluated. As a result, the study confirmed that prosumers were able to sell their surplus electricity, and consumers were able to preferentially purchase renewable energy when it was available. In addition, those trades were made economically. All trades were made automatically, and this efficiency allowed the users to continue using the P2P energy trading.

Keywords: distributed energy resources (DER); P2P energy trading; cooperative mechanism; renewable energy; multi agent system; blockchain

Citation: Takeda, Y.; Nakai, Y.; Senoo, T.; Tanaka, K. Designing a User-Centric P2P Energy Trading Platform: A Case Study—Higashi-Fuji Demonstration. *Energies* **2021**, *14*, 7289. https://doi.org/10.3390/en14217289

Academic Editor: Hongseok Kim

Received: 28 September 2021
Accepted: 26 October 2021
Published: 3 November 2021

Publisher's Note: MDPI stays neutral with regard to jurisdictional claims in published maps and institutional affiliations.

1. Introduction

The decarbonization of energy is accelerating to achieve the Paris Agreement's goal of limiting global warming to well below 2 (preferably 1.5) degrees Celsius, compared with pre-industrial levels [1]. The investment in renewable energy remains high in 2021, and the momentum is as strong as ever. This trend is expected to continue [2].

Despite national efforts, future population growth and the development of economic activities will create further demand for electricity. It will be more important to use energy efficiently and to promote the use of renewable energy [3]. Furthermore, the installation costs of renewable energy decrease year by year [4,5] thus, more renewable energy will be connected to the grid in the future.

However, the generation of renewable energy, such as photovoltaic (PV) or wind power, is highly weather-dependent [6], and can sometimes generate excessive amounts of power, which can adversely affect the quality of grid power if linked to the grid [7,8]. The key to solving this problem is in the technologies that mitigate rapid changes in power generation and high electricity demand. Battery energy storage systems (BESS) play a crucial role in this [9,10]. However, if BESS capacity reserved for grid operators is adjusted to when renewable energy sources generate the most power, the total usage rate of BESS will be reduced, and the performance will be lower [11]. In the end, this will cause a negative impact on the cost of BESS.

On the other hand, if existing assets, such as batteries installed in households and EVs, can be effectively used in addition to dedicated BESS for grid operations, it will be possible to increase the capacity of storage batteries while reducing costs. P2P energy trading is gaining attention for effectively utilizing installed assets [12,13], and is a mechanism to flexibly exchange surplus energy generated from distributed power resources (DER) among neighbors [14]. As the number of participants increases, the number of DER that can be handled will also increase. For that, it is necessary to provide users with incentives to participate in P2P energy trading.

In the first place, if the participants cannot perform the transactions they intend, the system may not be used. Several studies exist that reflect the preferences of participants in P2P energy trading. Reference [15] proposed an energy management method based on trading priorities that allow prosumers to trade energy as heterogeneous products in the P2P energy market. Reference [16] describes a method that reflects the ordering styles of participants with multiple parameters. Reference [17] shows a method for prioritizing transactions nearby.

Economic trading would be a clear incentive for P2P energy trading participants. There are several studies that discuss it from a market mechanism point of view. Reference [18] shows key indices for P2P market-clearing performance. Reference [19] compares several auction mechanisms and ordering strategies, then analyzes how they change the outcome in the P2P energy market. Reference [20] studies a multi-round double auction mechanism for local energy grids.

There are several P2P energy trading pilot projects. For example, the UK's Piclo [21] offers a market where consumers can choose a producer/generator, and Vandebron [22] offers a similar system in the Netherlands.

While there are a few pilot projects that require manual actions to do P2P energy trading, there are not many projects in which trading is done automatically, involve participants, and use hardware for measurement and control. One example that is already in operation is the Brooklyn Microgrid provided by Lo3, but it is not designed to predict the power usage of participants, and then order in advance [23].

This paper describes a mechanism that enhances users' participation in P2P energy trading by providing a user-centric cooperative mechanism. Here, orders can be conducted so as to reflect the trading intentions of participants, and energy will be secured in advance by predicting participants' energy demand and supply. The key components of the incentive for participants to engage in P2P energy trading are described and analyzed. A demonstration experiment conducted in the Higashi-Fuji area of Shizuoka, Japan, is verified as a case study. It is a joint project by Toyota Motor Corporation, the University of Tokyo, and TRENDE Inc. [24]. Volunteer participants were recruited in a total of 19 households, and 9 of them were each loaned a Toyota Prius, a plugin hybrid vehicle (PHV), for the duration of the experiment. In order to conduct this verification experiment, the following tasks were carried out: recruiting participants, procuring hardware (home energy management system (HEMS), PV, storage batteries, etc.), arranging for construction work, completing application procedures with the grid operator, dealing with hardware problems, managing supply and demand during the demonstration experiment period, and removing the equipment after the verification experiment was completed.

A one-week period of trading results was analyzed to validate whether the defined key components were satisfied or not.

2. User-Centric P2P Energy Trading Platform

2.1. Key Components

This study focuses on the following three key components, which incentivize participants to trade willingly in a user-centric P2P energy trading platform:

- The participants can reflect their preferences on the trade (K1);
- The participants can trade economically (K2);
- The participants can conduct P2P energy trading with lower effort (K3).

Regarding the first point (K1), electricity needs are becoming more diverse. For example, some people wish to purchase renewable energy preferentially. Therefore, it is important to specify the energy to be traded at the time of ordering. As for the second point (K2), even if participants have the means to conduct the desired trade, if it is not conducted economically, the trade will not be sustainable. Therefore, it is also important to measure whether the trade is economically viable. Finally, the third point (K3) is that if participants spend much effort to conduct K1 and K2, it is not feasible unless sufficient benefits are obtained. When dealing with inexpensive resources, such as electricity, it is not easy to obtain a benefit that exceeds the effort expended. Therefore, it is important to ensure that the trade can be processed with little effort.

If these three points are achieved, it will be possible to provide a P2P energy trading platform that incorporates a cooperative mechanism in which P2P energy trading participants with various energy demand characteristics can supplement each other's energy needs.

2.2. Trading Platform Design

The schematic diagram of the P2P energy trading platform is shown in Figure 1. The information layer in the figure represents the exchange of transaction information, and the physical layer represents the exchange of electricity. On the information layer in the figure, the virtual P2P energy exchange is performed by trading agents, and the information is treated as if the trading agents are exchanging power with each other, but the actual power exchange is performed using the existing power distribution network on the physical layer.

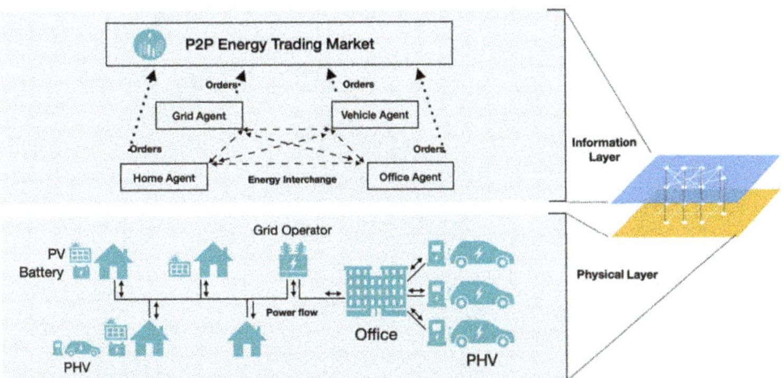

Figure 1. P2P Energy trading system.

In order to achieve K1, it is necessary to have a flexible ordering system. Therefore, in this project, in addition to the mandatory information, such as price and quantity, we will add tags as additional information to specify what type of electricity participants want to purchase and in which market they want to trade.

Next, for K2, we will use continuous double auction (CDA) as the market mechanism. CDA is widely used, not only for the financial sector, but also for energy trading all over the world [25]. There are some studies that utilize CDA to manage P2P energy trading. For example, [26] shows how P2P energy trading is done with neighbors using a CDA-based market. The use of CDA allows for price-first trading, where the execution price tends to be lower when there are many orders to sell in the market and higher when there are few. With this behavior, supply and demand are naturally adjusted.

The interaction between the P2P energy trading market and agents is shown in Figure 2. The P2P energy trading market is built on smart contracts using Ethereum Blockchain on a private network. Proof of authority was employed for the consensus algorithm. A block is generated in about 5 seconds. Ether, the virtual currency of Ethereum, is used for transactions exchanged on the P2P energy trading market.

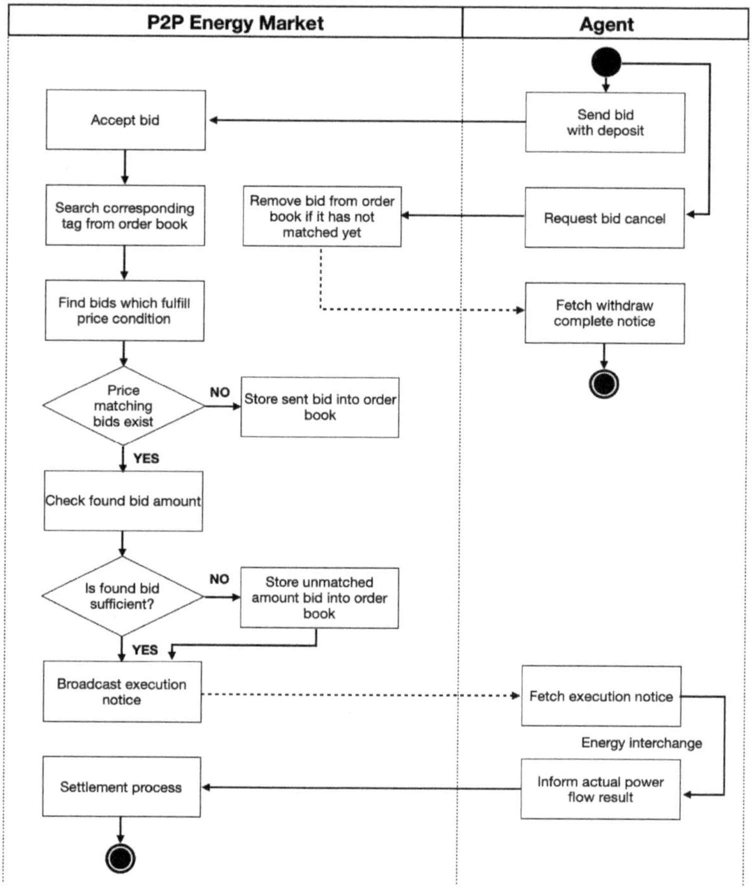

P2P Energy Market	Agent
Accept bid	Send bid with deposit
Search corresponding tag from order book	Request bid cancel
Remove bid from order book if it has not matched yet	
Find bids which fulfill price condition	Fetch withdraw complete notice
Price matching bids exist — NO → Store sent bid into order book	
Check found bid amount	
Is found bid sufficient? — NO → Store unmatched amount bid into order book	
Broadcast execution notice	Fetch execution notice
	Energy interchange
Settlement process	Inform actual power flow result

Figure 2. Activity chart between the P2P energy market and an agent.

Markets exist every 30 min; advance ordering can be conducted up to one day (48 slots) ahead of the real-time market. The real-time market gate closure is 10 min before the market ends.

Due to the nature of the blockchain, it is impossible to withdraw the executed price from the agent's cryptocurrency wallet later. Therefore, the participants must make a deposit when they submit an order. The amount of the deposit $P_{deposit}$ is shown in Equation (1)

$$P_{deposit} = (P_{max} + P_{transmission}) \cdot A_{order} \tag{1}$$

where P_{max} is the highest price on the market, which is the same as the price when buying from the grid agent. $P_{transmission}$ is the transmission cost, and A_{order} is the order amount. Both the seller and the buyer make deposits.

It is common for buyers to make a deposit, but sellers also need to do so in order to collect a fee from $P_{deposit}$ as a penalty if they make an order and fail to sell enough electricity. The penalty is calculated as the adjustment fee. If a deployed order is not executed, the $P_{deposit}$ is refunded to the agent. There is a different calculation method for the buyer and seller. First, the seller's adjustment fee is calculated as Equation (2).

$$P_{sell\_adj} = \begin{cases} -(A_{act\_sell} - A_{order\_sell}) \cdot P_{min} & (A_{act\_sell} >= A_{order\_sell}) \\ (A_{order\_sell} - A_{act\_sell}) \cdot P_{max} & (A_{act\_sell} < A_{order\_sell}) \end{cases} \tag{2}$$

A_{act_sell} is the actual amount of electricity sold, A_{order_sell} is the sold amount with the orders, and P_{max} refers to the highest price in the P2P electricity trading market. In case of excess sales, the grid purchases the excess at the lowest price P_{min} in the P2P energy trading market (with a post-transfer remittance).

Next, the buyer's adjustment fee is calculated by Equation (3). If the buying order amount A_{order_buy} is greater than the amount of actual electricity usage A_{act_buy}, the adjustment price becomes the difference between the execution price P_{exec} and P_{min}. This payment is made at the time of the completion of the energy interchange in Figure 2.

$$P_{buy\_adj} = \begin{cases} (A_{order\_buy} - A_{act\_buy}) \cdot (P_{exec} - P_{min}) & (A_{act\_buy} <= A_{order\_buy}) \\ (A_{act\_buy} - A_{order\_buy}) \cdot P_{max} & (A_{act\_buy} > A_{order\_buy}) \end{cases} \tag{3}$$

If a participant does not follow an execution result, their economic situation will worsen because of this adjustment mechanism. Thus, participants are encouraged to send a precise order and help the platform to remain stable. If the amount of electricity generated is less than the amount of orders submitted by the trading agent, the grid agent will compensate for that amount of electricity. The grid agent collects a fee from the trading agent as an adjustment according to the amount of compensation.

2.3. Trading Agent Design

The trading agent is software that performs P2P energy trading on behalf of the user. In order to achieve K3, it is desirable that all the necessary processing can be done without requiring any manual action by the user. The minimum required functions are listed below.

- Measurement;
- Prediction;
- Ordering.

As an example of the trading agent, the configuration of a home agent is described. The overall process of the home agent is shown in Figure 3.

The measurements vary depending on the user's assets; demand, PV power generation, and battery storage amount are recorded in real-time. The raw data are converted into a data format that can be processed by the agent by interpolating the missing data. Predictions of energy use are also made in real-time using user measurement data, weather forecasts, and the trading results are used to determine the amount of trading orders. To decide the ordering price, a fixed price table is used, as shown in Figure 4. The desirable price change in the buy order price is a lower price for future orders and a higher price for recent orders. This is because users can order more optimistically in the future. The sell order price is the opposite, with a higher price in the future market.

The home agent also has the ability to negotiate with the vehicle agent of the PHV that is associated with the same owner before ordering, but this is not covered in this paper.

Orders prioritize renewable energy, but will buy grid power if renewable energy cannot be purchased. In order for the trading agent to order on the smart contract in the blockchain, it uses a private key to sign the ordering transaction and deploy it on the blockchain. The deployed orders are constantly monitored, and when an execution occurs, the internal data are updated and reflected in the following order. It also cancels deployed orders if necessary. At the end of the transaction, the amount of power actually used for the execution result is measured, and the information is recorded on the blockchain. The detailed home agent settings are described in Section 3.2.

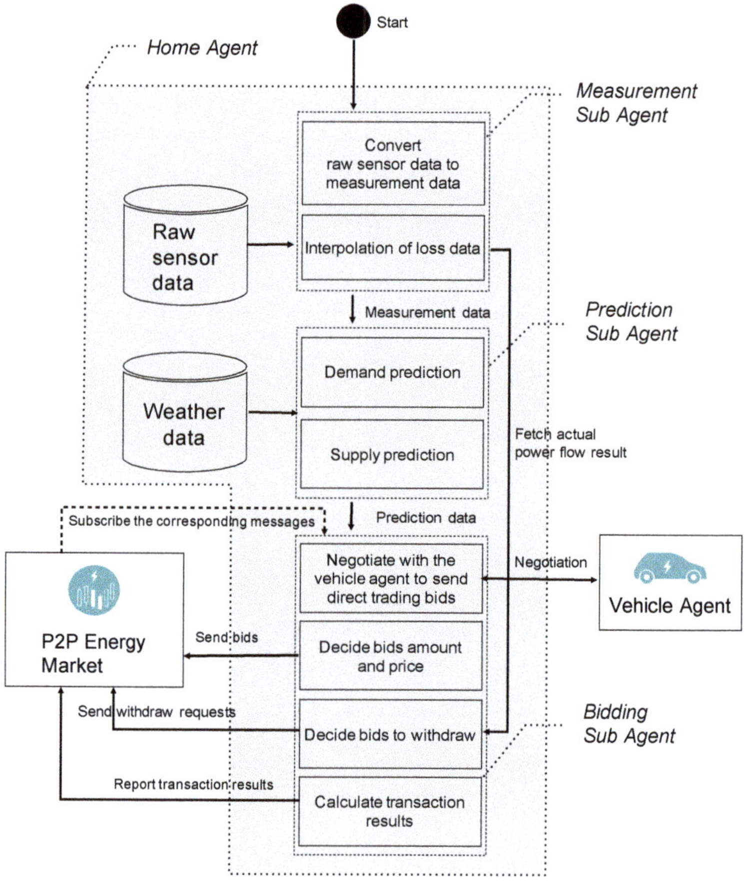

Figure 3. Home agent mechanism flow chart. The main features are divided into 3 components, measurement, prediction, and ordering.

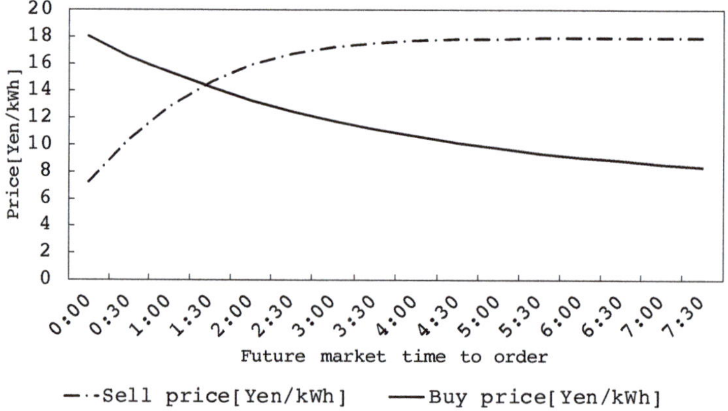

Figure 4. The ordering price change curves. The y-axis is the order price and the x-axis is the future market time as seen from the current time.

3. Case Study: Higashi-Fuji P2P Energy Trading Demonstration Experiment

The overall picture of this experiment is described in Figure 5. The process required to conduct this demonstration experiment is as follows.

- Recruit participants;
- Procure hardware (HEMS, PV, storage batteries, etc.);
- Arrange for construction work;
- Complete application procedures with the grid operator;
- Deal with hardware problems and manage supply and demand during the demonstration experiment period;
- Remove the equipment after the verification experiment is completed.

A total of nineteen households of volunteer participants were recruited. Initially, there was one more participant, but due to hardware trouble that could not be handled, the participant was excluded from this evaluation. Nine of them were loaned Toyota Prius (PHV) vehicles for the duration of the experiment (Figure 6). The hardware (HEMS, PV, storage batteries, and EV chargers) used by each participant was different. The details of the equipment for each participant are described in Section 3.2. The necessary construction work for the hardware and the application procedures for connecting the hardware to the grid was carried out before the start of the experiment. Actions were taken to resolve hardware problems that occurred during the demonstration experiment and daily balancing group [27] operations were also conducted. If the participants decided not to continue to use the equipment after the experiment, construction to remove the equipment was also carried out. The entire period of this demonstration experiment was from 17 June 2019 to 31 August, 2020. During this period, new functions of the trading agents were developed, and their bugs were fixed.

PHV charging was assumed to be done at the owner's house or office in this experiment. A grid agent sends an order that can be reliably executed if the other agent's energy supply and demand are unmet. It works as an energy retailer in the real world. The order price is assumed to be a minimum price for selling and a maximum price for buying. An office agent aims to reduce the peak energy purchase from the grid, and when it is possible, provides favorable orders to vehicle agents, which act as employee benefits.

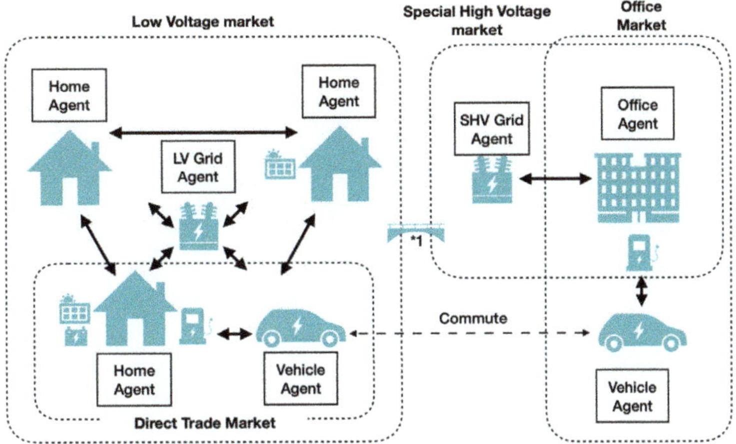

*1: LVM participants can send a sell order to SHVM but cannot send a buy order because they are connected with a low voltage grid line. Vice versa for SHVM participants.

Figure 5. Overall diagram of Higashi-Fuji P2P energy trading demonstration experiment.

Figure 6. A prosumer connects a Toyota Prius to a charger and conducts P2P energy trading.

3.1. Market Settings

In this demonstration experiment, each trading agent had a particular market to participate in. This allowed for flexibility to change the transmission cost for each market and limit participant types. Table 1 shows the correspondence between each market and its participants and transmission costs. The ordering price of the grid agent depends on the type of market (Table 2). The buy order by the grid agent is the lowest offered price for the agent in that market, and the sell order is the highest buy price for the agent.

Table 1. The relationship between agents, participating markets, and transmission costs. Check marks indicate the markets the agents can join. △ means only selling. The office cannot buy in LVM because it is not connected with a low voltage power line. The same thing can be said of the home agent and the vehicle agent in SHVM.

	Home Agent	Vehicle Agent	Office Agent	Grid Agent	Transmission Cost (Yen/kWh)
Low voltage market (LVM)	✓	✓	△	✓	8
Special high voltage market (SHVM)	△	△	✓	✓	4
Office market (OM)	-	✓	✓	-	0
Direct trade market (DTM)	✓	✓	-	-	0

Table 2. Buy or sell price depending on the target market of the grid agent. These prices are determined by referring to the grid tariffs for each voltage in Japan (not including transmission costs). The selling price here means the highest price in the market, and the buying price means the lowest price.

	Buy Price (Yen/kWh)	Sell Price (Yen/kWh)
LVM	7	18
SHVM	4	11

The office market (OM) does not have transmission costs because it uses a company-owned power line. Direct trade market (DTM), a private market for home agents and vehicle agents, is prepared for vehicle to home (V2H). Since the energy is exchanged over the home wiring, there is no transmission cost.

3.2. Home Agent Settings

Although vehicle agents and office agents participate in P2P energy tradings, this section describes the detail of the home agent that the authors worked on.

The home agent has several types depending on the assets it owns (Table 3). The details of the 19 home agents are shown in Table 4. All agents have smart meter and HEMS controller, which obtains energy usage on the smart meter via B-route and uploads the measured values to the cloud server using LTE. The B-route is a mechanism installed in smart meters in Japan that obtains electricity accumulation and instantaneous values using Wi-SUN (920 MHz wireless communication in compliance with IEEE 802.15.4 g) [28].

The agent that has all assets (P4) is shown in Figure 7. The communication between the PV system and the HEMS controller was conducted using the Echonet Lite protocol [29], but the battery system (9.8 kWh, OMRON) was the same. For deciding future market order amounts, predictions of power load and PV generation were made. In this case study, support vector regression (SVR) and moving average were adaptively used depending on the agent. Ordering was conducted twice, in 30 min. The markets for ordering were limited to markets that started 2.5 h ahead, including the real-time market.

Table 3. Correspondence table of the owned assets and agent types. Check marks indicate that the hardwares are owned by the agent types.

	Smart Meter and HEMS Controller	PV	Battery	PHV and EV Charger
C1	✓	-	-	-
C2	✓	-	-	✓
P1	✓	✓	-	-
P2	✓	✓	✓	-
P3	✓	✓	-	✓
P4	✓	✓	✓	✓

Table 4. Composition of the household participants.

Home Agent ID	Area	Agent Type	PV Capacity (W)	Battery Capacity (Wh)
HA_01	Susono	C1	-	-
HA_02	Susono	C1	-	-
HA_03	Mishima	C1	-	-
HA_04	Yokohama	C1	-	-
HA_05	Susono	C1	-	-
HA_06	Yamanakako	P3	5880	-
HA_07	Fuji	C2	-	-
HA_08	Mishima	C2	-	-
HA_09	Susono	C2	-	-
HA_10	Mishima	C2	-	-
HA_11	Odawara	C1	-	-
HA_12	Susono	C2	-	-
HA_13	Gotenba	P1	4200	-
HA_14	Gotenba	P1	4800	-
HA_15	Gotenba	P2	6000	9800
HA_16	Mishima	P2	5400	9800
HA_17	Fuji	P3	7200	-
HA_18	Gotenba	P4	7200	9800
HA_19	Mishima	C2	-	-

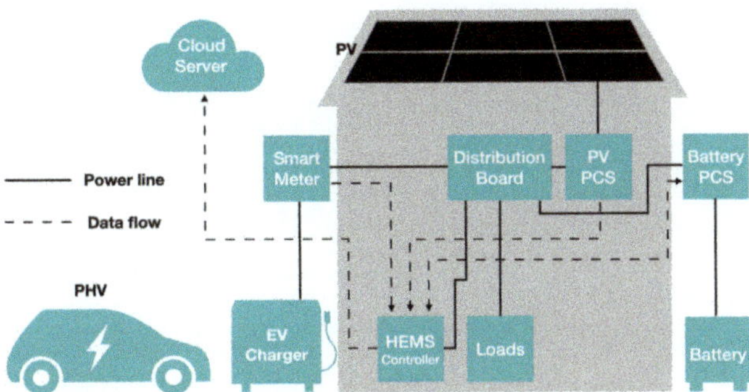

Figure 7. Hardwares installed in the P4 agent type home.

4. Demonstration Experiment Result

This section describes the results of transactions conducted between 25 and 31 August, during the entire experiment period.

A breakdown of the trading partners of the energy sold by prosumers (P1–P4) is shown in Figure 8. It shows that the transactions were concentrated between 8:00 and 19:00, with 15.4% of the unmatched energy sales absorbed by the grid agent and the remaining 84.6% of the energy sales purchased by other agents. The discharge of the battery made up the energy sold from 2:00 to 3:00. A breakdown of the sources of electricity purchased by consumers (C1–C2) is shown in Figure 9. It shows that between 8:00 and 18:00 , there were multiple purchases from prosumers and the percentage was 48.9%. These results indicate that redundant renewable energy is effectively absorbed within the P2P energy market. Purchases from the vehicle agents occurred between 18:30 and 7:30. The energy stored in the PHVs was purchased when no PV power was generated, indicating V2H behavior. There was a small amount of purchasing power from prosumers around 2:00 to 4:00. This was due to the discharge of the battery.

From these results, it was confirmed that the prosumer was able to sell surplus electricity, and the consumer was able to preferentially purchase renewable energy when it was available in the market. Therefore, it was confirmed that K1 was achieved.

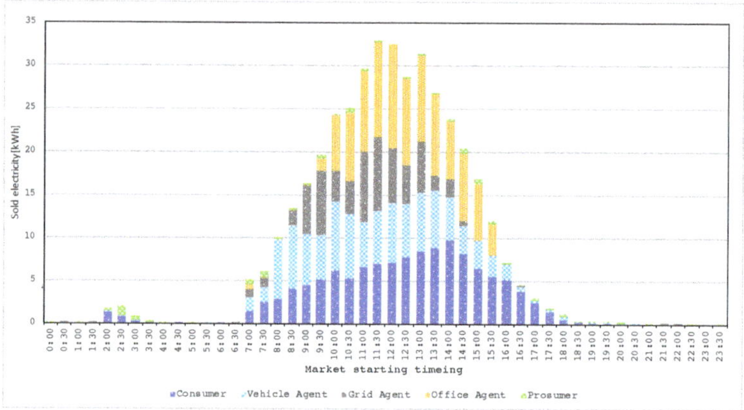

Figure 8. Execution result breakdown of where prosumer sold the electricity to.

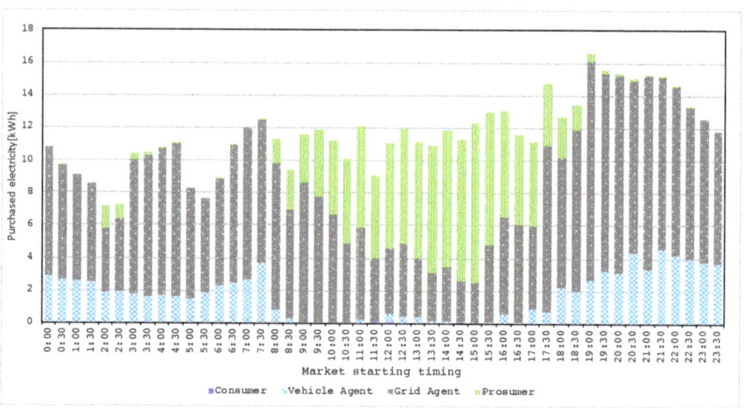

Figure 9. Execution result breakdown of the source of electricity purchased by consumers.

The economic results are shown in Table 5, with the price reduction rates comparing to when P2P energy trade was not used for transactions. The price reduction rates by agent type are 1.4% for C1, 2.2% for C2, 13.3% for P1, 60.3% for P2, 6.0% for P3, and 301.9% for P4. From this, it was confirmed that every agent type could get economic benefits. Meanwhile, P3 was lower than other prosumers although it was higher for agents with assets. Figure 10 shows the feed-in and feed-out amount measured by each agent's smart meter. It shows that P3 (HA_06, HA_17) had more feed-in amount than feed-out amount, and the amount sold was relatively low. That is the reason why the economic results were not as good as other prosumer agent types. P4 (HA_18) had the opposite result.

There was also an economic benefit for consumers without assets, but the reduction rate was about 2%; there is still room for improvement. One of the reasons why the consumer's reduction rate was low is that the ordering amount was intentionally reduced to avoid the negative adjustment fee. If the execution amount has excess, the adjustment result becomes a negative balance. It will be further improved if the prediction accuracy can be enhanced and increase the order amounts to the P2P market.

Table 5. Comparison of the P2P energy trading market and grid power trading prices for each agent. Sorted by agent type.

Home Agent ID	Agent Type	Incumbent Transaction (Yen)	P2P Transaction (Yen)	Difference (Yen)	Reduction Rate (%)
HA_01	C1	−2028.0	−2012.2	15.8	0.8
HA_02	C1	−2979.6	−2962.3	17.3	0.6
HA_03	C1	−3538.6	−3456.1	82.5	2.3
HA_04	C1	−2706.6	−2691.6	15.0	0.6
HA_05	C1	−3419.0	−3341.9	77.1	2.3
HA_11	C1	−2667.6	−2616.9	50.7	1.9
HA_07	C2	−1369.6	−1339.0	30.5	2.2
HA_08	C2	−4023.5	−3943.0	80.5	2.0
HA_09	C2	−3461.1	−3358.3	102.8	3.0
HA_10	C2	−1651.8	−1658.4	−6.6	0.4
HA_12	C2	−4791.8	−4647.4	144.4	3.0
HA_19	C2	−2390.4	−2333.0	57.4	2.4
HA_13	P1	−568.1	−464.5	103.6	18.2
HA_14	P1	−1250.4	−1146.8	103.6	8.3
HA_15	P2	−274.9	−28.0	246.9	89.8
HA_16	P2	−684.0	−473.5	210.5	30.8
HA_06	P3	−2539.8	−2433.1	106.7	4.2
HA_17	P3	−2621.8	−2417.5	204.3	7.8
HA_18	P4	42.6	171.2	128.6	301.9

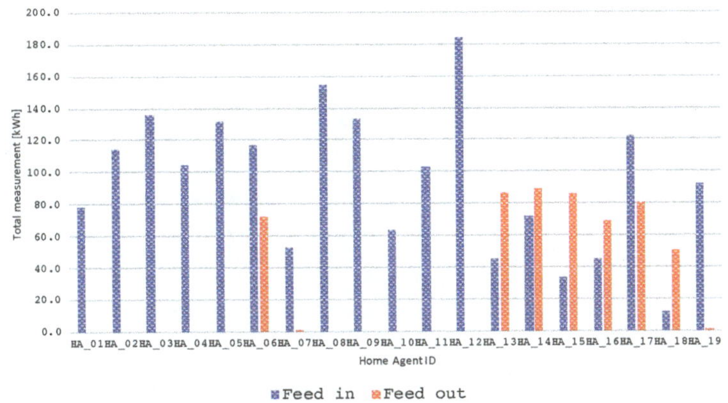

Figure 10. The total feed-in and out amount measured by each agent's smart meter.

A boxplot of LVM executed prices for the P2P energy trading market is shown in Figure 11. This shows that the nighttime price is almost the same as the grid price, but the price drops around 7:30 a.m., and around noon the average price is about 5 Yen/kWh lower than the grid price. This is because many orders are received during the daytime when PV generation is high and cheaper energy is available. From these results, it was confirmed that economic transactions (K2) could be conducted.

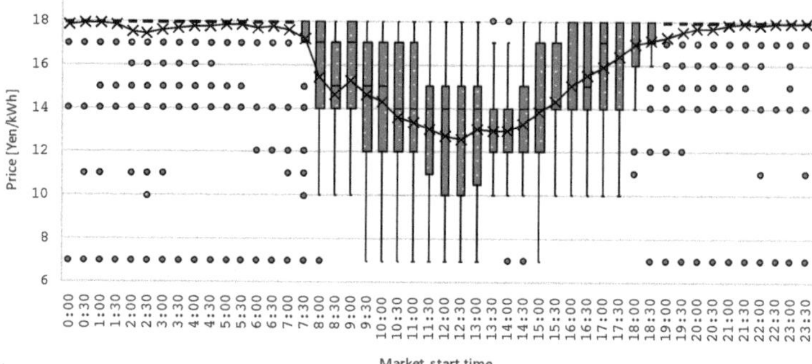

Figure 11. Boxplots of executed prices in LVM. The circle means outlier and the x means mean value.

All transactions were done automatically by the trading agents, so participants did not have to perform manual actions to conduct P2P energy trading. Therefore, it was confirmed that P2P energy trading could be used by real users and that K3 was achieved. Participants provided comments on the demonstration experiment such as, "The reassurance of safety and security against disasters was a major attraction", and, "The exchange of electricity also provided an opportunity to think about energy".

5. Conclusions

This paper describes a mechanism that enhances users' participation in P2P energy trading by providing a user-centric cooperative mechanism. Three key components were defined for evaluating the system as follows: reflect user preferences on the trade (K1); trade economically (K2); conduct P2P energy trading with low effort (K3).

In the proposed platform, tags are added to the orders to express users' ordering intentions for K1, and continuous double auction (CDA) is incorporated as a market mechanism for K2. For K3, it is designed to execute orders automatically without any manual actions by the user.

As a case study, we verified a demonstration experiment consisting of the proposed contents. This experiment was conducted in Higashi-Fuji, Japan. Volunteer participants were recruited in a total of nineteen households, and nine of them were each loaned a Toyota Prius, a plugin hybrid vehicle (PHV), for the duration of the experiment.

As a result, the study confirmed that prosumers were able to sell their surplus electricity, and consumers were able to preferentially purchase renewable energy when it was available. In addition, those trades were made economically; the average price in the P2P energy trading market was about 27% (about 5 Yen) lower than the grid price when PV power was generated. Furthermore, every trade was made automatically, and this efficiency allowed the users to continue using the P2P energy trading. From all of this, K1, K2, and K3 were achieved.

The results show that if the prediction accuracy of user demand and supply can be improved, more orders can be placed in the P2P energy trading market, leading to improved economic efficiency. Therefore, future improvements to the prediction accuracy are needed. In addition, there were a few times when the connections of the HEMS devices were unstable, and measurement information could not be acquired. It is necessary to consider a hardware configuration that will enable a more stable connection.

In this demonstration experiment, the PHV charging locations were limited to homes and the office. This is because it was necessary to use chargers that could be controlled by the vehicle agent. In order to provide more general service, other locations, such as quick charging stations, should be considered. In addition, because the evaluation was conducted by actual participants, it was not possible to compare the results with existing P2P energy trading methods. For the comparison, it is necessary to set up an evaluation method in advance and work on verification.

In a future study, it may be necessary to work on the analysis of vehicle to home (V2H) control performed through direct trades between home and vehicle agents.

Author Contributions: Conceptualization, T.S. and K.T.; methodology, Y.T.; software, Y.T.; validation, Y.T., Y.N.; formal analysis, Y.T.; investigation, Y.T. and Y.N.; data curation, Y.T.; writing—original draft preparation, Y.T.; writing—review and editing, K.T.; visualization, Y.T.; supervision, T.S. and K.T.; project administration, T.S. and K.T. All authors have read and agreed to the published version of the manuscript.

Funding: This research received no external funding

Acknowledgments: This project was made possible by the generous support of Toyota Motor Corporation in terms of resources and technology. We would like to express our sincere gratitude to them. The fourth author is supported by Grant-in-Aid for Scientific Research (A) 20H00285.

Conflicts of Interest: The authors declare no conflict of interest.

References

1. Unitated Nation. The Paris Agreement. Available online: https://unfccc.int/sites/default/files/english_paris_agreement.pdf (accessed on 30 October 2021).
2. International Energy Agency. World Energy Investment 2021. Available online: https://iea.blob.core.windows.net/assets/5e6b3821-bb8f-4df4-a88b-e891cd8251e3/WorldEnergyInvestment2021.pdf (accessed on 30 October 2021).
3. International Energy Agency. Net Zero by 2050 A Roadmap for the Global Energy Sector. Available online: https://iea.blob.core.windows.net/assets/deebef5d-0c34-4539-9d0c-10b13d840027/NetZeroby2050-ARoadmapfortheGlobalEnergySector_CORR.pdf (accessed on 30 October 2021).
4. Bloomberg New Energy Finance. New Energy Outlook 2020-Executive Summary. Available online: https://assets.bbhub.io/professional/sites/24/928908_NEO2020-Executive-Summary.pdf (accessed on 30 October 2021).
5. International Renewable Energy Agency. Renewable Power Generation Costs in 2019. Available online: https://www.irena.org/-/media/Files/IRENA/Agency/Publication/2020/Jun/IRENA_Power_Generation_Costs_2019.pdf (accessed on 30 October 2021).
6. Bremen, L.V. Large-Scale Variability of Weather Dependent Renewable Energy Sources. In *Management of Weather and Climate Risk in the Energy Industry*; Troccoli, A., Ed.; Springer: Amsterdam, The Netherlands, 2010; pp. 189–206.
7. Chawda, G.S.; Shaik, A.G.; Shaik, M.; Padmanaban, S.; Holm-Nielsen, J.B.; Mahela, O.P.; Kaliannan, P. Comprehensive Review on Detection and Classification of Power Quality Disturbances in Utility Grid with Renewable Energy Penetration. *IEEE Access* **2020**, *8*, 146807–146830. [CrossRef]
8. Mahela, O.P.; Khan, B.; Alhelou, H.H.; Tanwar, S. Assessment of power quality in the utility grid integrated with wind energy generation. *IET Power Electron.* **2020**, *13*, 2917–2925. [CrossRef]
9. Diouf, B.; Pode, R. Potential of lithium-ion batteries in renewable energy. *Renew. Energy* **2015**, *76*, 375–380. [CrossRef]
10. Keck, F.; Lenzen, M.; Vassallo, A.; Li, M. The impact of battery energy storage for renewable energy power grids in Australia. *Energy* **2019**, *173*, 647–657. [CrossRef]
11. Yang, Y.; Bremner, S.; Menictas, C.; Kay, M. Battery energy storage system size determination in renewable energy systems: A review. *Renew. Sustain. Energy Rev.* **2018**, *91*, 109–125. [CrossRef]
12. International Renewable Energy Agency. Peer-to-Peer Electricity Trading Innovation Landscape Brief. Available online: https://irena.org/-/media/Files/IRENA/Agency/Publication/2020/Jul/IRENA_Peer-to-peer_trading_2020.pdf (accessed on 30 October 2021).
13. Soto, E.A.; Bosman, L.B.; Wollega, E.; Leon-Salas, W.D. Peer-to-peer energy trading: A review of the literature. *Appl. Energy* **2021**, *283*, 116268. [CrossRef]
14. Zhang, C.; Wu, J.; Zhou, Y.; Cheng, M.; Long, C. Peer-to-Peer energy trading in a Microgrid. *Appl. Energy* **2018**, *220*, 1–12. [CrossRef]
15. Morstyn, T.; McCulloch, M.D. Multiclass Energy Management for Peer-to-Peer Energy Trading Driven by Prosumer Preferences. *IEEE Trans. Power Syst.* **2019**, *34*, 4005–4014. [CrossRef]
16. Takeda, Y.; Tanaka, K. Bidding Agent Model for P2P Energy Trading. *IEEJ Trans. Ind. Appl.* **2020**, *140*, 738–745. [CrossRef]
17. Guerrero, J.; Sok, B.; Chapman, A.C.; Verbič, G. Electrical-distance driven peer-to-peer energy trading in a low-voltage network. *Appl. Energy* **2021**, *287*, 116598. [CrossRef]
18. Khorasany, M.; Mishra, Y.; Ledwich, G. Design of auction-based approach for market clearing in peer-to-peer market platform. *J. Eng.* **2019**, *2019*, 4813–4818. [CrossRef]
19. Lin, J.; Pipattanasomporn, M.; Rahman, S. Comparative analysis of auction mechanisms and bidding strategies for P2P solar transactive energy markets. *Appl. Energy* **2019**, *255*, 113687. [CrossRef]

20. Flikkema, P.G. A multi-round double auction mechanism for local energy grids with distributed and centralized resources. In Proceedings of the IEEE International Symposium on Industrial Electronics, Santa Clara, CA, USA, 8–10 June 2016; pp. 672–677. [CrossRef]
21. Mujeeb, A.; Hong, X.; Wang, P. Analysis of Peer-to-Peer (P2P) Electricity Market and Piclo's Local Matching Trading Platform in UK. In Proceedings of the 2019 3rd IEEE Conference on Energy Internet and Energy System Integration: Ubiquitous Energy Network Connecting Everything, EI2 2019, Changsha, China, 8–10 November 2019; pp. 619–624. [CrossRef]
22. Vandebron. 2021. Available online: https://vandebron.nl/ (accessed on 30 October 2021).
23. Mengelkamp, E.; Gärttner, J.; Rock, K.; Kessler, S.; Orsini, L.; Weinhardt, C. Designing microgrid energy markets. A case study: The Brooklyn Microgrid. *Appl. Energy* **2017**, *210*, 870–880. [CrossRef]
24. Toyota. The University of Tokyo, Toyota, and TRENDE to Begin Testing of Next-Generation Electricity System. 2020. Available online: https://global.toyota/en/newsroom/corporate/28231367.html (accessed on 30 October 2021).
25. Shah, D.; Chatterjee, S. A comprehensive review on day-ahead electricity market and important features of world's major electric power exchanges. *Int. Trans. Electr. Energy Syst.* **2020**, *30*, e12360. [CrossRef]
26. Ilic, D.; Da Silva, P.G.; Karnouskos, S.; Griesemer, M. An energy market for trading electricity in smart grid neighbourhoods. In Proceedings of the IEEE International Conference on Digital Ecosystems and Technologies, Campione d'Italia, Italy, 18–20 June 2012; pp. 1–6. [CrossRef]
27. Dr. Patrick Matschoss (IZES), K.N.u. Analysis of Framework Conditions for Founding of Green Retailers in Japan. Available online: https://www.izes.de/sites/default/files/publikationen/EM_18_014.pdf (accessed on 30 October 2021).
28. Harada, H.; Mizutani, K.; Fujiwara, J.; Mochizuki, K.; Obata, K.; Okumura, R. IEEE 802.15.4g based Wi-SUN communication systems. *IEICE Trans. Commun.* **2017**, *100*, 1032–1043. [CrossRef]
29. Matsumoto, S. Echonet: A home network standard. *IEEE Pervasive Comput.* **2010**, *9*, 88–92. [CrossRef]

Article

Feasibility Conditions for Demonstrative Peer-to-Peer Energy Market

Reo Kontani [1,*], Kenji Tanaka [1] and Yuji Yamada [2]

[1] Department of Technology Management for Innovation, The University of Tokyo, Tokyo 113-8656, Japan; tanaka@tmi.t.u-tokyo.ac.jp

[2] Faculty of Business Sciences, University of Tsukuba, Tokyo 112-0012, Japan; yuji@gssm.otsuka.tsukuba.ac.jp

* Correspondence: kontani@ioe.t.u-tokyo.ac.jp

Abstract: Distributed energy resources (DERs) play an indispensable role in mitigating global warming. The DERs require flexibility owing to the uncertainty of their power output when connected to the power grid. Recently, blockchain technology has actualized peer-to-peer (P2P) energy markets, promoting efficient and resilient flexibility in the power grid. This study aimed to extract insights about the contribution of the P2P energy markets to ensuring flexibility through analyzing transaction data. The data source was a demonstration project regarding the P2P energy markets conducted from 2019 to 2020 in Urawa-Misono District, Japan. The participants in the project were photovoltaic generators (PVGs), convenience stores (CSs), and residences equipped with battery storage as the only flexibility in the market. We quantitatively analyzed the prices and volumes ordered or transacted by each participant. The execution prices purchased by the residences were lower than those purchased by CSs; the differences between execution prices and order prices of the residences were narrower than those of PVGs and CSs; the lower state-of-charge (SoC) in the storage battery induced the higher purchasing prices. Thus, P2P energy markets, where holding flexibility resulted in the advantageous position, can promote installing flexibility through market mechanisms.

Keywords: peer-to-peer energy trading; distributed energy resources; microgrid; blockchain; digital grid; bidding strategy

Citation: Kontani, R.; Tanaka, K.; Yamada, Y. Feasibility Conditions for Demonstrative Peer-to-Peer Energy Market. *Energies* **2021**, *14*, 7418. https://doi.org/10.3390/en14217418

Academic Editor: Ricardo J. Bessa

Received: 10 October 2021
Accepted: 2 November 2021
Published: 8 November 2021

1. Introduction

Decarbonization is an essential process for mitigating global warming, which causes natural disasters, such as lethal heatwaves and extreme precipitation [1]. Nowadays, various pathways to reduce carbon dioxide emissions are being explored, and distributed energy resources (DERs) are attracting significant attention [2]. However, DERs have several disadvantages, including power grid disturbances, such as duck curves and dark doldrums, stimulating the need for flexibility [3–5].

Blockchain technology is expected to develop efficient and resilient flexibility and contribute to the realization of peer-to-peer (P2P) energy markets [6]. In these markets, generators and consumers recognize each other, conduct one-to-one power transactions, and act as aggressive players in the power distribution network. The P2P energy markets enable participants to trade directly without mediation and seek a better outcome, that is, buying sides can save costs while selling sides profit in trading electricity [7].

Research on P2P energy markets has been conducted from various perspectives [8]. Transmission system operators in Japan and system integrators have entered into strategic partnerships, such as the Tokyo Electric Power Company Holdings, Inc. and Innogy (currently integrated into RWE AG [9]) [10]; the Chugoku Electric Power Co., Inc. and IBM Japan, Ltd. [11]; Tohoku Electric Power Co., Inc. and Toshiba Corporation [12]; Kansai Electric Power Co., Inc. (KEPCO) and Power Ledger Pty. Ltd. [13]; and Shikoku Electric Power Co., Inc. and LO3 energy, Inc. [14]. Several energy management system providers in Japan, such as ENERES Co., Ltd. [15], UPDATER, Inc. (formerly Minna-Denryoku, Inc.) [16], and

Hitachi, Ltd. [17,18] have commercialized blockchain-based systems to actualize and trace non-fossil fuel energy values. In addition to tracing power on blockchain-based databases, several studies have attempted to seamlessly access devices [19] and control electricity delivery according to P2P-matched results [20,21]. Moreover, demonstration projects implementing the P2P energy market on real power grids have gradually increased. One of the earliest projects was the Brooklyn Microgrid [22], operated by LO3 Energy, Inc. [23], and projects in Japan include the DC-based Open Energy System initiated by Sony Computer Science Laboratories, Inc. and Okinawa Institute of Science and Technology [24,25]; the EV charging project co-conceptualized by the Chubu Electric Power Co., Inc., Nayuta, Inc., and Asteria Corporation (formerly Infoteria Corporation) [26]; the Tatsumi Research Center Project demonstrated by KEPCO and Nihon Unisys, Ltd. [27]; and the Higashi-Fuji project carried out by the Toyota Motor Corporation, TRENDE, Inc. and the University of Tokyo [28].

This study focuses on a demonstration project conducted by Digital Grid Co., in Urawa-Misono District, Japan, from August 2019 to March 2020, financially supported by the Ministry of the Environment, Japan [29–32] (see Acknowledgments). This project featured a digital grid and blockchain-based platform, in which the digital grid allows for the acceptance of the high penetration of DERs [33]. Controlling power flow [34,35] and routing algorithms [36] with a digital grid have been researched previously.

This study aims to analyze the transaction data recorded in the Urawa-Misono demonstration project and extract insights about the contribution of P2P to ensuring flexibility. It represents the first attempt to quantitatively investigate the details of transaction data after the project completion. Various attributes of transaction data enable us to comprehend the benefits of holding flexibility for each facility in the market, which addresses information regarding the order, executions, and the timing of P2P electricity trading in the project as follows. The order data specify the facilities responsible for posting the order data, order volumes, order prices, keys to link to the executed data, and time intervals of electricity delivery. The executed data specify the suppliers, consumers, executed volumes, executed prices, keys to link to the order data, and time intervals of electricity delivery.

The purpose of this study is to analyze the extent to which flexible facilities are advantageous in the market. The featuring indicators were the volume-weighted averages of the order and execution prices, associated with the time intervals of electricity delivery, durations of electricity delivery, and the state-of-charge (SoC) of battery storage.

The remainder of this paper is organized as follows: In Section 2, we provide an overview of the project; in Section 3, we depict the overall trends among the executed and order volumes and prices, associating them with the time intervals of electricity delivery and the duration of electricity delivery; in Section 4, we summarize the findings and limitations.

2. Overview of the Demonstration Project

This section describes the Urawa-Misono demonstration project [31,32].

2.1. Location of the Demonstration Project

Figure 1 shows the location of the Urawa-Misono District. The district, a metropolitan suburb of Japan situated in the southeastern part of Saitama Prefecture, is built around Urawa-Misono Station, operated by the Saitama Railway Corporation.

The climate in the Saitama Prefecture is regulated by the Pacific Ocean [37]. In the winter, the air is dry, with numerous sunny days due to the northwest monsoon. Figure 2 shows the weather conditions (precipitation, temperature, and daylight duration) during the demonstration project period in the city of Saitama [38]. Daylight duration is defined as the duration of direct solar radiation of 0.12 kW/m^2 or more [39]. In particular, from February to March, the weekly total precipitation was below 50 mm, the weekly mean temperature was approximately 9 °C, and the weekly daylight duration was more than 20 h.

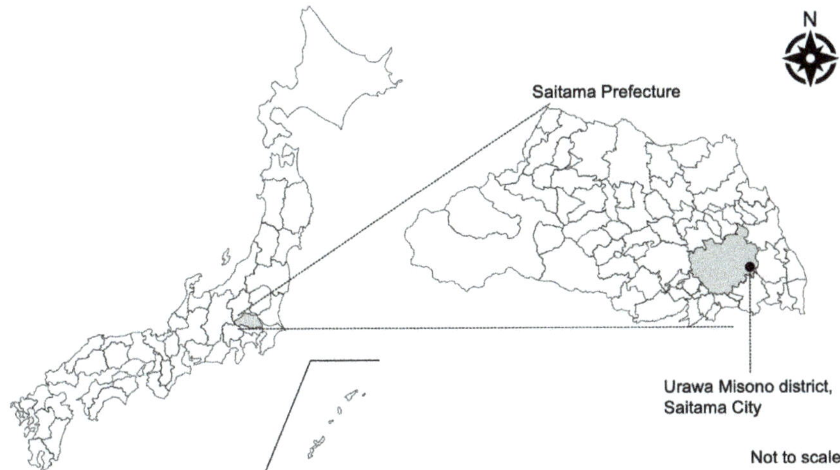

Figure 1. Location of the demonstration project in Urawa-Misono District.

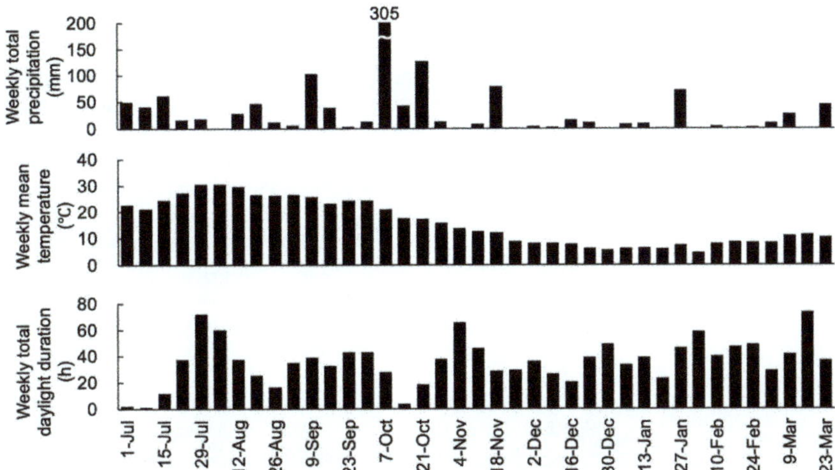

Figure 2. Weather conditions during the time of the demonstration project.

2.2. Demonstration Project System

Figure 3 shows the system applied for the demonstration project. Digital Grid, Co. (Tokyo, Japan), designed the facility configurations [31]. Tanaka conceptualized the electricity trading market and bidding algorithms in each agent [32]. The project participants were individual power producers (suppliers), consumers (demanders), and prosumers (both producers and consumers), represented by actual infrastructural facilities: three photovoltaic generators (PVGs), four convenience stores (CSs), and four residences. All the PVGs, equipped with 18 kW panels and located on the roof of the same shopping center, behaved as a supplier in the market. The CSs were separated from each other and behaved as consumers in the market. Incidentally, a convenience store in Japan is a retail open almost 24 h a day, seven days a week, and sells a large variety of food and daily sundries. The residences were situated in the same block, contained ≤5.5 kW panels and 12 kWh lithium-ion batteries, and behaved as prosumers.

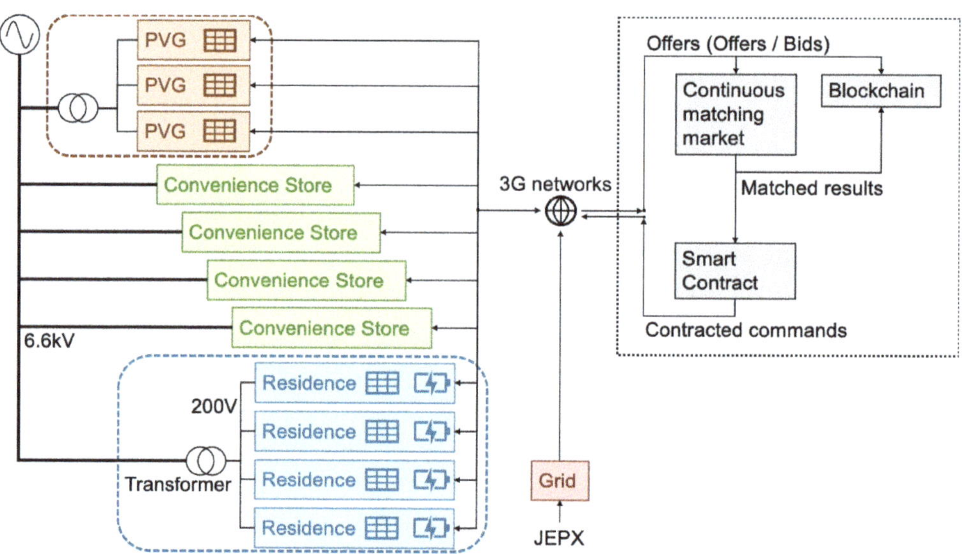

Figure 3. System used for the demonstration project.

The PVGs and CSs were connected to a 6.6 kV distribution line operated by TEPCO Power Grid, Inc. [40]. The residences were connected to the distribution line via a transformer and a 200 V private line operated by Digital Grid, Co. [41].

The main procedure for order and execution was as follows [31,32]: First, each facility individually posted orders on the market. Each facility was equipped with a digital grid controller (DGC) [42]. DGCs are programmable devices for reading smart meters and communicating with the market via 3G networks, which can forecast the demand and supply for the facilities and schedule orders. The order schedules of DGCs for the PVGs and CSs were based on the forecasted demand and supply, while those for the residences were based on the SoC of their battery storage. Orders were either offers (selling orders) or bids (purchasing orders). The Ethereum blockchain preserved all the posted requests. Attributes of the posted requests included the time at which the requests are posted, volumes and prices, and the requesting facility. In addition to the existing facilities, it was imperative that the power grid agent offer electricity at a price of 30 JPY/kWh or higher than the JEPX [43], which is the wholesale electricity exchange market in Japan.

Second, each market was kept open 24 h before the start time of electricity delivery and accepted orders in continuous sessions. The participants placed orders in the books executed according to the principle of time and price priority. The market detected an offer coupled with a bid on the condition that the bid price is greater than the offered price. All offers with prices exceeding those of bids and all bids with prices lower than those of the offers remained on the board. If the offered volumes and bid volumes differed, the market would set the matched volumes to a smaller one. If multiple requests with the same price existed, the earliest request was given priority. The Ethereum blockchain preserved all matched results. Attributes of the matched results included times when the results were matched, matched volumes and prices, start times and end times of electricity delivery, and a matched supplier and consumer pair.

Third, the market notified each facility when its orders were executed. The market converted each of the matched results into a smart contract, an Ethereum-based program that is automatically executed on the occurrence of a defined event and transmits these results to each of the facilities.

Fourth, each facility supplied or consumed power according to the contract. In addition to a DGC, each of the PVGs and residences was equipped with a Digital Grid

Router (DGR), a multifunctional inverter capable of measuring and controlling power according to the attached DGC; at the time of electricity delivery, the DGRs installed in the PVGs supplied as much power as the matched volumes, even if a room capable of generating more power than the matched volumes existed.

3. Analysis of Traded Data

This section mentions the results of quantitative analysis of the traded data. The quantitative analysis was implemented in the Python programming language and numerical libraries such as pandas [44].

3.1. Selection of Target Period

The number of participants was not always the same, meaning several participants withdrew in the middle of the demonstration project. Here, we will focus on the data period from 17 February to 22 March (target period), during which all the participants were involved in the demonstration project. Figure 4 shows the weekly total executed volumes, where the horizontal axes represent the weekly periods starting from these dates. The average total executed volume in the target period was approximately 25 MWh per week, higher than in the first half of the demonstration period.

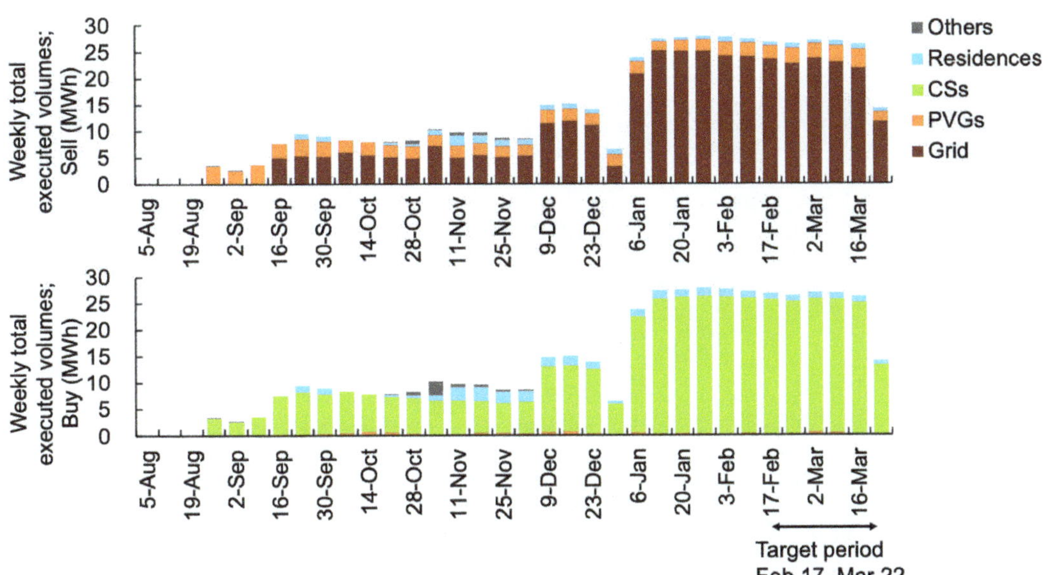

Figure 4. Selection from the target period for trade data analysis.

3.2. Analysis of Executed Volumes and Prices in the Target Period

3.2.1. Executed Volume Shares among Each Agent

Figure 5 illustrates the share of the executed volumes of each agent with eight Sankey diagrams. The left side of each Sankey diagram represents the share of the supplied volumes, and the right side represents the share of the purchased volumes during the time interval shown at the bottom of the diagram, for example, "00:00–03:00" indicates the executed volumes from 0:00 to 3:00 a.m. in the target period. "PVGs", "CSs", and "Residences" each represent the total volumes executed by the respective facilities, for example, "PVGs" are the total volumes executed by the three PVGs.

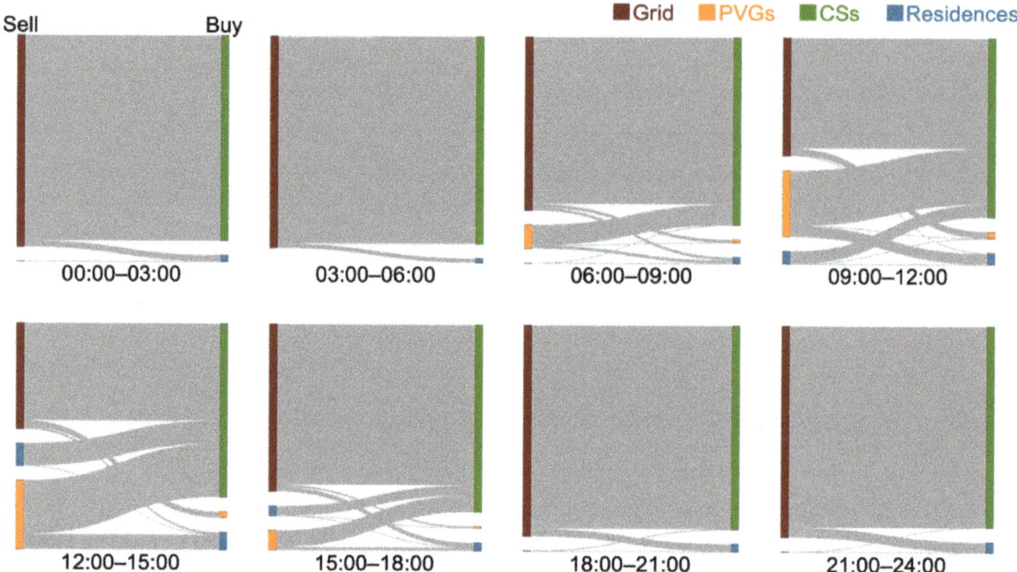

Figure 5. Share of the executed volumes among each agent in the target period.

From Figure 5, it can be seen that each Sankey diagram pertains to a different time interval of electricity delivery. At night (00:00–03:00, 03:00–06:00, 18:00–21:00, and 21:00–24:00), the grid was dominant on the supply side and the CSs were dominant on the purchase side. During the daytime (09:00–12:00 and 12:00–15:00), the PVGs accounted for a large share (34% at 09:00–12:00 and 35% at 12:00–15:00). Moreover, the PVGs accounted for 31% of the total volumes purchased by the CSs in the daytime, and 93% at 09:00–12:00 and 86% at 12:00–15:00 of the total volumes purchased by the residences.

3.2.2. Executed Volumes and Prices Associated with Electricity Delivery Time Intervals

Figure 6 shows the total executed volumes for each time interval of electricity delivery in the target period, where the upper panel represents the breakdown of the executed volumes on the supplying side, and the lower panel indicates the breakdown of the executed volumes on the purchasing side. The horizontal axes denote the time intervals of electricity delivery, for example, "00–01" represents the total volumes transacted from 0:00 to 1:00 in the target period.

Figure 7 shows the mean price, that is, the weighted average of the executed prices by the executed volumes for each time interval of electricity delivery in the target period, where the upper and lower panels represent the supplied and purchased price, respectively, and the horizontal axes denote the time intervals of electricity delivery, similar to Figure 6.

Several observations were made based on Figure 6. Here, the trend of the total executed volumes was approximately steady at 5.3–6.0 MWh and the mean executed prices supplied by the grid were steady at 36–39 JPY/kWh throughout the day. Moreover, during the daytime (9:00–10:00, 10:00–11:00, 11:00–12:00, 12:00–13:00, and 14:00–15:00), the mean executed prices supplied by the PVGs were steady at 27–29 JPY/kWh, while the mean executed prices supplied by the residences were steady at 29–32 JPY/kWh. The mean executed prices purchased by the CSs were steady at 34–39 JPY/kWh throughout the day, and in the daytime, the mean executed prices purchased by the residences were 9–19 JPY/kWh, significantly lower than those purchased by the CSs.

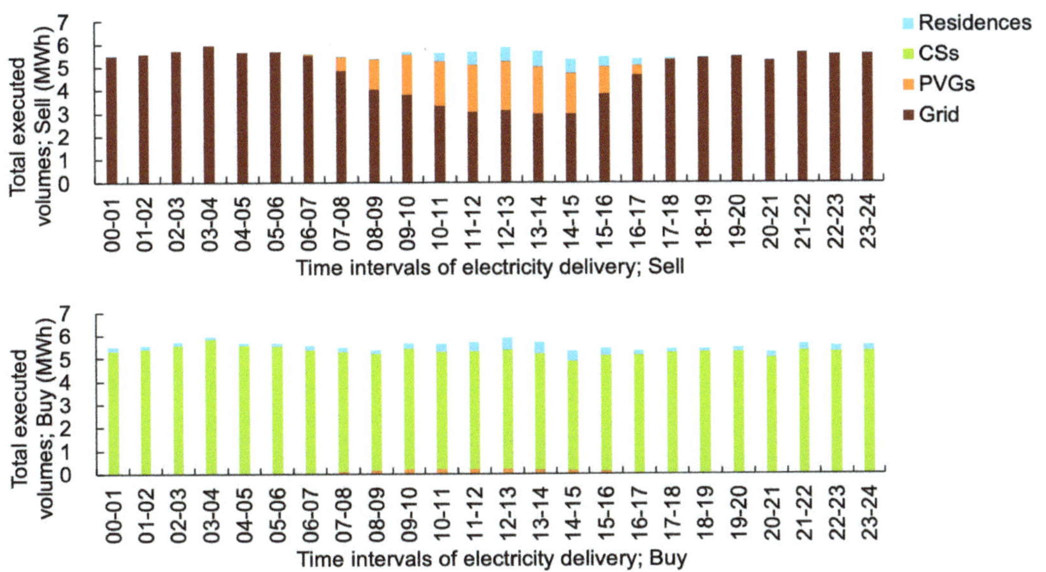

Figure 6. Total executed volumes for each time interval of electricity delivery in the target period.

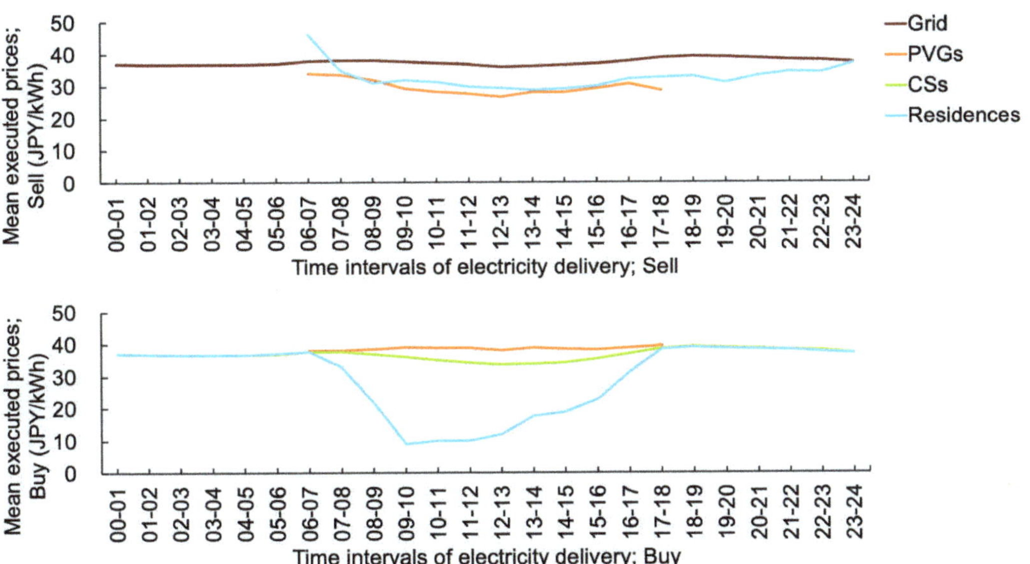

Figure 7. Mean executed prices for each time interval of electricity delivery in the target period.

3.2.3. Executed Volumes and Prices Associated with the Durations of Electricity Delivery

Next, we compared the durations of electricity delivery with the total executed volumes and mean executed prices. The upper panel in Figure 8 represents the breakdown of the executed volumes on the supplying side and the lower panel represents the purchasing side. The horizontal axis in each plot denotes the duration of the electricity delivery. Note that the duration period is indicated in decreasing order; therefore, "24–21" represents the duration from 24 to 21 h before delivery.

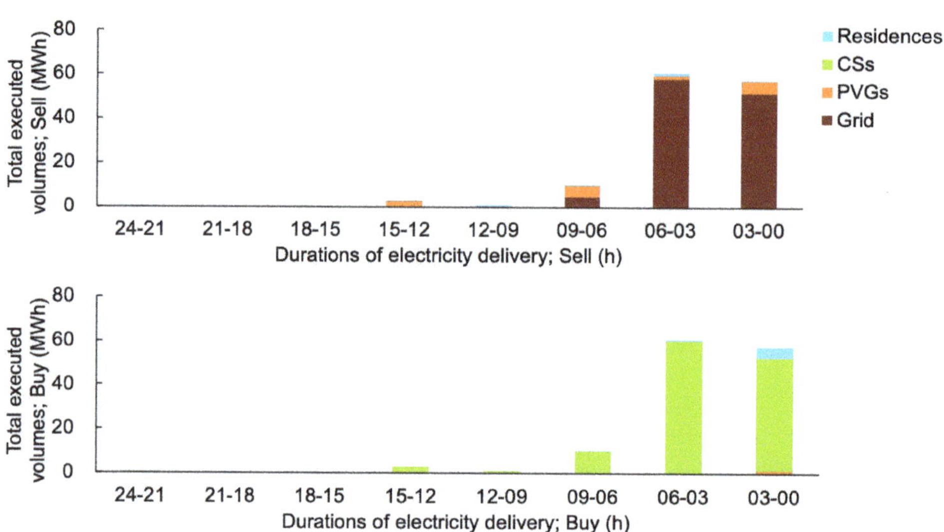

Figure 8. Total executed volumes for each duration of the electricity delivery in the target period.

Figure 9 shows the mean executed prices for each duration period, where the mean prices are the weighted averages of the executed prices by the executed volumes. The upper panel represents each supplied price, and the lower panel represents each purchased price. The horizontal axes represent the durations of electricity delivery, similar to Figure 8.

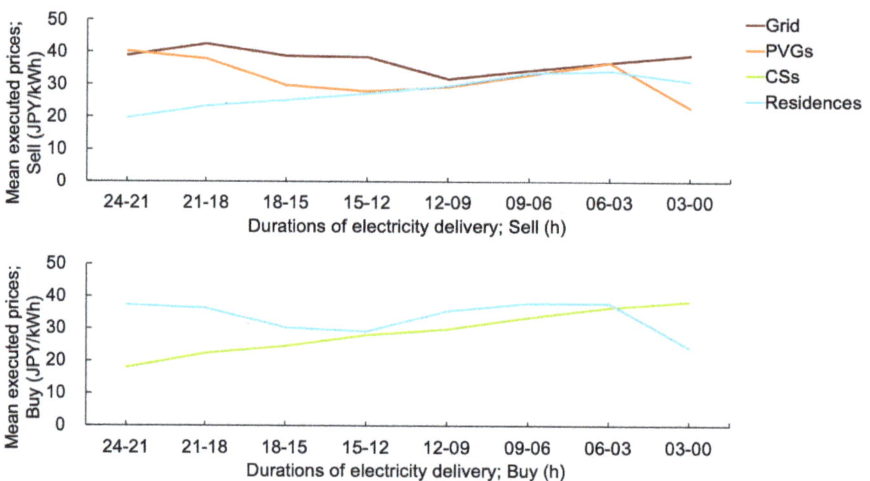

Figure 9. Mean executed prices for each duration of electricity delivery in the target period.

The executed volume within the last 6 h period (i.e., 06–03 and 03–00) covers the highest volume of electricity delivery. The mean executed prices at 06–03 were approximately the same, whereas the mean executed prices at 03–00 varied. The mean executed price supplied by the PVGs (23 JPY/kWh) was significantly lower than that supplied by the residences (31 JPY/kWh) at 03–00. Moreover, the mean executed price purchased by the CSs (39 JPY/kWh) was significantly higher than that purchased by the residences (24 JPY/kWh) at 03–00.

3.3. Analysis of Order Volumes and Prices during the Target Period

3.3.1. Order Volumes and Prices for Electricity Delivery

Figure 10 shows the total order volume for each time interval of electricity delivery, where the upper panel represents the offered volumes, and the lower panel represents the bid volumes. The horizontal axes denote the time intervals of electricity delivery. For example, "00–01" represents the total order volumes delivered from 0:00 to 1:00. The total offers from the grid, which were 70 GWh at each bar and significantly higher than the other offers, are intentionally invisible. Note that the volumes and prices here are based on all orders accepted for the given time intervals of electricity delivery.

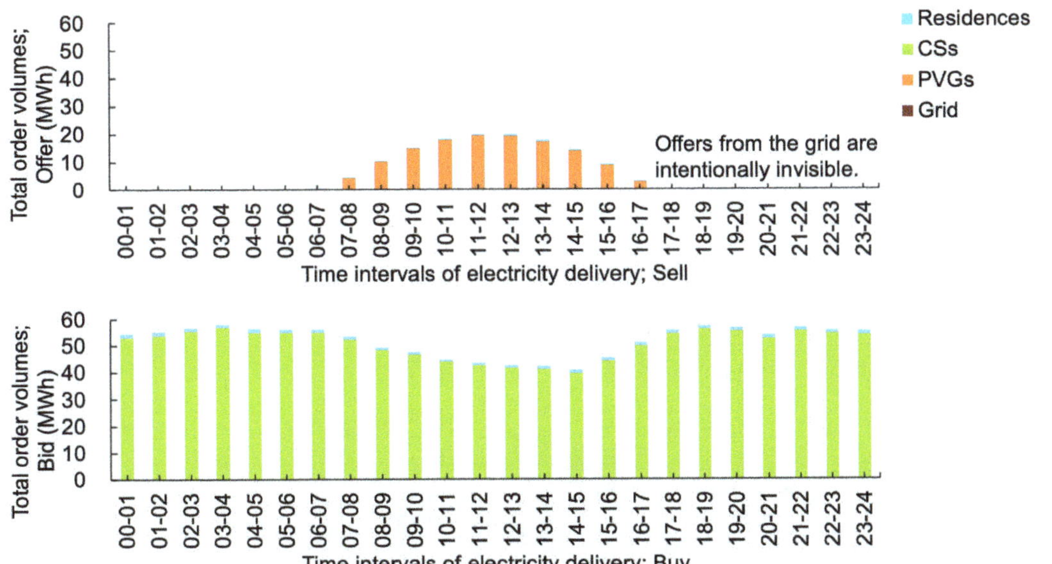

Figure 10. Total order volumes for each time interval in the target period.

Figure 11 shows the mean order prices for the time intervals of electricity delivery. The upper panel represents the mean offered prices, and the lower panel represents the mean bid prices. The horizontal axes represent the time intervals of electricity delivery, similar to Figure 10. In this figure, the mean prices denote the averages of the order prices weighted by the order volumes.

Comparing the mean order prices depicted in Figure 11 with the mean executed prices in Figure 7. The order volumes of the PVGs were higher than those supplied. For example, at 12–13 (ordered between 12:00 and 13:00), the order volumes were 19 MWh, while the executed volumes were 2 MWh. Moreover, the CSs' order volumes were higher than those purchased. For example, at 12–13, the order volume was 42 MWh, while only 5 MWh was executed.

The mean order prices of the PVGs and CSs tended to deviate from the executed prices. For example, the mean prices offered by the PVGs (40–42 JPY/kWh) were significantly higher than the mean executed prices supplied by the PVGs (27–29 JPY/kWh). Moreover, the mean order prices bid by CSs (26–28 JPY/kWh) were significantly lower than the mean executed prices purchased by CSs (34–39 JPY/kWh).

The mean order prices for residences were relatively similar to those executed, at least during the daytime (from 9:00 to 15:00). The mean order prices offered by residences were between 28 and 31 JPY/kWh, whereas the mean executed prices supplied by the residences were between 29 and 32 JPY/kWh, as shown in Figure 7. Moreover, the mean order prices bid by the residences were between 6 and 11 JPY/kWh, and the mean executed prices were between 9 and 19 JPY/kWh (see Figure 7).

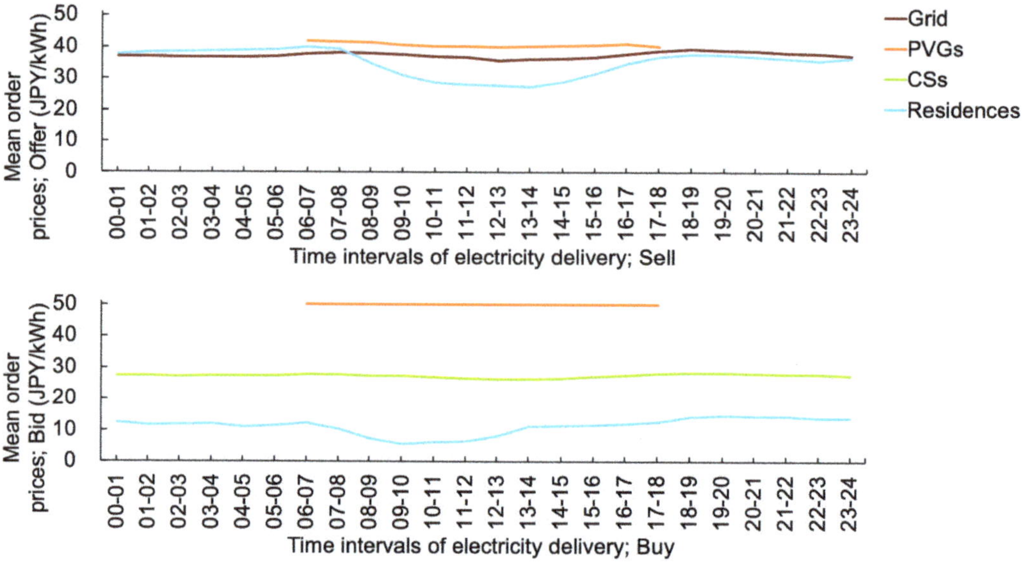

Figure 11. Mean order prices for each time interval in the target period.

3.3.2. Order Volume and Price as Compared with Electricity Delivery Duration

Figure 12 shows the total order volumes for durations of electricity delivery; the upper panel represents the breakdown of the offered volumes, and the lower panel represents the breakdown of the bid volumes. Here, the total volumes offered from the grid, which are significantly higher than the other offers, are invisible.

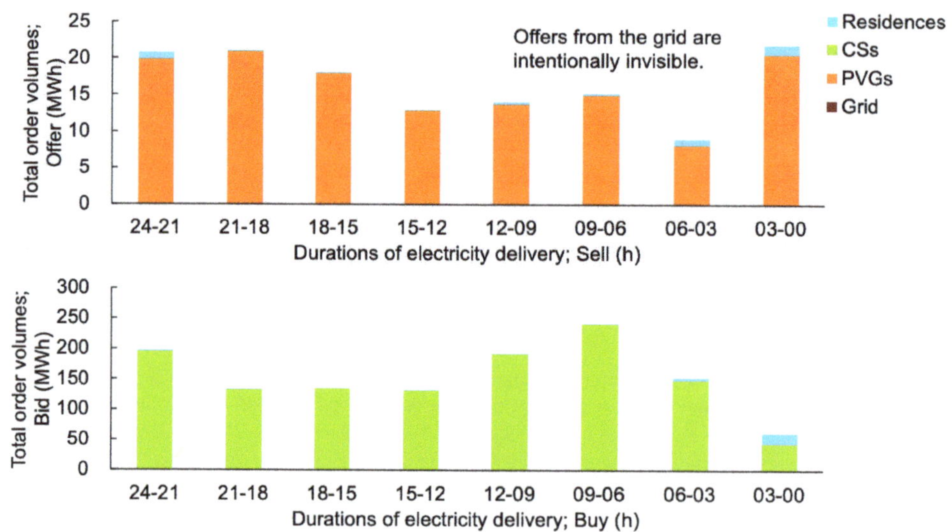

Figure 12. Total order volumes for each electricity delivery duration in the target period.

Figure 13 shows the relationship between the mean order prices and electricity delivery duration, where the upper and lower panels represent the offered and bid prices, respectively. Similar to Figure 12, the mean prices are the average of the order prices weighted by the order volumes.

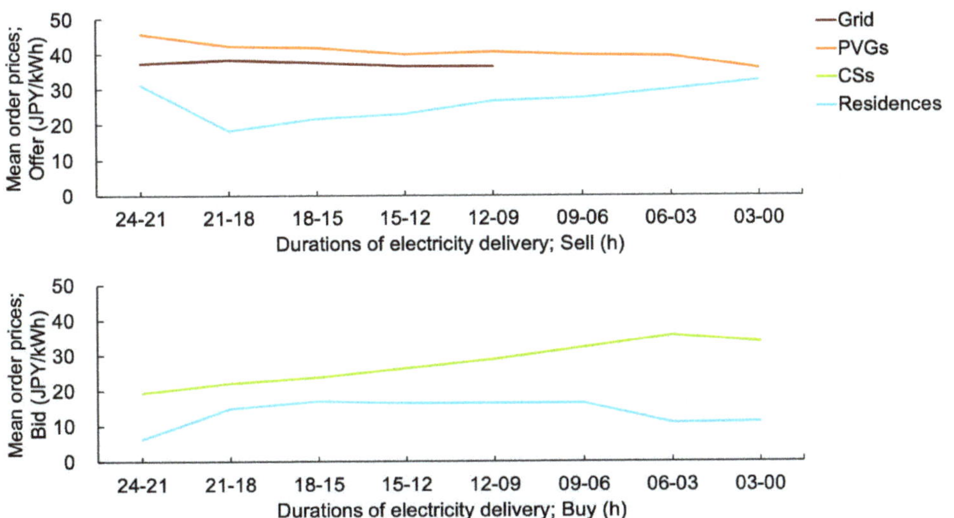

Figure 13. Mean order prices for each electricity delivery duration in the target period.

The total bid volumes (131–241 MWh) were more abundant than the executed ones (0–10 MWh) until six hours before delivery (24–21, 21–18, 18–15, 15–12, 12–09, and 09–06), as shown in Figure 8. The trend in order prices for duration diverges. The prices offered by the PVGs steadily decreased, whereas those offered by the residences gradually increased. Moreover, the prices bid by the CSs steadily increased, whereas those bid by residences gradually decreased.

3.3.3. Relation between Order Volumes/Prices and State-of-Charge (SoC) of Battery Storages

Each of the residences analyzed in this study had a lithium-ion battery. The SoC is the ratio of the residual charges remaining in the battery storage to the battery capacity. The SoC is expressed in percentage points, where 0% indicates an empty SoC and 100% indicates a full SoC. Figure 14 shows the appearance frequencies of each SoC during the target period. Moreover, residences were distinguished using DGR identifiers (24, 28, 29, and 35). Overall, the trends were similar; for example, mid-level SoC (31–40 and 41–50) was a frequent occurrence, whereas high SoC (91–100) and low SoC (0 and 1–10) occurrences were rare.

Figure 15 shows the total order volumes for each SoC during the target period. The upper panel represents the total offer volumes for each SoC, and the lower panel represents the total order volumes for the bids.

Figure 16 shows the mean order prices for each SoC in the target period, where the upper panel represents the mean offer prices for each SoC, and the lower panel represents those for the bids. The mean order prices are the averages weighted by the order volumes. Both horizontal axes represent the ranges of SoC; for example, "00–20" indicates that the SoC was between 0% and 20% when the order was placed. Overall, we observed four different volumes and prices because each residence posted orders individually.

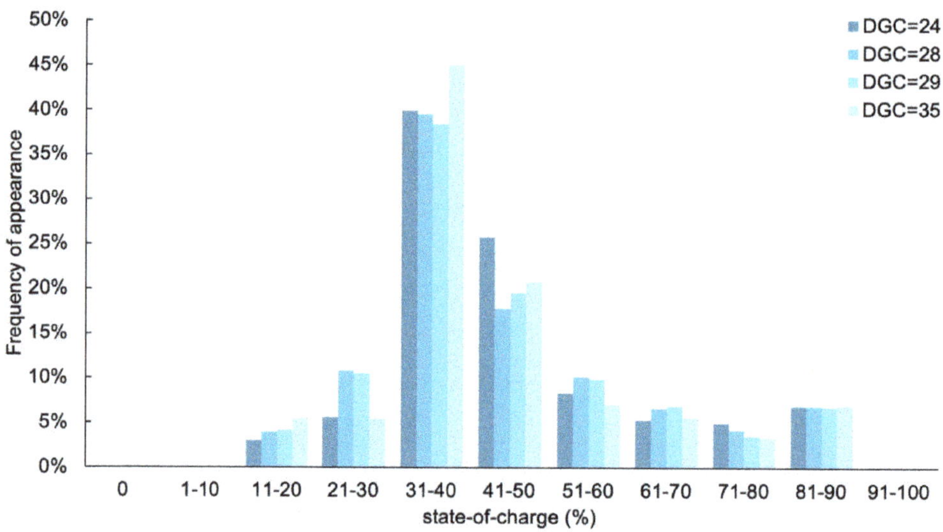

Figure 14. The appearance frequencies of each SoC in the target period.

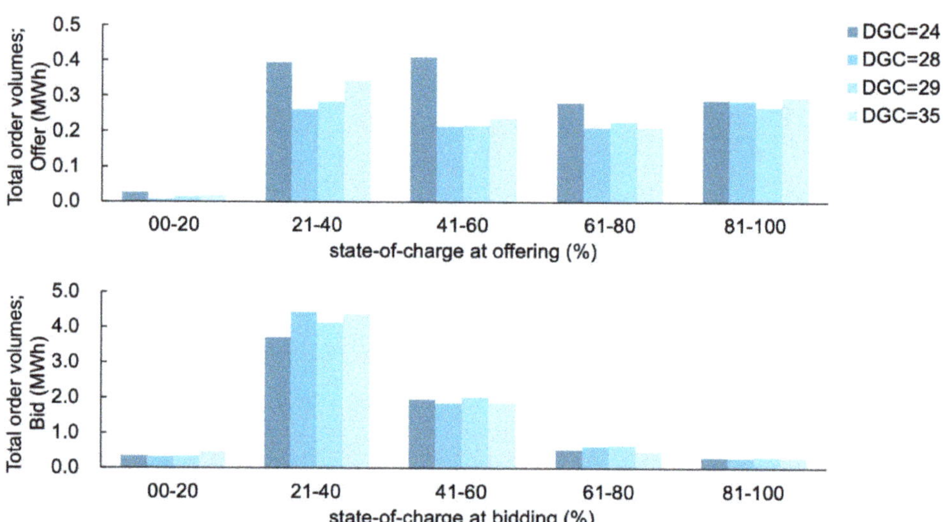

Figure 15. Total order volumes for each (SoC) in the target period.

The total offer volumes were small when the SoC at the time of the ordering offers was low (00–20), and the total offer volumes were steady at 0.2–0.4 MWh when the SoC was higher than 20% (21–40, 41–60, 61–80, and 81–100). The total bid volumes were also small when the SoC at the time of the ordering bids was low (00–20) and gradually declined when the SoC was higher (21–40, 41–60, 61–80, and 81–100). The total bid volumes were more numerous than the offer volumes. Moreover, the differences were more significant when the SoC values were in the middle (21–40 and 41–60), and less significant when the SoC values were high (61–80 and 81–100; offer volumes were between 0.27 and 0.30 MWh, whereas the bids volumes were between 0.30 and 0.33 MWh).

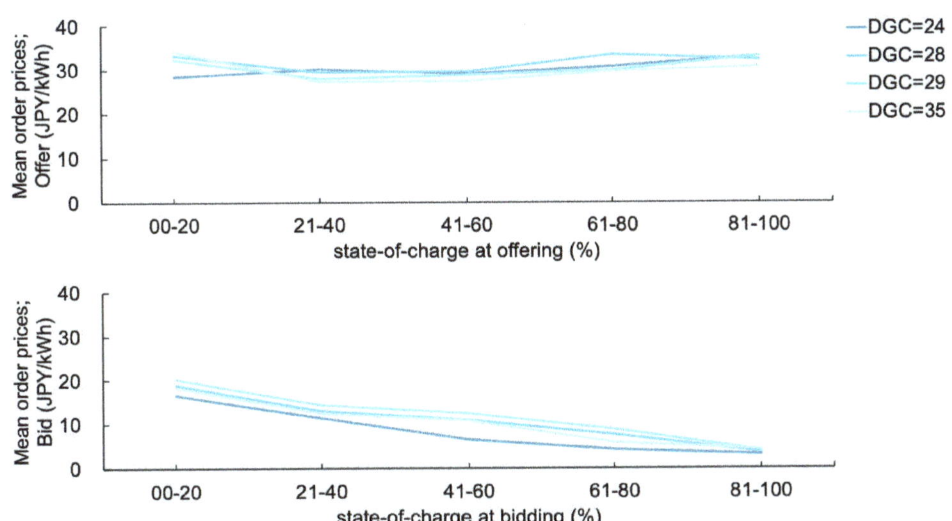

Figure 16. Mean order prices for each SoC in the target period.

Compared to the bid prices, the offered prices were relatively steady at 27–34 JPY/kWh. Moreover, the overall trend for the bid prices was: the higher the SoC values, the lower the price (for example, 12–14 JPY/kWh at 21–40 and 3–4 JPY/kWh at 81–100).

3.4. Chronological Analysis of Unexecuted/Uncancelled Volumes and Prices

This section analyzes the variations in the unexecuted/uncancelled volumes and prices in the order book at certain time intervals during the target period. Figure 17 shows the unexecuted/uncancelled volumes and prices for the delivery period of 12:00 to 12:30 on 26 February 2020. The left column depicts the variations in such volumes, while the right column denotes the mean prices during this period. The mean prices are the weighted averages of unexecuted/uncancelled volumes. The rows indicate the grid offers, PVG offers, residence offers, CS bids, and residence bids.

The volumes offered by the grid were significantly higher than the other requests, and the prices offered by the grid were flat. The volumes offered by the PVGs gradually decreased, and the mean prices offered by the PVGs declined with fluctuations as the time to electricity delivery approached zero. The volumes offered by the residences oscillated significantly, whereas the mean prices they offered were steady. The volumes bid by the CSs gradually decreased, and the mean prices bid by the CSs gradually increased until 9:00, three hours before electricity delivery was expected. The volumes and mean prices bid by the residences rapidly increased after 9:00, and the mean prices bid by the residences were lower than those bid by the grid and PVGs.

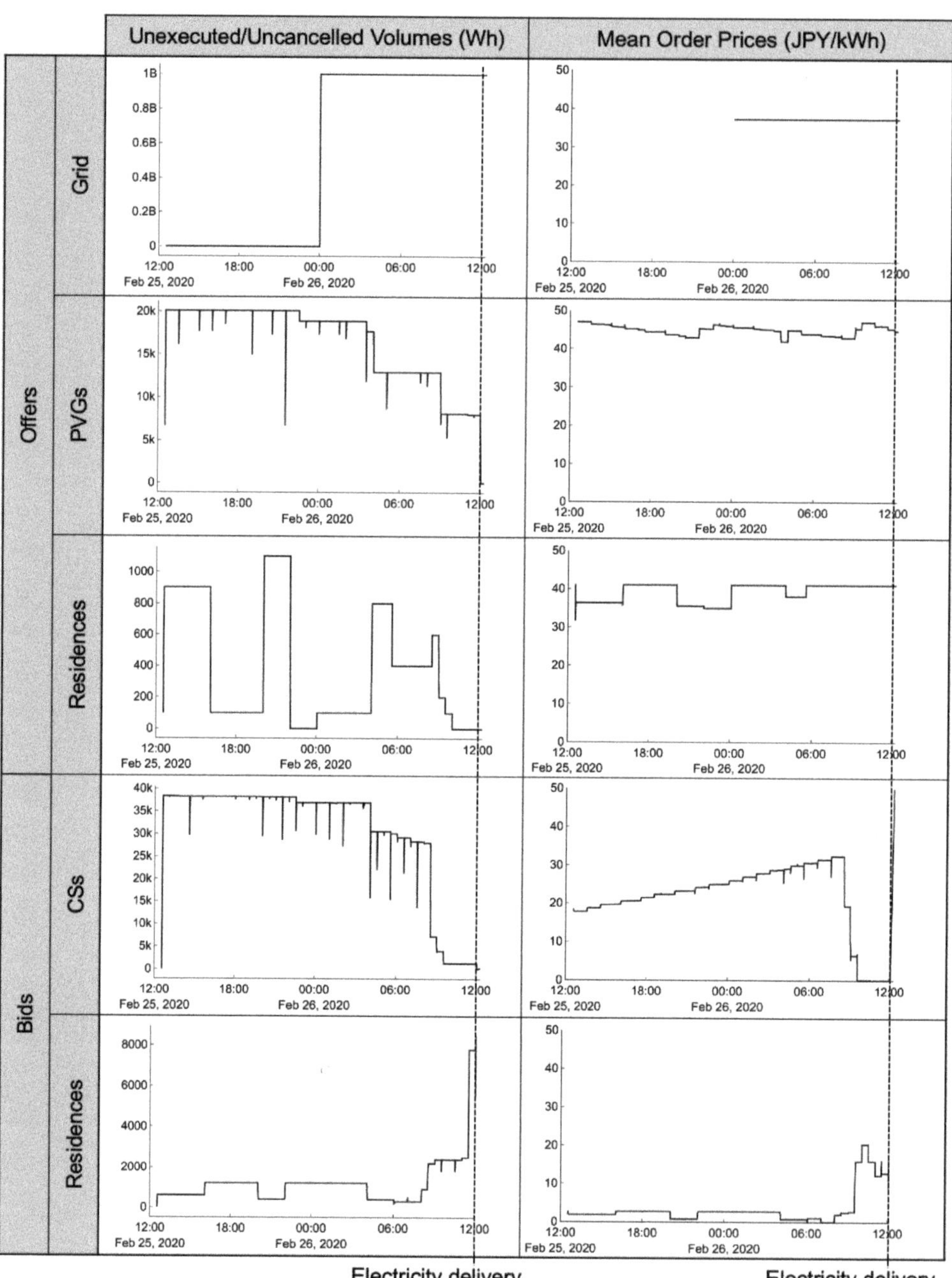

Figure 17. Offers and bids for electricity delivery from 12:00 to 12:30 on 26 February 2020.

4. Discussion

4.1. Principal Findings

P2P energy markets, which prepare the opportunity for buying sides to purchase electricity at lower prices, and the chances for selling sides to supply electricity at higher prices, are expected to enhance the penetration of DERs [7]. Recently, demonstration projects featuring P2P energy markets on the physical power grids have increased [22–28]. One of them was the project conducted in Urawa-Misono District, Japan [29,30]. Although some studies have described the concept and the facilities of the project, few of them have quantitatively analyzed the prices and volumes transacted in the project [31,32]. Based on the results of our analysis, the following primary observations were noted:

- Since the main power source for the supply participants (expect grid) in the P2P electricity market is solar power, offer volumes tended to increase predominantly during the day (Figure 10). Therefore, the demand participants, such as the CSs and residences, probably expected to procure electricity at a lower price during the daytime. However, the residences were more successful in purchasing electricity at reasonable prices than CSs. Moreover, the order prices bid by the residences were lower than those bid by the CSs, regardless of the time intervals of electricity delivery (Figure 11). In the daytime, most of the volumes purchased by the residences were supplied by the PVGs (Figure 5), and the mean prices purchased by the residences were lower than those purchased by the CSs (Figure 7).
- The residences appeared to continue executing more advantageous prices than the PVGs and CSs with the onset of the electricity delivery time. Moreover, the mean prices supplied by the residences were higher than those supplied by the PVGs, while the mean prices purchased by the residences were lower than those purchased by the CSs when the time of electricity delivery approached (Figure 9).
- Even though the holding storages were advantageous, a lower SoC induced higher bidding prices, that is, the decline in flexibility resulted in disadvantageous conditions (Figure 16).

We noted that battery storage was, in fact, the only flexible factor in this market; therefore, the residences were able to execute more advantageously than the facilities without flexibility, that is, the PVGs and CSs. Moreover, the facilities with flexibility could execute more reasonable prices (i.e., supply at higher prices and purchase at lower prices) than those without flexibility. Even when facilities posted requests individually without considering the possibility of grid disturbances, the market mechanism, in which more flexibility is likely to increase the profit, implicitly enhanced the flexibility.

4.2. Limitations

Several limitations were noted during this study. First, because the analysis depended on limited transaction data in the demonstration projects, different configurations may have altered the overall trends; however, this change was small. Such a variation may occur with other types of DERs, such as wind turbines, micro-hydro plants, and biomass-fueled co-generation systems, and other types of consumers, such as schools and offices, whose energy consumption in the daytime differs from that in the nighttime. Second, prices may change when non-fossil energy resources and fossil energy resources are traded in different markets. Eco-friendly consumers may prefer non-fossil energy resources and willingly pay significantly higher prices than fossil energy resources. Lastly, even though holding flexibility, such as battery storage, enables its owner to earn a profit on the market, it does not necessarily guarantee that it is sufficient to cover the initial cost of installing the flexibility.

4.3. Recommendations for Future Research

This study focused thoroughly on the Urawa-Misono project, where battery storage is the only type of flexibility. In the future, we will investigate other types of flexibility and risk management techniques, such as demand responses, vehicle-to-grid with electric vehicles, fuel cells, and weather derivatives, which may have affected the results.

As the orders of the grid agent were derived from JEPX, the executed prices and volumes were partly affected by a conventional power grid. A further comprehensive comparison between the conventional power grid and peer-to-peer electricity trading will be a future issue.

Further investigation may be required to provide a more efficient bidding strategy based on flexibility. Battery storages in this study ordered their bids/offers according to their SoC. Maintaining the SoC within an adequate range—high enough to discharge and low enough to charge—is necessary. In addition to battery storage, even agents without battery storage can hold flexibility; for example, consumers can achieve flexibility through the demand response, which is the concept of controlling their consumption. Additionally, because offering and bidding prices and volumes are partially based on the weather forecast, repositioning prices and volumes according to the changes in weather forecast may be addressed.

4.4. Implications

P2P energy markets enable DERs to trade without the mediation of the incumbents such as conventional transmission system operators. Conversely, the incumbents may resist P2P energy markets because an increase in the P2P energy markets may take away their role in the power grids from them. However, the regulatory authority should rationalize the existing regulations and restrictions related to P2P energy markets. Since the high penetration of DERs stimulates unprecedented needs for flexibility, any available mechanisms to increase flexibility should be addressed. As the technological components of P2P energy markets have matured, and P2P energy markets contribute to installing flexibility, the regulatory obstacles may require mitigation.

4.5. Summary

Blockchain technologies and P2P energy markets are believed to realize efficient and resilient flexibility in power grids. This study focused on a demonstration project conducted in Urawa-Misono District, Japan. The project featured multifunctional inverters called Digital Grid Routers, a continuous matching market, and an automatic execution system implemented with blockchain to actualize P2P energy markets. The analysis of the traded results quantitatively revealed that holding flexibility results in advantageous conditions for the market; thus, P2P energy markets have the potential to induce flexibility through market mechanisms.

Author Contributions: Conceptualization, R.K., K.T. and Y.Y.; methodology, R.K.; software, R.K.; validation, R.K.; formal analysis, R.K.; investigation, R.K.; resources, K.T.; data curation, K.T.; writing—original draft preparation, R.K.; writing—review and editing, R.K. and Y.Y.; visualization, R.K.; supervision, K.T. and Y.Y.; project administration, K.T. and Y.Y.; funding acquisition, Y.Y. All authors have read and agreed to the published version of the manuscript.

Funding: The third author was supported by a Grant-in-Aid for Scientific Research (A) 20H00285.

Acknowledgments: The demonstration project, "Digital grid router (DGR) and power interchange settlement system to accelerate the introduction of renewable energy," was financially supported by the Ministry of the Environment, Government of Japan's Low Carbon Technology Research, Development and Demonstration Program. We express our sincere gratitude to Digital Grid Co., and other related companies.

Conflicts of Interest: The authors declare no conflict of interest.

References

1. IPCC, Intergovernmental Panel on Climate Change. Headline Statements from the Summary for Policymakers. In *Climate Change 2021: The Physical Science Basis, the Working Group I Contribution to the Sixth Assessment Report*; IPCC: Geneva, Switzerland, 2021. Available online: https://www.ipcc.ch/report/ar6/wg1/downloads/report/IPCC_AR6_WGI_Headline_Statements.pdf (accessed on 15 August 2021).
2. IEA, International Energy Agency. Net Zero by 2050. Available online: https://www.iea.org/reports/net-zero-by-2050 (accessed on 15 August 2021).
3. CAISO, California Independent System Operator. Fast facts: What the Duck Curve Tells us about Managing a Green Grid. Available online: https://www.caiso.com/documents/flexibleresourceshelprenewables_fastfacts.pdf (accessed on 15 August 2021).
4. Matsuo, Y.; Endo, S.; Nagatomi, Y.; Shibata, Y.; Komiyama, R.; Fujii, Y. Investigating the economics of the power sector under high penetration of variable renewable energies. *Appl. Energy* **2020**, *267*, 113956. [CrossRef]

5. Ogimoto, K.; Wani, H. Making Renewables Work: Operational Practices and Future Challenges for Renewable Energy as a Major Power Source in Japan. *IEEE Power Energy Mag.* **2020**, *18*, 47–63. [CrossRef]

6. Ahl, A.; Yarime, M.; Goto, M.; Chopra, S.S.; Kumar, N.M.; Tanaka, K.; Sagawa, D. Exploring blockchain for the energy transition: Opportunities and challenges based on a case study in Japan. *Renew. Sustain. Energy Rev.* **2020**, *117*, 109488. [CrossRef]

7. IRENA, International Renewable Energy Agency. Innovation Landscape Brief: Peer-to-Peer Electricity Trading. Available online: https://irena.org/-/media/Files/IRENA/Agency/Publication/2020/Jul/IRENA_Peer-to-peer_trading_2020.pdf (accessed on 15 August 2021).

8. Tushar, W.; Saha, T.K.; Yuen, C.; Smith, D.; Poor, H.V. Peer-to-peer trading in electricity networks: An overview. *IEEE Trans. Smart Grid* **2020**, *11*, 3185–3200. [CrossRef]

9. RWE AG. Press Release (1 July 2020): RWE Closes Deal with E.ON. Available online: https://jp.rwe.com/en/press/2020-07-01-rwe-closes-deal-with-eon (accessed on 1 September 2021).

10. Tokyo Electric Power Company Holdings, Incorporated. Press Release (10 July 2017): Launching a Direct Power Trading Platform Business in Cooperation with Germany's Leading Power Company Innogy Leveraging Advanced IT to Develop Power Trading Business in Germany. Available online: https://www.tepco.co.jp/press/release/2017/1443908_8706.html (accessed on 1 September 2021). (In Japanese).

11. Chugoku Electric Power Company, Incorporated. Press Release (25 April 2019): Implementation of a Demonstration Test on Electricity Flexibility Using Blockchain Technology. Available online: https://www.energia.co.jp/press/2019/11766.html (accessed on 1 September 2021). (In Japanese).

12. Tohoku Electric Power Company, Incorporated. Press Release (26 April 2019): Conclusion of Contract for Joint Research on Direct Power Trading Using Distributed Power Sources (P2P Power Trading). Available online: https://www.tohoku-epco.co.jp/pastnews/normal/1201039_1049.html (accessed on 1 September 2021). (In Japanese).

13. Kansai Electric Power Company, Incorporated. Press Release (9 December 2019): Commencement of a Demonstration Study of Environmental Value Trading Using Blockchain Technology by Power Ledger Australia. Available online: https://www.kepco.co.jp/corporate/pr/2019/1209_1j.html (accessed on 1 September 2021). (In Japanese).

14. Shikoku Electric Power Company, Incorporated. Press Release (12 March 2021): Investment in U.S. Venture Company LO3 Energy. Available online: https://www.yonden.co.jp/press/2020/__icsFiles/afieldfile/2021/03/12/pr002.pdf (accessed on 1 September 2021). (In Japanese).

15. ENERES Company, Limited. Press Release (4 December 2020): Launch of a Demonstration Project to Build a P2P Power Trading Platform for Next-Generation Power Systems—Issuing Blockchain-Based Environmental Value Tokens to Examine a New Renewable Energy Trading Scheme for RE100 Companies. Available online: https://www.eneres.co.jp/news/release/20201204.html (accessed on 1 September 2021). (In Japanese).

16. UPDATER, Incorporated. Press Release (29 June 2021): Launching Sales of P2P Power Tracking System to Other Companies, Enabling the Designation of Power Plants When Purchasing Power—We Aim to Spread the Use of Origin-Guaranteed Renewable Energy with our SaaS. Available online: https://minden.co.jp/news/2021/06/29/4456 (accessed on 1 September 2021). (In Japanese).

17. Hitachi, Limited. Press Release (22 January 2021): Development of a System to Visualize the Use of Renewable Energy for Each Facility and Service for the Realization of a Decarbonized Society—Operation of a System to Certify the Use of 100% Renewable energy as Powered by Renewable Energy Started. Available online: https://www.hitachi.co.jp/New/cnews/month/2021/01/0122.html (accessed on 1 September 2021). (In Japanese).

18. Hitachi, Limited. Powered by Renewable Energy. Available online: https://www.powered-by-re.com/ (accessed on 1 September 2021). (In Japanese).

19. Nakamura, Y.; Zhang, Y.; Sasabe, M.; Kasahara, S. Exploiting smart contracts for capability-based access control in the internet of things. *Sensors* **2020**, *6*, 1793. [CrossRef] [PubMed]

20. Tokyo Institute of Technology. Press Release (18 January 2021): Development of Blockchain Technology to Optimize P2P Electricity Trading According to the Trading Needs of Consumers—Contributing to the Effective Use of Surplus Electricity by Realizing a Flexible Trading Environment. Available online: https://www.titech.ac.jp/news/2021/048738 (accessed on 1 September 2021). (In Japanese).

21. Nagatsuka, T.; Sano, M.; Yamaguchi, N. Decentralized Transaction System of Surplus PV Output Using Blockchain. In Proceedings of the Grand Renewable Energy Proceedings Japan council for Renewable Energy (2018), Yokohama, Japan, 20–22 June 2018; Japan Council for Renewable Energy: Tokyo, Japan, 2018; p. 304. [CrossRef]

22. Brooklyn Microgrid. Available online: https://www.brooklyn.energy/ (accessed on 1 September 2021).

23. LO3 Energy, Incorporated. Available online: https://lo3energy.com/ (accessed on 1 September 2021).

24. Werth, A.; Kitamura, N.; Tanaka, K. Conceptual Study for Open Energy Systems: Distributed Energy Network Using Interconnected DC Nano Grids. *IEEE Trans. Smart Grid* **2015**, *6*, 1621–1630. [CrossRef]

25. Sony Computer Science Laboratories, Incorporated. Press Release (1 December 2020): Microgrid System Core Module (Power-Interchange Management Software) Going Open Source—Promote Locally Produced & Consumed Renewable Energy, Work towards a Sustainable Society. Available online: https://www.sonycsl.co.jp/press/prs20201201/ (accessed on 1 September 2021).

26. Chubu Electric Power Company, Incorporated. Press Release (1 March 2018): Implementation of a Demonstration Experiment for a New Service for Recharging Electric Vehicles and Other Vehicles Using Blockchain. Available online: https://www.chuden.co.jp/publicity/press/3267230_21432.html (accessed on 1 September 2021). (In Japanese).

27. Kansai Electric Power Company, Incorporated. Press Release (9 December 2019): Launch of a Demonstration Project on Environmental Value Trading Using Blockchain Technology, Including the Determination of Trading Prices for Environmental Values. Available online: https://www.kepco.co.jp/corporate/pr/2019/1209_2j.html (accessed on 1 September 2021). (In Japanese).

28. Toyota Motor Corporation. Press Release (13 November 2020): Joint Demonstration Experiment on P2P Power Trading System Confirms Its Effectiveness—Exploring Further Partner Collaboration to Provide New Value. Available online: https://www.toyota.co.jp/jpn/tech/partner_robot/news/20201113_01.html (accessed on 1 September 2021). (In Japanese).

29. Ministry of the Environment. Brief summary: Digital grid router (DGR) and Power Interchange Settlement System to Accelerate the Introduction of Renewable Energy. Available online: http://www.env.go.jp/earth/ondanka/cpttv_funds/pdf/db/202.pdf (accessed on 15 August 2021). (In Japanese)

30. Ministry of the Environment. Low Carbon Technology Research, Development and Demonstration Program. Available online: https://www.env.go.jp/earth/ondanka/cpttv_funds/pdf/pamph_e2019.pdf (accessed on 15 August 2021).

31. Wencong, S.; Huang, A. (Eds.) *The Energy Internet: An Open Energy Platform to Transform Legacy Power Systems into Open Innovation and Global Economic Engines*; Woodhead Publishing: Cambridge, UK, 2018; pp. 241–264. [CrossRef]

32. Tanaka, K.; Abe, R.; Nguyen-Van, T.; Yamazaki, Y.; Kamitamari, T.; Sako, K.; Koide, T. A Proposal on an Electricity Trading Platform Using Blockchain. In *Transdisciplinary Engineering Methods for Social Innovation of Industry 4.0*; IOS Press: Amsterdam, The Netherlands, 2018; pp. 976–983. [CrossRef]

33. Abe, R.; Taoka, H.; McQuilkin, D. Digital grid: Communicative electrical grids of the future. *IEEE Trans. Smart Grid* **2011**, *2*, 399–410. [CrossRef]

34. Nguyen-Van, T.; Abe, R.; Tanaka, K. Digital Adaptive Hysteresis Current Control for Multi-Functional Inverters. *Energies* **2018**, *11*, 2422. [CrossRef]

35. Hayashi, K.; Kato, R.; Torii, R.; Taoka, H.; Abe, R. Bi-directional power flow through a digital grid router. *J. Int. Counc. Electr. Eng.* **2015**, *5*, 42–46. [CrossRef]

36. Shibano, K.; Kontani, R.; Hirai, H.; Hasegawa, M.; Aihara, K.; Taoka, H.; McQuilkin, D.; Abe, R. A Linear programming formulation for routing asynchronous power systems of the Digital Grid. *Eur. Phys. J. Spec. Top.* **2014**, *223*, 2611–2620. [CrossRef]

37. Kumagaya Meteorological Office. Characteristics of the Climate in Saitama. Available online: https://www.jma-net.go.jp/kumagaya/shosai/chishiki/tokutyou.html (accessed on 15 August 2021). (In Japanese).

38. Japan Meteorological Agency. Historical Weather Observations. Available online: https://www.data.jma.go.jp/obd/stats/etrn/index.php?prec_no=43&block_no=0363 (accessed on 15 August 2021). (In Japanese)

39. Japan Meteorological Agency. Frequently Asked Questions about Humidity, Barometric Pressure, and Daylight Duration. Available online: https://www.jma.go.jp/jma/kishou/know/faq/faq5.html (accessed on 15 August 2021). (In Japanese)

40. TEPCO Power Grid, Incorporated. Available online: https://www.tepco.co.jp/pg/ (accessed on 15 August 2021). (In Japanese).

41. Digital Grid, Corporation. Available online: https://www.digitalgrid.com/ (accessed on 15 August 2021). (In Japanese).

42. ECHONET Consortium. Digital Grid Controller | ECHONET. Available online: https://echonet.jp/introduce_en/qz-000069/ (accessed on 15 September 2021). (In Japanese).

43. JEPX, Japan Electric Power Exchange. Available online: http://www.jepx.org/ (accessed on 15 August 2021). (In Japanese).

44. McKinney, W. Data structures for statistical computing in python. In Proceedings of the 9th Python in Science Conference, Austin, TX, USA, 28 June–3 July 2010; Volume 445, p. 51.

energies

Article

Demonstration of Blockchain Based Peer to Peer Energy Trading System with Real-Life Used PHEV and HEMS Charge Control

Yuki Matsuda [1,*], Yuto Yamazaki [2], Hiromu Oki [2], Yasuhiro Takeda [1,3], Daishi Sagawa [1] and Kenji Tanaka [1]

1 School of Engineering, The University of Tokyo, Tokyo 113-8654, Japan; yasu@g.ecc.u-tokyo.ac.jp (Y.T.); sagawa@ioe.t.u-tokyo.ac.jp (D.S.); tanaka@tmi.t.u-tokyo.ac.jp (K.T.)
2 USD Co., Ltd., Tokyo 141-0032, Japan; yamazaki@us-design.co.jp (Y.Y.); hiromu_oki@us-design.co.jp (H.O.)
3 Trende Inc., Tokyo 101-0031, Japan
* Correspondence: matsuda@ioe.t.u-tokyo.ac.jp

Abstract: To further implement decentralized renewable energy resources, blockchain based peer-to-peer (P2P) energy trading is gaining attention and its architecture has been proposed with virtual demonstrations. In this paper, to further socially implement this concept, a blockchain based peer to peer energy trading system which could coordinate with energy control hardware was constructed, and a demonstration experiment was conducted. Previous work focused on virtually matching energy supply and demand via blockchain P2P energy markets, and our work pushes this forward by demonstrating the possibility of actual energy flow control. In this demonstration, Plug-in Hybrid Electrical Vehicles(PHEVs) and Home Energy Management Systems(HEMS) actually used in daily life were controlled in coordination with the blockchain system. In construction, the need of a multi-tagged continuous market was found and proposed. In the demonstration experiment, the proposed blockchain market and hardware control interface was proven capable of securing and stably transmitting energy within the P2P energy system. Also, by the implementation of multi-tagged energy markets, the number of transactions required to secure the required amount of electricity was reduced.

Keywords: blockchain; peer to peer energy market; hardware control; demonstration experiment; home energy management systems; electric vehicles

Citation: Matsuda, Y.; Yamazaki, Y.; Oki, H.; Takeda, Y.; Sagawa, D.; Tanaka, K. Demonstration of Blockchain Based Peer to Peer Energy Trading System with Real-Life Used PHEV and HEMS Charge Control. *Energies* **2021**, *14*, 7484. https://doi.org/10.3390/en14227484

Academic Editor: Yuji Yamada

Received: 30 September 2021
Accepted: 3 November 2021
Published: 9 November 2021

Publisher's Note: MDPI stays neutral with regard to jurisdictional claims in published maps and institutional affiliations.

1. Introduction

To further implement decentralized renewable energy resources [1], blockchain based peer to peer (P2P) energy trading is a promising concept [2]. The realization of P2P energy trading has high affinity with implementation of renewable energy, for it is capable of empowering small energy producers, and diversify energy production profiles quickly [3,4]. In the realization of P2P energy trading, blockchain is a promising concept, with its benefits in data security, immutability, and higher efficiency in administrative processes [5,6].

The architecture and concrete viable algorithms of blockchain based P2P energy markets have been designed and proposed in-depth [7,8]. Also, many researchers have proposed and virtually verified blockchain system architectures, which could possibly realize this concept [9–14]. This trend is backed up with extensive consideration on social acceptance and policy adjustments as energy distribution being a social infrastructure which involves several stakeholder responses when trying to be updated [15–19].

Building up on this trend of P2P blockchain energy system implementation, in this paper, a unified demonstration experiment of both the virtual layer of blockchain systems, and the physical layer of energy charge/consumption hardware, was conducted. Previous work has focused on virtually matching supply and demand of existing energy flow, by measuring the values via smart meters. Our demonstration pushes this forward by enabling control of energy flow, and thus charge/discharge hardware, according to energy transactions on the blockchain market. This demonstration experiment aims to confirm that blockchain systems could realize energy flow control, and clarify issues regarding stable operation of the system.

In the demonstration experiment conducted, PHEV and HEMS actually used in daily life by users were connected to the blockchain system, and energy matchmaking was done in a blockchain continuous energy market. Based on the secured matches, actual energy transmission was controlled, transmitting the transacted energy through existing power transmission infrastructure. In construction of this market, additional system requirements for future energy procurement user needs and system stability were clarified, and a multi-tagged continuous market was introduced for the market algorithm.

The main idea of this paper is to control real-life P2P energy trading hardware based on matching results of the blockchain energy system in order to demonstrate the possibility of energy flow control using blockchain based energy markets.

Our contributions are as follows.

- Implement a blockchain based P2P energy trading system, which could coordinate with real-life used energy charge/discharge control hardware (Section 3).
- Propose a multi-tagged energy market algorithm, which was required to further match energy market participant requirements, and simultaneously stabilize the system. This algorithm was also implemented (Section 4).
- Conduct a demonstration experiment for one year with real-life participants and energy charge hardware to confirm blockchain system stability and also define further blockchain system implementation requirements (Section 5).

2. Related Work

2.1. Blockchain Based Peer to Peer Energy Markets

Various approaches have been taken to implement the power trading functions on blockchain as a P2P power trading market.

Mengelkamp et al. built a blockchain-based microgrid energy market called The Brooklyn Microgrid, which introduced seven market components. A case study was conducted to show its effectiveness and future work [9]. Green et al. measured the electricity consumption of a family of two adults and two children in Perth, Australia. The house was named 'Josh's House' and was equipped with a 3 kW photovoltaic system. It was concluded that citizen-based distributed power systems and conventional integrated power grids need to coordinate combining the use of storage batteries [10]. Janusz et al. selected the machine to machine (M2M) power market as a model for developing blockchain-based applications for Industry 4.0, and implemented it on MultiChain [11]. Further, addressing the issue of system delay in the usage of blockchain in actual energy trade, blockchain network algorithms have been proposed to create secure and minimum latency communications [14].

Also, as an initiative of the Japanese government, at Urawa Misono Saitama Prefecture, the Ministry of the Environment created a blockchain platform to transact electric power between the photovoltaic (PV) system installed in a shopping mall, the PV/batteries installed in five buildings in the subdivision (three in the central housing area), and five convenience stores [12,13].

The blockchain energy market platform constructed in this study is an updated version of the platform constructed by the Ministry of the Environment, designed to control real-life used HEMS and PHEVS (Section 3) and further realize efficient matchmaking via the multi-tagged continuous market (Section 4).

2.2. Peer to Peer Energy Transmission Control

As structured in [20,21] and modeled in [22], peer to peer energy trading requires fine management and execution of bilateral energy transmission.

In terms of virtual energy management of peers, Erol-Kantarci et al. proposed a home energy system (HEMS) method that was based on appliances, a smart meter, and storage units; a convenient time to execute participant demand was obtained [23]. To create a sensor based network of participants, Han et al. used IEEE 802.15.4 and zigbee [24], to further realize smart home energy management [25]. In the electric vehicle (EV) sector, to ensure accurate information synchronization, Hussain et al. proposed a communication network architecture based on IEC 61850-7-420 logical nodes [26,27]. Furthermore, to ensure charge completion of EVs within uncertainties and grid load restrictions, EV behaviour modelling and management methods have been proposed [28,29].

For physical energy transmission, Abe et al. proposed the digital grid [30], where the power system is separated into asynchronously connected grids by hardware named the digital grid router, a multi-legged ac/dc/ac converter. There is no need for additional placement of transmission lines, for energy is sent through the existing transmission lines in a cost efficient way [31]. The router is capable of bi-directional power flow, with the direction of current flow controlled through the leg and the value of current by hysteresis control [32,33].

2.3. Social Acceptance of Peer to Peer Energy Trading

As energy being a social infrastructure with regulations and a vast number of stakeholders participating, legal foundations and participant incentives are being studied further. In a review by Ahl et al. [15], future possibilities of regulations and standards were pointed out. Cali et al. analyzed in-depth incentive mechanisms of the market in order to support policy makers in preparing relevant energy policies [19]. Issues in the business domain were discussed by Hanna et al. [16]. There is currently a gap between technological advances in blockchain P2P energy trading and its social standards, but efforts are underway to address this.

In terms of participant understanding, simulational studies have been conducted, with each participant's utilities numerically analyzed, with the overall convergence of the system also taken into account [17,18]. Other approaches could be taken to analyze each participant's behavioural data. In the context of energy disaggregation, prediction models of energy consumption behaviour is being proposed [34,35]. Also, studies to gain characteristics of EVs through behaviour data clustering have been conducted [36].

3. Constructed Blockchain Based Peer to Peer Trading System

3.1. Overall Architecture

The overall architecture of the constructed blockchain-based peer-to-peer trading system is shown in Figure 1.

The blockchain system was constructed using Ethereum with private blockchain, andpProof of authority as the consensus algorithm. There are two types of nodes, named the fullNode and the authority node. The duties of the nodes are further described in Section 3.2.

Through an cloud-deployed API, each market participant makes bids to the energy market, confirms the contracted bids to control energy flow, and reports actual executed energy transactions. These actions account for "Bid Flow", "Energy Control Flow", and "Execution Done Report Flow" in Figure 1, respectively. Bids creation is assumed to be done by bidding agents representing each participant's energy procurement requirements. Contract confirmation /energy hardware control / energy transaction reporting is assumed to be done by a client system, and energy transaction measurement is assumed to be done by smart meters. This enables an end to end machine to machine (M2M) control of energy transaction, without any human intervention. This could lower market participation barriers from the viewo of each participant, and also increase the security of the overall system. Further information and energy flow will be described in Section 3.3.

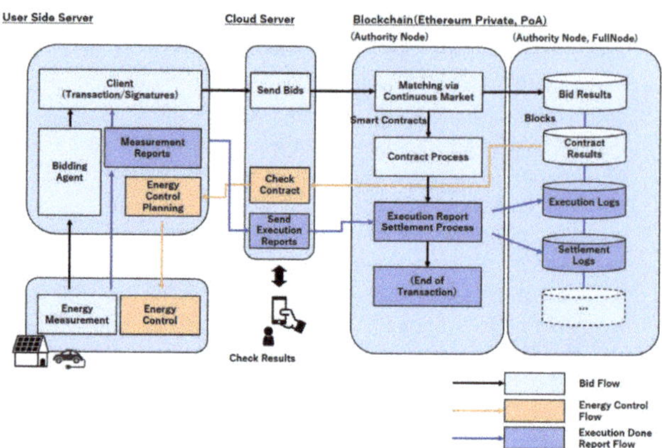

Figure 1. overall architecture.

3.2. Blockchain Energy Market System Composition

Ethereum [37] was selected in order to build the blockchain energy market. This is mainly because Ethereum is capable of using proof of authority (PoA) [38] as the consensus algorithm, which enables high-speed transaction processing.There exists a tradeoff between transaction broadcast speed and security, and proof of authority via Ethereum was selected as the balance point [39].

The structure and transaction flow of the market is shown in Figure 2. Two types of nodes, the fullNode and authority, were designed as the building blocks of blockchain. The fullNode is responsible for accepting transactions and referencing data, while the authority is responsible for transaction processing and block generation. The user cannot directly access the authority, thus minimizing external influence.

The user sends power buy and sell bid transactions to the fullNode via the agent program, which forwards the received transactions to the authority. Also, when the fullNode receives a reference process such as data acquisition, it refers to its own ledger and returns the result to the agent. When the authority recieves a transaction, it conducts a calculation and logging process to log the transaction results to its ledger. Since the fullNode synchronizes its data with authority, the fullNode's ledger can be kept up-to-date.

In this composition, scalability of the system could be designed according to user requirements. By increasing the number of authority nodes, the reliability of transaction processing could be improved. Also, by increasing the number of fullNodes, the amount of possible concurrent transactions and data reference processing could be increased.

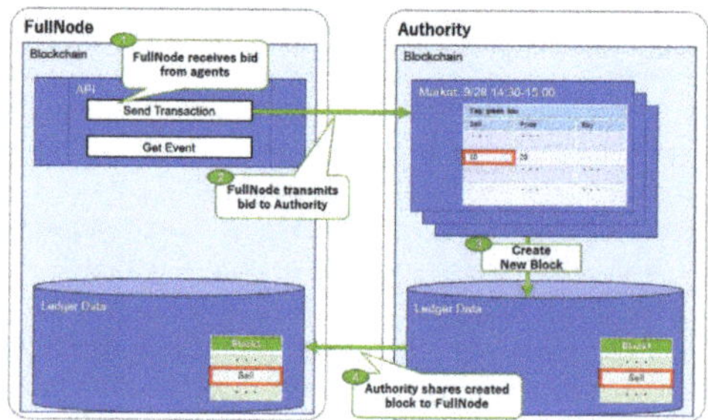

Figure 2. Blockchain system composition.

3.3. System Interface

The types of requests designed are shown in Table 1. Using this interface, the participants could make bids to the market, obtain current market contractions, and report energy transaction. From the participant point of view, using these request structures automatically creates unforgeable energy transaction certificates, which is fundamental in a reliable market.

The bid structure to the blockchain system is shown in Table 2. Following this structure, each market participant generates bids expressing its energy trading requirements. In doing so, the market window tags and energy feature tags were introduced to further enable trading requirement expression. For example, energy procurement requirements, such as those listed below, could be expressed with low cost.

- We want to buy a total of 10kWh of energy in the cheapest way, within 10:30 a.m.–11:30 a.m., which overlaps two market time windows of 10:30 a.m.–11:00 a.m. and 11:00 a.m.–11:30 a.m.
- We want to buy renewable energy, even if it costs a little higher than fossil-based energy.

  This concept and its merits will be discussed further in Section 4.

After bids are contracted, the contracts are notified to the market participant systems, and actual energy charge/discharge is executed. The execution reports are sent to the blockchain system in the structure shown in Table 3.

Table 1. Request types designed.

Type Number	Transaction Type
1	sell bid
2	buy bid
3	cancel
4	report energy transaction executed
5	obtain market status

Table 2. Bid Structure.

Code Based Bid Information	Explanation
bytes32 bidId	Unique id for the bid
BidType btype	Sell or Buy type information of transaction
address payable addr	Bidder's blockchain wallet address
unit32 amount	Bid amount of energy
uint64[] times	All bid market time windows tags for the id amount in a list
btypes32[] tags	All energy feature tags for the bid amount in a list
uint[] prices	All energy bid prices per Wh for each market and tag combination, in a list

Table 3. Energy Transaction Execution Report Structure.

Code Based Bid Information	Explanation
bytes32 execId	Contract id notified when bid was contracted
unit32 amount	Actually transacted amount of energy
uint32 chargeId	ID of energy charger used in transaction

4. Multi-Tagged Continuous Market

4.1. Market and Bid Structure Design

As briefly stated in Section 3.3, in construction of the P2P blockchain energy market, a market that adds a tagging element to normal electricity trading was constructed. The overall bid structure resulting from this was shown in Table 2.

Tags are additional attributes of energy that are added to the market, such as renewable energy and fossil fuelled power. Market participants could express their willingness to pay for these additional attributes by setting separate prices for each tags. Market time windows could be also expressed. In former non-tagged market structures, in order to express these features, additional energy markets needed to be newly constructed. This promotes market fragmentation, which leads to an increase in user stress and system load. This will be further mentioned in Section 4.2.

4.2. System Design Intention and Assumed Market Participants

General system requirements of a continuous market are listed below, and implementing these functions using blockchain are the fundamentals for continuous markets.

- User bid account management
- First come first serve matching
- Bid price and amount order book management

Additional to this, through user feedback and literature reviews of ways to further engage P2P market participation [40], the requirements listed below require addressing .

1. Enable the feature expression of energy, which is currently transacted as a commodity. Examples of features could be energy generation method, energy generation location, etc.
2. Reduce the possibility of over-contraction when bidding in numerous market time windows. For example, when a user wants to obtain 10Wh of energy between 4:00 p.m.–5:00 p.m., the market may be split into 30 min time windows of 4:00 p.m.–4:30 p.m. and 4:30 p.m.–5:00 p.m., resulting in bidding 10 Wh to both 30 min markets with a total of 20 Wh bids in the market. If these bids simultaneously contract, the userends up obtaining unneeded energy.

The installation of tags in the bid market and energy type is capable of overcoming these issues without raising the blockchain system load. Separate prices could be set for each tag combination. Examples of the usage of tags to tackle the issues is further described in Section 4.3.

4.3. Example Usage of Tags

In the proposed bid structure, by setting the bid amount of energy in a single value, while setting the corresponding market time/energy feature tag/each bid price in numerous options, a wide range of bids can be made and weighted according to taste.

For example, in the case of bidding for 50 Wh of electricity, a parallel bid as shown in Table 4 could be generated, expressing the will to purchase renewable energy ("green" tag) at a slightly higher price than fossil based energy ("brown" tag).

Table 4. Example of bid expressing needs for renewable energy purchase.

Item	Tag Combination 1	Tag Combination 2
time	14:00–14:30	14:00–14:30
feature tag	green	brown
price	$1	$0.8
amount	50 Wh	

Another example is shown in Table 5. In this case, a time-based parallel bidding is used. This type of strategy could be used when the market participant is only connected to energy charge/discharge devices at a range of time windows, and thus wants to set a range of bids to acquire the necessary and sufficient amount of energy within the range.

Table 5. Example of bid expressing needs for time based bidding.

Item	Tag Combination 1	Tag Combination 2
time	14:00–14:30	18:00–18:30
feature tag	green	green
price	$1	$1
amount		50 Wh	

This bid structure fixes quantity over a range of markets and makes it is possible to prevent the excessive procurement or sales of electricity, while bidding to multiple market time windows at the same time. For example, in the bid in Table 4, if 30 Wh of energy is obtained from 2:00 p.m.–2:30 p.m. in tag combination 1, the total bidded amount of energy would automatically be reduced to 20 Wh by the market system, preventing excessive procurement. This is not the case if there exist multiple energy markets for each time window or tag, where the market participant needs to actively detect energy contraction, and quickly adjust its market bidding position over multiple markets. This raises participant stress and system load, for the number of requests needed to be sent to the blockchain system rises.

5. Demonstration Experiment Settings and Results

5.1. Demonstration Experiment Settings

In order to verify the effectiveness of the platform created, a demonstration experiment was conducted from 17 June 2019 to 31 August 2020, in Higashifuji, Shizuoka-Prefecture of Japan.

Demonstration experiment participants are shown in Table 6.

Table 6. Demonstration Experiment Participants.

Participant Type	Hardware Owned				Number
	PV	Battery	PHEV	PHEV Charger	
household	x	x	x	x	6 households
household	x	x	o	o	6 household
household	o	x	x	x	2 households
household	o	o	x	x	3 household
household	o	x	o	o	2 household
household	o	o	o	o	1 household
company office	o	o	x	o	1 office

An example image of batteries and EV chargers installed at demonstration experiment participant houses are shown in Figure 3. Also, an image of the bird's eye view of the company office and installed PHEV charger is shown in Figures 4 and 5.

Demonstration experiment market settings are shown in Table 7. The market time window was set according to Japan Electric Power Exchange(JPEX) settings. Energy type tags were designed according to business energy transmission needs.

One fullNode and one authority was set up for the blockchain system. As stated above, in future systems, by increasing the number of authority nodes, the reliability of transaction processing could be improved. Also by increasing the number of full nodes, the amount of concurrent transactions can be increased, and the amount of data reference processing can be increased.

Figure 3. Hardware installed at participant homes.

Figure 4. Birds eye view of company office participant.

Figure 5. Hardware installed at participant office.

Table 7. Demonstration Experiment Market Settings.

Setting Items	Setting
market window	30 min
energy type tags	"green": Renewable Energy "brown": Other Energy
market type tags	"Low Voltage Market(LVM)" "Special High Voltage Market(SHVM)" "Direct Markets": For priority contracting for specific agents

5.2. Results

5.2.1. System Performance

The system performance was verified by first checking that daily transactions were executed without any issues, and second analyzing issues in blockchain system operation. Previous work [9,11] demonstrated that blockchain systems are capable of stable energy matchmaking in the virtual layer. The performance mentioned here additionally has the scope of physical energy charge/discharge control and its following execution reports.

Overall, the system was capable of handling and executing the bid transactions from the participants without any faults. The data of the number of transactions handled are shown in Table 8. One block accounts for 5 s.

Table 8. System Performance Measurements(from 17 June 2019 to 31 August 2020).

Item	Data
Total Transactions	7,861,004
Server Maximum Permissible Transactions	1000 per block
Actual Maximum Transactions	805 per block

In the blockchain system management operation, the matters shown below had to be dealt with in the following manner.

1. Forced reboot of fullnode
   Issue: occurs when the participating agent systems connected to the blockchain system tries to fetch hundreds of thousands of blocks worth of information at once. The fullnode runs out of memory and is forced to restart.
   Handled: fix the system connecting to blockchain
2. Insufficient disk space
   Issues: occurs due to increase in data storage.
   Handled: both the authority and fullnode storage was increased from 50 GB to 100 GB to 150 GB accordingly.

The former issue is due to extending the scope of the blockchain system from the virtual layer to the physical layer, and future work should address this issue in system design. Future system implications from these matters will be discussed in Section 6.

5.2.2. Effect of Multi-Tagged Continuous Market

The market performance was compared to the performance of the Urawa Misono project of the Ministry of the Environment [12,13]. The proposed market in this paper was implemented based on the market in the Urawa Misono project, which makes this a reasonable comparison. The evaluation results are summarized in Table 9.

First, in comparison of one energy market type (i.e., brown energy market only), a reduction in the number of transactions per user, compared to the existing method, was confirmed. Installing tags allows bidding to multiple markets simultaneously while executing only the required amount, which lead to this reduction. This effect was further confirmed for two energy market types (i.e., green and brown energy market).

This reduction effect is due to the feature whereby the proposed multi-tagged bid structure could express more information in a single transaction. The previous structure required the creation of a transaction per energy market type and per market time window, resulting in fragmentation of bids, and also many bid cancellations. The proposed structure could express this information in a single transaction, and thus improve system efficiency.

In addition, through participant transaction and requirement analysis, it was confirmed that the over-execution of transactions was suppressed.

Table 9. Average transactions per contract.

Number of Market Types	Urawa Misono Project	Ours
1 market	16	4
2 market	32	4

6. Discussion

6.1. Further Usage of Tags

Introducing the concept of tags enabled the expression of additional characteristics and values of energy. This paper mainly mentioned the RE value of energy, but the possibilities tag usage is more extensive.

For example, different voltage levels could be expressed using tags as well. Energy is transacted at different voltage levels according to consumer requirements. Using previous non-tagged markets to express this difference results in the number of required transactions increasing, as shown in Table 9.

Another usage is to express additional emerging energy values, such as local energy consumption [41,42]. From the power transmission system point of view, promoting local energy production and consumption is environmentally friendly, reducing system load. Also, local energy consumption could be used as marketing tools for companies, for this consumption implies that the company is restoring earned cash to the local economy, and thus further activating it.

The concept of tags allows for the flexible updating of the P2P energy market according to the distributed needs of each participant, which is a distinct aspect for distributed energy markets to have an attraction compared to conventional centralized energy distribution.

6.2. Blockchain System Operations

As stated above, in a blockchain system operation from 17 June 2019 to 31 August 2020, there were matters in access load control from external systems, and node server storage.

The former matter should be avoided in the future by creating access load limits to APIs, offered to the external systems. Further management should be done by setting access limits according to the participant agent type in order to balance agent system execution and blockchain system stability. The latter matter should be avoided by setting server storage alerts and actively raising the number of nodes connected to the blockchain system. This scalability is an advantage of using blockchain, and further leverage of this is expected.

7. Conclusions

In this paper, to further socially implement blockchain based P2P energy trading, a blockchain based P2P energy trading system which could coordinate with energy control hardware was constructed, and a demonstration experiment was conducted. Previous work focused on virtually matching energy supply and demand via blockchain P2P energy markets, and our work advances this forward by demonstrating the possibility of actual energy flow control. In the demonstration, PHEVs and HEMS actually used in daily life were controlled in coordination with the blockchain system. In doing so, the need of a multi-tagged continuous market was found and proposed.

The blockchain system was constructed using Ethereum with private blockchain,with Proof of Authority as the consensus algorithm. Through a cloud-deployed API, each market participant makes bids to the energy market, confirms the contracted bids to control energy flow, and reports actual executed energy transactions. The processes are automated, enabling an end-to-end, machine-to-machine (M2M) control of the energy transaction without human intervention.

The multi-tagged continuous market adds a tagging element to normal electricity trading. The inclusion of tags in the bid information allows users to express individual features of energy, which is currently transacted as commodities. It is also capable of reducing over contraction when bidding over numerous market time windows.

In the demonstration experiment, the proposed blockchain market and hardware control interface was proven capable of securing and stably transmitting energy within the P2P energy system. Also, by the implementation of multi-tagged energy markets, the number of transactions required to secure the required amount of electricity was reduced. In terms of blockchain system operation, matters in the external system requested handling, and the system storage was activated. Based on these issues, implications to future blockchain system implementation were given.

Blockchain based P2P energy transaction is a promising concept in energy decentralization, and its feasibility is being proven. The next step is to make distinct its difference compared to conventional centralized energy distribution from the user's point of view. The usage of tags could express a vast variety of energy value, and the low cost and the ability to design the types of tags, as well as to measuring the effect on market participant utility, is future work that would build on this base.

Author Contributions: Conceptualization, K.T.; methodology, K.T., Y.Y., Y.T., D.S. and Y.M.; software, Y.Y., H.O., Y.T. and D.S.; validation, Y.Y., H.O., Y.M. and K.T.; investigation, Y.M., Y.T. and D.S.; writing—original draft preparation, Y.M. and H.O.; writing—review and editing, H.O., Y.Y. and Y.M.; visualization, Y.M. and H.O.; supervision, K.T.; All authors have read and agreed to the published version of the manuscript.

Funding: This research received no external funding.

Data Availability Statement: Data available on request due to restrictions. The data presented in this study are available on request from the corresponding author. The data are not publicly available due to privacy.

Acknowledgments: This research is based on the joint demonstration project with Toyota Motor Corporation, the University of Tokyo, and TRENDE Inc.

Conflicts of Interest: The authors declare no conflict of interest.

References

1. Energy—United Nations Sustainable Development. Available online: https://www.un.org/sustainabledevelopment/energy/ (accessed on 26 September 2021).
2. Siano, P.; De Marco, G.; Rolán, A.; Loia, V. A survey and evaluation of the potentials of distributed ledger technology for peer-to-peer transactive energy exchanges in local energy markets. *IEEE Syst. J.* **2019**, *13*, 3454–3466. [CrossRef]
3. Zhang, C.; Wu, J.; Zhou, Y.; Cheng, M.; Long, C. Peer-to-Peer energy trading in a Microgrid. *Appl. Energy* **2018**, *220*, 1–12. [CrossRef]
4. Sousa, T.; Soares, T.; Pinson, P.; Moret, F.; Baroche, T.; Sorin, E. Peer-to-peer and community-based markets: A comprehensive review. *Renew. Sustain. Energy Rev.* **2019**, *104*, 367–378. [CrossRef]
5. Khatoon, A.; Verma, P.; Southernwood, J.; Massey, B.; Corcoran, P. Blockchain in energy efficiency: Potential applications and benefits. *Energies* **2019**, *12*, 3317. [CrossRef]
6. Erturk, E.; Lopez, D.; Yu, W.Y. Benefits and risks of using blockchain in smart energy: a literature review. *Contemp. Manag. Res.* **2019**, *15*, 205–225. [CrossRef]
7. Vangulick, D.; Cornélusse, B.; Ernst, D. Blockchain for peer-to-peer energy exchanges: design and recommendations. In Proceedings of the 2018 Power Systems Computation Conference (PSCC), Dublin, Ireland, 11–15 June 2018; pp. 1–7.
8. Monroe, J.G.; Hansen, P.; Sorell, M.; Berglund, E.Z. Agent-based model of a blockchain enabled peer-to-peer energy market: Application for a neighborhood trial in Perth, Australia. *Smart Cities* **2020**, *3*, 1072–1099. [CrossRef]
9. Mengelkamp, E.; Gärttner, J.; Rock, K.; Kessler, S.; Orsini, L.; Weinhardt, C. Designing microgrid energy markets: A case study: The Brooklyn Microgrid. *Appl. Energy* **2018**, *210*, 870–880. [CrossRef]
10. Green, J.; Newman, P. Citizen utilities: The emerging power paradigm. *Energy Policy* **2017**, *105*, 283–293. [CrossRef]
11. Sikorski, J.J.; Haughton, J.; Kraft, M. Blockchain technology in the chemical industry: Machine-to-machine electricity market. *Appl. Energy* **2017**, *195*, 234–246. [CrossRef]
12. Tanaka, K.; Abe, R.; Nguyen-Van, T.; Yamazaki, Y.; Kamitamari, T.; Sako, K.; Koide, T. A Proposal on an Electricity Trading Platform Using Blockchain. In *Transdisciplinary Engineering Methods for Social Innovation of Industry 4.0*; IOS Press: Amsterdam, The Netherland, 2018; pp. 976–983.
13. Ahl, A.; Yarime, M.; Goto, M.; Chopra, S.S.; Kumar, N.M.; Tanaka, K.; Sagawa, D. Exploring blockchain for the energy transition: Opportunities and challenges based on a case study in Japan. *Renew. Sustain. Energy Rev.* **2020**, *117*, 109488. [CrossRef]

14. Shukla, S.; Thakur, S.; Hussain, S.; Breslin, J.G. A Blockchain-Enabled Fog Computing Model for Peer-To-Peer Energy Trading in Smart Grid. In *Blockchain and Applications*; Prieto, J., Partida, A., Leitão, P., Pinto, A., Eds.; Springer International Publishing: Cham, Switzerland, 2022; pp. 14–23.

15. Ahl, A.; Yarime, M.; Tanaka, K.; Sagawa, D. Review of blockchain-based distributed energy: Implications for institutional development. *Renew. Sustain. Energy Rev.* **2019**, *107*, 200–211. [CrossRef]

16. Hanna, R.; Ghonima, M.; Kleissl, J.; Tynan, G.; Victor, D.G. Evaluating business models for microgrids: Interactions of technology and policy. *Energy Policy* **2017**, *103*, 47–61. [CrossRef]

17. Zhou, Y.; Wu, J.; Long, C. Evaluation of peer-to-peer energy sharing mechanisms based on a multiagent simulation framework. *Appl. Energy* **2018**, *222*, 993–1022. [CrossRef]

18. Long, Y.; Wang, Y.; Pan, C. Incentive mechanism of micro-grid project development. *Sustainability* **2018**, *10*, 163. [CrossRef]

19. Cali, U.; Çakir, O. Energy policy instruments for distributed ledger technology empowered peer-to-peer local energy markets. *IEEE Access* **2019**, *7*, 82888–82900. [CrossRef]

20. Fang, X.; Misra, S.; Xue, G.; Yang, D. Smart grid—The new and improved power grid: A survey. *IEEE Commun. Surv. Tutor.* **2011**, *14*, 944–980. [CrossRef]

21. Tushar, W.; Saha, T.K.; Yuen, C.; Smith, D.; Poor, H.V. Peer-to-peer trading in electricity networks: An overview. *IEEE Trans. Smart Grid* **2020**, *11*, 3185–3200. [CrossRef]

22. Khorasany, M.; Mishra, Y.; Ledwich, G. A decentralized bilateral energy trading system for peer-to-peer electricity markets. *IEEE Trans. Ind. Electron.* **2019**, *67*, 4646–4657. [CrossRef]

23. Erol-Kantarci, M.; Mouftah, H.T. Wireless sensor networks for cost-efficient residential energy management in the smart grid. *IEEE Trans. Smart Grid* **2011**, *2*, 314–325. [CrossRef]

24. Ergen, S.C. ZigBee/IEEE 802.15. 4 Summary. *UC Berkeley Sept.* **2004**, *10*, 11.

25. Han, D.M.; Lim, J.H. Smart home energy management system using IEEE 802.15. 4 and zigbee. *IEEE Trans. Consum. Electron.* **2010**, *56*, 1403–1410. [CrossRef]

26. Hussain, S.; Mohammad, F.; Kim, Y.C. Communication Network Architecture based on Logical Nodes for Electric Vehicles. In Proceedings of the 2017 International Symposium on Information Technology Convergence, Shijiazhuang, China, October 2017.

27. Cleveland, F. IEC 61850-7-420 communications standard for distributed energy resources (DER). In Proceedings of the 2008 IEEE Power and Energy Society General Meeting-Conversion and Delivery of Electrical Energy in the 21st Century, Pittsburgh, PA, USA, 20–24 July 2008, pp. 1–4.

28. Chaudhari, K.; Kandasamy, N.K.; Krishnan, A.; Ukil, A.; Gooi, H.B. Agent-based aggregated behavior modeling for electric vehicle charging load. *IEEE Trans. Ind. Inform.* **2018**, *15*, 856–868. [CrossRef]

29. Hussain, S.; Lee, K.B.; Ahmed, M.A.; Hayes, B.; Kim, Y.C. Two-stage fuzzy logic inference algorithm for maximizing the quality of performance under the operational constraints of power grid in electric vehicle parking lots. *Energies* **2020**, *13*, 4634. [CrossRef]

30. Abe, R.; Taoka, H.; McQuilkin, D. Digital grid: Communicative electrical grids of the future. *IEEE Trans. Smart Grid* **2011**, *2*, 399–410. [CrossRef]

31. Shibano, K.; Kontani, R.; Hirai, H.; Hasegawa, M.; Aihara, K.; Taoka, H.; McQuilkin, D.; Abe, R. A Linear programming formulation for routing asynchronous power systems of the Digital Grid. *Eur. Phys. J. Spec. Top.* **2014**, *223*, 2611–2620. [CrossRef]

32. Hayashi, K.; Kato, R.; Torii, R.; Taoka, H.; Abe, R. Bi-directional power flow through a digital grid router. *J. Int. Counc. Electr. Eng.* **2015**, *5*, 42–46. [CrossRef]

33. Nguyen-Van, T.; Abe, R.; Tanaka, K. Digital Adaptive Hysteresis Current Control for Multi-Functional Inverters. *Energies* **2018**, *11*, 2422. [CrossRef]

34. Kelly, J.; Knottenbelt, W. Neural nilm: Deep neural networks applied to energy disaggregation. In Proceedings of the 2nd ACM International Conference on Embedded Systems for Energy-Efficient Built Environments, Seoul, Korea, 4–5 November 2015; pp. 55–64.

35. Brown, J.; Abate, A.; Rogers, A. Disaggregation of household solar energy generation using censored smart meter data. *Energy Build.* **2021**, *231*, 110617. [CrossRef]

36. Crozier, C.; Apostolopoulou, D.; McCulloch, M. Clustering of usage profiles for electric vehicle behaviour analysis. In Proceedings of the 2018 IEEE PES Innovative Smart Grid Technologies Conference Europe (ISGT-Europe), Sarajevo, Bosnia and Herzegovina, 21–25 October 2018; pp. 1–6.

37. Wood, G. Ethereum: A secure decentralised generalised transaction ledger. *Ethereum Proj. Yellow Pap.* **2014**, *151*, 1–32.

38. Openethereum/Parity-Ethereum: The Fast, Light, and Robust Client for Ethereum-Like Networks. Available online: https://github.com/openethereum/parity-ethereum (accessed on 26 September 2021).

39. Suciu, G.; Nădrag, C.; Istrate, C.; Vulpe, A.; Ditu, M.C.; Subea, O. Comparative analysis of distributed ledger technologies. In Proceedings of the 2018 Global Wireless Summit (GWS), Chiang Rai, Thailand, 25–28 November 2018; pp. 370–373.

40. Morstyn, T.; Farrell, N.; Darby, S.J.; McCulloch, M.D. Using peer-to-peer energy-trading platforms to incentivize prosumers to form federated power plants. *Nat. Energy* **2018**, *3*, 94–101. [CrossRef]

41. Van Der Schoor, T.; Scholtens, B. Power to the people: Local community initiatives and the transition to sustainable energy. *Renew. Sustain. Energy Rev.* **2015**, *43*, 666–675. [CrossRef]

42. Local Energy Supply Solution for Realizing Distributed Power Supply for Local Production and Consumption: Hitachi Review. Available online: https://www.hitachi.com/rev/archive/2020/r2020_04/04b04/index.html (accessed on 26 September 2021).

Article

Bidding Agents for PV and Electric Vehicle-Owning Users in the Electricity P2P Trading Market

Daishi Sagawa [1,*]**, Kenji Tanaka** [1]**, Fumiaki Ishida** [2]**, Hideya Saito** [2]**, Naoya Takenaga** [3]**, Seigo Nakamura** [3]**, Nobuaki Aoki** [3]**, Misuzu Nameki** [3] **and Kosuke Saegusa** [3]

[1] School of Engineering, The University of Tokyo, Tokyo 113-8656, Japan; tanaka@tmi.t.u-tokyo.ac.jp
[2] The Kansai Electric Power Co., Inc., Osaka 530-8270, Japan; ishida.fumiaki@b5.kepco.co.jp (F.I.); saito.hideya@d5.kepco.co.jp (H.S.)
[3] Nihon Unisys, Ltd., Tokyo 135-8560, Japan; naoya.takenaga@unisys.co.jp (N.T.); seigo.nakamura@unisys.co.jp (S.N.); nobuaki.aoki@unisys.co.jp (N.A.); misuzu.nameki@unisys.co.jp (M.N.); kosuke.saegusa@unisys.co.jp (K.S.)
* Correspondence: sagawa@ioe.t.u-tokyo.ac.jp

Abstract: As the world strives to decarbonize, the effective use of renewable energy has become an important issue, and P2P power trading is expected to unlock the value of renewable energy and encourage its adoption by enabling power trading based on user needs and user assets. In this study, we constructed a bidding agent that optimizes bids based on electricity demand and generation forecasts, user preferences for renewable energy (renewable energy-oriented or economically oriented), and owned assets in a P2P electricity trading market, and automatically performs electricity trading. The agent algorithm was used to evaluate the differences in trading content between different asset holdings and preferences by performing power sharing in a real scale environment. The demonstration experiments show that: EV-owning and economy-oriented users can trade more favorably in the market with a lower average execution price than non-EV-owning users; forecasting enables economy-enhancing moves to store nighttime electricity in batteries in advance in anticipation of future power generation and market prices; EV-owning and renewable energy-oriented users can trade more favorably in the market with other users. EV-owning and renewable energy-oriented users can achieve higher RE ratios at a cost of about +1 yen/kWh compared to other users. By actually issuing charging and discharging commands to the EV and controlling the charging and discharging, the agent can control the actual use of electricity according to the user's preferences.

Keywords: P2P energy trading; bidding agent; electric vehicle

Citation: Sagawa, D.; Tanaka, K.; Ishida, F.; Saito, H.; Takenaga, N.; Nakamura, S.; Aoki, N.; Nameki, M.; Saegusa, K. Bidding Agents for PV and Electric Vehicle-Owning Users in the Electricity P2P Trading Market. *Energies* **2021**, *14*, 8309. https://doi.org/10.3390/en14248309

Academic Editor: Yuji Yamada

Received: 20 October 2021
Accepted: 30 November 2021
Published: 9 December 2021

1. Introduction

1.1. Background

The current electricity network is undergoing a major transformation with the introduction of renewable energy. The European Union (EU) has set a goal to increase the share of renewable energy to at least 32% by 2030 and to reduce greenhouse gas emissions by 40% compared to 1990 levels [1,2]. However, in order to increase the proportion of power sources that are decentralized and whose output is affected by weather conditions, such as renewable energy, a mechanism is needed to ensure that supply and demand are coordinated to make effective use of renewable energy. In this context, there is a growing need for supply and demand adjustment and energy storage through distributed power networks in order to cope with decentralized power sources.

In particular, peer-to-peer (P2P) energy trading is expected to be a promising model for the future power system, which consists of energy buyers, sellers, and their matching mechanisms, and is expected to enable users to match each other's needs. P2P transactions are expected to enable matching according to the needs of users, and have been the subject of extensive research in recent years [3]. The significance of P2P trading is that it allows consumers, who are passive in the existing system, to trade while taking into account prices and their own preferences. As a result, when power generation is low and the price of electricity is high, consumers are expected to move their use of electricity to other times of the day or discharge electricity from storage batteries, and when the price is low, they are expected to store electricity or run heat pumps. Through these actions, the uncertainty of renewable energy generation is expected to be absorbed by the demand side by shifting their own demand as much as possible through prices and by using storage facilities. This will contribute to improving the balance of supply and demand, not at the micro level of frequency adjustment, but at the macro level of shifting demand and storing electricity.

1.2. Related Work

The forms of P2P transactions can be broadly classified into three categories: full P2P markets, community-based markets, and hybrid P2P markets [4]. In the full P2P market, peers negotiate directly with each other in order to buy and sell electric energy. An example of this is the study of bilateral method matching [5,6]. A relaxed consensus + innovation (RCI) approach in P2P market structure based on the multi-bilateral economic dispatch (MBED) method [6] has been proposed. It was shown that the MBED approach can effectively produce optimal market outcomes in terms of maximizing social welfare while respecting consumer preferences. In these studies of full P2P markets, the challenge is to reduce communication as peers interact with each other. Community-based markets are more structured, with a community manager to manage trading activities within the community and an intermediary between the community and the rest of the world. Mengelkamp et al. showed that both buyers and sellers of energy can benefit from P2P trading by harnessing excess renewable energy in the Brooklyn Microgrid experiment and reaching mutually satisfactory prices and quantities [7]. Another example showed that P2P markets can balance the local energy supply and demand and reduce energy transmission losses [8]. Hybrid P2P markets are proposed as a hybrid of full P2P markets and community-based markets, where transactions between peers are hierarchically defined in a model. An example is a study that aims to minimize the overall energy cost and the loss of P2P energy sharing in a distribution network consisting of multiple MGs [9].

P2P energy trading has been studied not only from the perspective of efficiency, but also from the perspective of fulfilling user preferences. While the increase in willingness to pay (WTP) for renewable energy has been reported in various countries [10–13], not everyone necessarily has the same WTP, and there is a need to realize transactions that meet the needs of individual users, and there are expectations for P2P transactions to meet these needs.

One study of the user preference perspective is [14], in which the authors proposed a P2P market based on multi-class energy management that respects user preferences and assumes that individual users work to maximize overall utility, rather than to maximize their own profit or utility. There are also studies that consider preferences for renewable energy within a community [15,16], but they have practical issues in that they are not optimized based on predictions of electricity demand and generation. To make effective use of the fluctuating output of renewable energies, an approach of sequential optimization while predicting the output of power generation and the power demand of consumers is necessary, and thus research on optimization based on prediction is needed.

In addition, in order to satisfy the needs of users, individual users themselves may utilize storage batteries, electric vehicles, and other energy storage equipment, as well as electric water heaters and other equipment that can shift demand. By conducting P2P energy transactions while optimizing the operation of such equipment according to the economic perspective and personal preferences of individual users, the system as a whole is expected to absorb the uncertainty of renewable energy and make more effective use of renewable energy. Kobashi et al. conducted a techno-economic analysis of an urban-scale energy system with rooftop solar PV, batteries, and electric vehicles, and showed that rooftop solar PV could be popularized at a significantly lower cost by actively introducing electric vehicles and using electric vehicles as energy storage devices [17]. In addition to improving the economics, these facilities could also be used to increase the percentage of renewable energy. The system itself, which assumes consumers who own storage batteries and electric vehicles automatically trade electricity, has already been proposed as previously introduced, and research on bidding strategies for aggregators who bundle peers [18–21] can be cited. However, all of them aim at maximizing profits for the entire community, and not for individuals to maximize their own profits or satisfy their own preferences. Therefore, bidding for satisfaction of individual preferences is not taken into account. This study differs from existing studies in this respect, as it gives bidding strategies for individuals to trade while taking their own preferences into account as they aim to maximize their own profits through market principles. In addition, although the studies in [19,20] take user preferences into account, the user preferences there are like the "departure SoC" for EV usage. These studies do not optimize bids with the goal of satisfying individual consumers' preferences for renewable energy usage. In contexts other than P2P electricity trading, although there are studies [22] that reveal the economic benefits of owning EVs by optimizing EV battery operation, they do not consider the needs of users other than economic benefits.

1.3. Contribution

In this research, we adopt an electricity P2P trading market where consumers and generators trade with each other and develop an agent system that automatically trades electricity on behalf of users in the market. The system not only makes bids based on the user's assets, such as electric vehicles and solar power generators, and the user's demand for electricity, but also makes bids based on the user's preferences for renewable energy, enabling the trading of electricity according to the circumstances of the individual user. There are two major novelties in our research.

① Individuals optimize their bidding by using energy storage facilities in order to maximize profits and satisfy their own preferences for renewable energy.
② By developing a bidding strategy that considers the individual's preference for renewable energy, we have achieved both economic efficiency and satisfaction of the individual's preference for renewable energy.

The agent system is used to conduct a demonstration experiment of P2P electricity trading. This agent system not only bids on the electricity market, but also plays a role in optimizing the use of electricity by users by controlling the charging and discharging of electric vehicles based on the results of bid execution. This research is also unique in that the P2P power market and its surrounding systems work together with physical objects such as electric vehicles. There are only a few studies on this topic. Through demonstration experiments, this study shows that this agent system and the P2P electricity market mechanism enable the effective use of electricity through the use of assets such as electric vehicles, while taking into account users' costs and preferences regarding renewable energy.

2. Overall Picture of the Demonstration Experiment

Figure 1 shows the overall picture of the demonstration experiment. User agents bid into the blockchain market on behalf of consumers and generators. Each user agent represents a household and conducts electricity transactions according to its preferences and the availability of EVs. In this experiment, user agents interact with each other through a retail power provider. User agents can also purchase electricity from the retail power provider. The blockchain market is implemented using Ethereum smart contracts to process the bids of user agents. The actual demand and power generation are also sent from the smart meter to the blockchain through an Internet gateway (GW) and recorded. Since it takes time to recall past demand data and past generation data from the blockchain, the information recorded in the blockchain is also synchronized in the RDBMS, and data are referenced from the RDBMS. In this experiment, there is only one EV, and the EV is treated as belonging to HOME1; the user agent in HOME1 plans the electricity usage, including charging and discharging the EV. The user agent of HOME1 plans the electricity usage including the charging and discharging of the EV. The other HOME2~4 do not own any EVs. As for the power generation side, there is only one PV, and there is a user agent that bids in the market on behalf of the PV. Therefore, there is one user agent for PV and four for HOME1~4, for a total of five user agents. The agent in HOME1 can issue charging and discharging commands to EVs via REST/HTTPS through the Internet gateway (EV GW) and can also obtain state of charge (SOC) from EVs. In addition, the measurement data from the smart meter are received by the Internet gateway (GW) via Wi-SUN and sent from the GW to the blockchain server via REST/HTTPS.

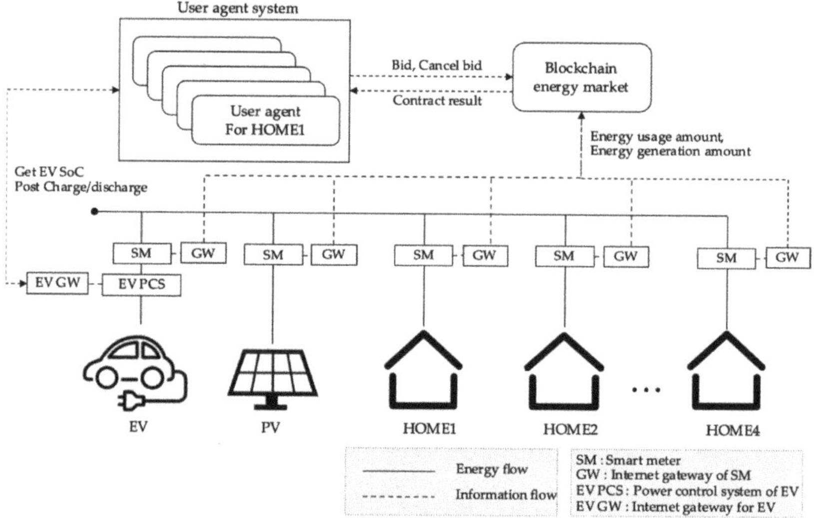

Figure 1. Diagram of the demonstration experiment.

3. Functions of User Agents

The user agent determines the amount and price of bids according to the user's equipment and demand, aiming to maximize the profit of each consumer or prosumer. The user agent makes decisions about when and how much to sell (or buy) and at what price, based on demand and power generation forecasts, and executes bids to the blockchain market. In the case of P2P transactions at the individual level, it is unrealistic for electricity consumers to constantly monitor the market just like day traders in the stock market, calculate the amount of electricity they need, and place bids. Therefore, we need such an agent module that automatically procures the amount of electricity consumers need from the market.

Bidding agents are required to take into account the various needs of consumers and power generators and automatically execute transactions in accordance with their preferences. The purpose of the bidding agent is to realize various needs, such as the financial need to purchase cheap electricity anyway, and the environmental value need to use renewable energy as much as possible. In the development of bidding agents, the assets they own are also an important factor. By optimizing the charging and discharging of EVs, consumers who own EVs can be expected to enjoy cost advantages, such as procuring more electricity from the market when electricity prices are low, storing it in batteries, and discharging it from batteries when electricity prices in the market are rising. In addition, optimizing the charging and discharging of EVs is an important factor not only in terms of cost, but also in terms of satisfying the RE preferences of individual users, as it can be expected to increase the RE ratio at a low cost by charging EVs when surplus inexpensive

RE is generated. When EVs and storage batteries are not owned, the amount of electricity demanded is a constraint on electricity transactions, but when EVs and storage batteries are owned, it is possible to reduce costs and increase the RE ratio by recharging and discharging at appropriate times.

The processing flow of the user agent is shown in Figure 2. In the electricity demand forecasting function, electricity demand forecasting is performed based on consumer demand data and weather data. In the case of PV power generation, the PV power generation forecast is based on past power generation data and weather data. Here, the solar radiation forecasting API of the Meteorological Engineering Center [23] is used to create a machine learning model using random forest [24] that learns the relationship between actual PV power generation values and forecasted solar radiation values to make forecasts. The bid creation function creates bids specifying the time frame, amount of electricity, and price based on the trading mode (green mode or economy mode) set by the user, the forecast results, and the SOC of the EV, and the bid execution function puts the created bids into the energy market. The details of the bid creation method will be described in the following sections. The execution result acquisition function acquires a record of the executed bids in the energy market, and the results are submitted again to the bid creation function to recalculate a new bid. The energy market trades electricity in 30-min increments, bids can be submitted from 24 h before the actual electricity fusion to one hour before the fusion, and user agents change their bids for the same market every 30 min. At that time, the bid cancellation function is a function that sends a command to the market to cancel the old bids from the past. The entire process from forecasting to bidding is repeated for each agent at 30-min intervals until one hour before the market closes. The EV charging/discharging command function actually issues charging/discharging commands to EVs through the EV PCS API based on the calculated EV charging/discharging plan once the target market has been closed. The EV charging/discharging plan is calculated in the optimization calculation in the bid creation.

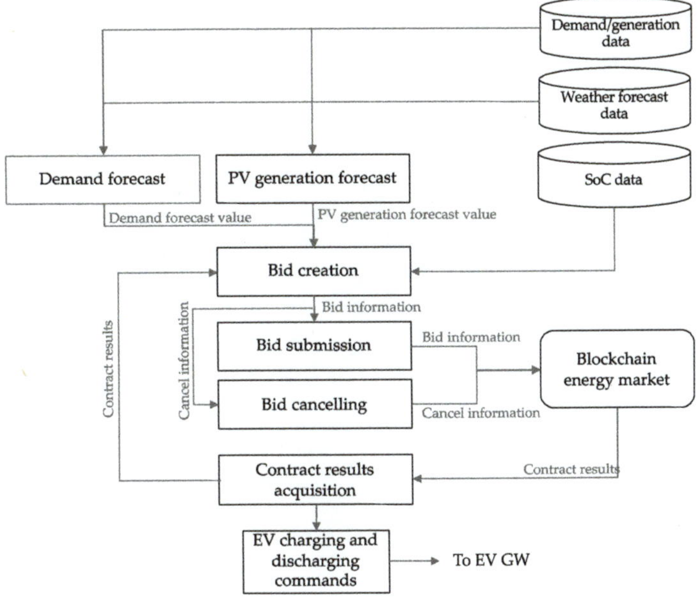

Figure 2. Calculation flow of user agent.

4. User Agent Bidding Modes and Bid Optimization

The bid creation function optimizes the bidding to the market and the charging and discharging of EVs based on the forecasted amount of demand, the forecasted amount of power generation, the SOC value, the expected market contract price, and the retail price of electricity. Two types of bid creation modes have been established: the economy mode and the green mode. In the economy mode, optimization is performed with the objective function of minimizing costs, including electricity sales revenue. It aims to maximize profits (minimize costs) by adjusting the timing and amount of procurement from the market and the grid, and by controlling the charging and discharging of its own EVs. The green mode is optimized by minimizing the cost, including the revenue from electricity sales, as the objective function while placing the constraint of meeting the target RE ratio set by the user. The objective is to maximize profits (or minimize costs) by adjusting the timing and amount of procurement from the market and grid, and by controlling the charging and discharging of its own EVs, while meeting the desired renewable energy consumption ratio (target RE ratio).

Equations (1)–(9) show the optimization equation for the economy mode. Each agent optimizes its own bid using this optimization equation. The objective function, Equation (1), to be minimized is the cost of procurement from the market (including revenue from electricity sales) + the cost of procurement from the grid + a penalty term, each of which is the sum of the values from the market one hour ahead to the market 48 h ahead of the target bid. The penalty term is expected to have the effect of preventing unnecessary trading from occurring, for example, buying 100 kWh and selling 99 kWh at the same price when one wants to buy 1 kWh. The variables to be optimized are B_t^m, S_t^m, B_t^g, C_t, and D_t and they are optimized by the calculations in Equations (1)–(9). In other words, we optimize the values from 1 h ahead to 49 h ahead for these variables. Each of these variables represents the amount of electricity purchased in the market, the amount of electricity sold in the market, the amount of electricity purchased from the retail business, and the amount of charging and discharging of the EV's battery. In addition to optimizing the charging and discharging of the EV's battery, the amount of electricity bought and sold in the market and the amount of electricity bought and sold from the retail business are simultaneously optimized. Since $Charge_t$ is the amount of charge for a certain 30 min, the upper limit of C_t is the maximum charging speed of the battery (C_{max}) [kW] multiplied by 0.5. Similarly, the value obtained by multiplying D_{max} [kW] by 0.5 is the upper limit of discharge (D_t).

$$Minimize. \quad \sum_{t=n}^{n+48*2} \left[P_t^m (B_t^m - S_t^m) + P_t^g B_t^g + C(B_t^m + S_t^m) \right] \tag{1}$$

$$Subject\ to.$$
$$B_t^m \geq 0 \tag{2}$$

$$S_t^m \geq 0 \tag{3}$$

$$\frac{C_{max}}{2} \geq C_t \geq 0 \tag{4}$$

$$\frac{D_{max}}{2} \geq D_t \geq 0 \tag{5}$$

$$A_t^d - B_t^m - \left(A_t^p - S_t^m \right) + C_t - D_t - B_t^g = 0 \tag{6}$$

$$E_t \geq E_{ll} \tag{7}$$

$$E_t \leq E_{hl} \tag{8}$$

$$E_{t+1} = \begin{cases} \frac{E_t E_{cap} + C_t R_c - D_t R_d}{E_{cap}} & (if\ V_t = False) \\ E_t - F_t & (if\ V_t = True) \end{cases} \tag{9}$$

Each variable is defined as follows.

B_t^m	Amount of electricity to be purchased in the market at time t [kWh] (Optimization target)
S_t^m	Amount of electricity to be sold in the market at time t [kWh] (Optimization target)
B_t^g	Amount of electricity to be purchased from electricity retailers at time t [yen/kWh] (Optimization target)
C_t	Amount of charge to the battery at time t [kWh] (Optimization target)
D_t	Amount of discharge from the battery at time t [kWh] (Optimization target)
P_t^m	Expected price at time t [yen/kWh] (estimated by each agent based on expected power generation)
P_t^g	Retail price of electricity at time t [yen/kWh] (defined in advance)
A_t^d	Expected demand at time t [kWh] (calculated by demand forecast)
A_t^p	Expected power generation at time t [kWh] (calculated by power generation forecast)
E_t	Percentage of remaining charge of the battery at time t [%]
C_{max}	Maximum charging output of the battery [kW] (6.7 [kW])
D_{max}	Maximum discharge output of the battery [kW] (6.0 [kW])
E_{ll}	Lower limit of SOC [%] (set to 20 percent)
E_{hl}	Upper limit of SOC [%] (set to 90%)
E_{cap}	Rated capacity of battery [kWh] (40 [kWh] was set.)
R_c	Battery charging efficiency [%] (set to 86.6%, so that CHARGE_RATE*DISCHARGE_RATE=75%)
R_d	Discharge efficiency of the battery [%] (set to 86.6%, the same as CHARGE_RATE)
F_t	Expected energy consumption by driving at time t [kWh]. This is always set to 0 because the EV is not running in this demonstration experiment.
V_t	The bool value indicating whether or not the EV is running at time t. It is always set to "false" because it is not run in this verification experiment.

P_t^m is the expected price in the market at time t. This expected price is calculated by each agent based on the weather information of the target day to predict the PV power generation on that day, and the expected price is calculated based on the power generation rate, which is the predicted PV power generation divided by the rated maximum output. Since the only electricity to be sold in the market in this case study is PV-derived, we believe it is a reasonable approach to forecast the PV power generation and predict the price according to the amount. The formula for calculating the expected price from the generation rate p_t is defined in Equation (2). Figure 3 plots the relationship between the power generation rate defined in Equation (2) and the expected market price. As the power generation rate p increases, the price approaches D = 5. In addition, when $p = 0$, the price is C + D = 28. In this demonstration experiment, we have taken the approach of calculating the price based on the expected amount of electricity generated. However, if such a trading market has actually been in operation for some time and sufficient data have been accumulated, a better method would be to create a regression model to predict the price using past contract prices and the weather conditions of the target market.

$$P_t^m = C * exp\left(-A * p_t^{\,B}\right) + D \tag{10}$$

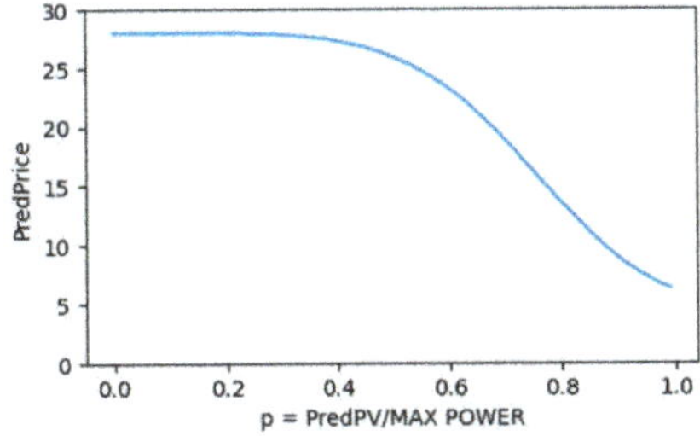

Figure 3. Relationship between generation rate and predicted price.

P_t^g gives the price list for each time. For the price list, we used the pay-as-you-go rates of the price table of "Hapi-e-time R" of Kansai Electric Power Co., Osaka, Japan [25]. This price list is shown in Table 1.

Table 1. Price table of $GridPrice_t$.

Hour	Price [Yen/kWh]
7:00–10:00	22.89
10:00–17:00	26.33
17:00–23:00	22.89
23:00–7:00	15.20

A_t^d is the agent's prediction of its own demand. The temperature and time information of the weather forecast data are used as explanatory variables, and a regression by random forest is conducted to make predictions. The predictions are made for 96 frames in 30-min increments for 48 h from 1:00 to 49:00 on the previous day.

A_t^p is the agent's prediction of its own photovoltaic power generation. The prediction is made by using the predicted solar radiation and time information as explanatory variables and conducting a regression by random forest. The predictions are made for 96 frames in 30-min increments for 48 h from 1:00 to 49:00 on the previous day.

E_t is the percent [%] of remaining charge of the battery at time t. The current battery state is obtained from the EV GW, and it is given as the initial state, but the subsequent times are calculated in the optimization according to the amount of charge and discharge, so it can be said that it is also optimized as a result.

Equations (11)–(20) show the optimization equation for green mode. The fact that the objective function (11) to be minimized is the cost of procurement from the market (including the revenue from electricity sales) + the cost of procurement from the grid + the penalty term is the same as in the economy mode, but the condition that the ratio of RE to the electricity consumed by the user should exceed the target RE ratio (R_{re}) has been added to the constraints (12). This allows us to plan the bidding to the market and the charging and discharging of the EVs so that the target RE ratio is exceeded. It should be noted that there may be cases where no solution exists due to this constraint condition. If a solution does not exist, the target RE ratio will be temporarily lowered by 5% in stages until a solution is found.

$$Minimize. \quad \sum_{t=n}^{n+48*2} \left[P_t^m (B_t^m - S_t^m) + P_t^g B_t^g + C(B_t^m + S_t^m) \right] \tag{11}$$

Subject to.

$$\sum_t \left(A_t^p - S_t^m + B_t^m \right) \geq R_{re} \sum_t \left[A_t^d + F_t + C_t(1 - R_c) + D_t(1 - R_d) \right] \tag{12}$$

$$B_t^m \geq 0 \tag{13}$$

$$S_t^m \geq 0 \tag{14}$$

$$\frac{C_{max}}{2} \geq C_t \geq 0 \tag{15}$$

$$\frac{D_{max}}{2} \geq D_t \geq 0 \tag{16}$$

$$A_t^d - B_t^m - \left(A_t^p - S_t^m \right) + C_t - D_t - B_t^g = 0 \tag{17}$$

$$E_t \geq E_{ll} \tag{18}$$

$$E_t \leq E_{hl} \tag{19}$$

$$E_{t+1} = \begin{cases} \frac{E_t E_{cap} + C_t R_c - D_t R_d}{E_{cap}} & (if\ V_t = False) \\ E_t - F_t & (if\ V_t = True) \end{cases} \tag{20}$$

Each variable is defined as follows.

R_{re} Target RE ratio (set by user between 0~100%)

The other items are the same as in (1)–(9).

In the case that the user does not own the EV, among the variables related to the EV (E_t, E_{ll}, E_{hl}, C_t, D_t, E_{cap}) in Equations (1)–(9) and Equations (11)–(20), respectively, all variables other than E_{cap} are set to 0. *BatteryCap* can be any real number other than 0 since it can be the denominator in the constraint.

Next, in the bid submission section, among the results calculated by the above optimization, B_t^m and S_t^m are bid into the blockchain market as the purchase and sales amount, respectively, and the unit price as P_t^m. Bidding is done for 48 markets every 30 min for the next 24 h. Here, optimization is performed until 48 h in the future, aiming to calculate the charging and discharging strategies for the last 24 h in a way that takes into account the future from 24 to 48 h in the future. If only the last 24 h are taken into account for optimization, even if the next two days are sunny and inexpensive electricity is supplied in abundance during the daytime, it is possible to store a lot of electricity in the batteries, so that when you try to store inexpensive electricity the next two days, the batteries are too full to store it. Therefore, the optimization is conducted for a longer period of time than the actual bidding.

In the contract results acquisition section, the contract status of the bids is obtained. Bids that have not yet been contracted are submitted under new conditions after optimization calculations. In this case, the existing bids are cancelled, and new bids are made.

This process of contract results acquisition, bid creation, bid cancelling, and bid submission is repeated every 30 min, and a time-evolving bidding experiment is conducted.

Regarding the optimization calculation of bidding agents, a single optimization calculation of an agent itself takes only a few seconds, and the calculation time increases linearly as the number of agents increases. Since the agents do not share information with each other, parallel computation is possible, and the problem can be solved by preparing multiple servers for computation.

5. About the Demonstration Experiment

5.1. Configuration of the Demonstration Experiment

The demonstration experiment was conducted with the following two main objectives.

- Confirmation that electricity costs can be further reduced when EVs are owned in economy mode
- Confirmation that the target RE ratio can be achieved at a relatively low cost when EVs are owned in green mode.

In the demonstration experiment, electricity trading was conducted in a P2P market with the participation of four consumers and one PV power generator as shown in Table 2.

This experiment was conducted over a period of two weeks, from 22 February 2021 to 7 March 2021. The settings were as follows.

Week 1 (22–28 February)

Setting: All in economy mode.

Objective: To confirm that the procurement costs of consumers who own EVs are lower than those of other consumers.

Week 2 (1–7 March)

Setting: Green mode for only EV owning consumers, economy mode for other users. In green mode, the setting is to conduct transactions aiming for a RE ratio of 40% or higher.

Objective: To confirm that EV-owning consumers can achieve a high RE ratio.

Table 2. Composition of consumers and generators in the demonstration experiment.

Type	User	EV(40 kWh)	Data Description
Demand	Home1	○	The electric load was reproduced in the experimental environment with a load device based on actual household demand data.
	Home2	-	An almost constant demand pattern was generated in the experimental environment.
	Home3	-	An almost constant demand pattern was generated in the experimental environment.
	Home4	-	An irregularly fluctuating demand pattern was generated in the experimental environment.
Supply	PV1	-	A PV system placed in the experimental environment was actually generating power.

5.2. Results of the Demonstration Experiment

5.2.1. Results and Discussion of the First Week

Table 3 below shows a summary of the trading results for the first week, showing that consumers with EVs (Home1) were able to trade more favorably in the market with a lower average trading price of 11.22 yen compared to the other consumers who traded at around 18–20 yen.

Table 3. Summary of trading results for the first week.

User	Mode	Average Contracted Price [Yen/kWh]	Contracted Amount [kWh]	Demand [kWh]	RE Rate [%]
Home1	Economy	11.22	29.6	84.8	34.9
Home2	Economy	20.13	9.4	16.6	56.6
Home3	Economy	18.75	32.5	70.6	46.0
Home4	Economy	20.32	38.2	102.6	37.2

Figure 4 shows the transition of the contract price and retail price for each user and time period. Here, the blue dots are the contract prices for users who do not own EVs, and the orange dots are the contract prices for users who own EVs. The gray line shows the retail price. From this figure, we can see that the contracted price is lower than the retail price for each corresponding time period, which means that agents were able to procure electricity more economically than purchasing electricity from retail. We can also see that most of the orange dots are distributed in the range of 5–15 yen, which is cheaper than the cheaper nighttime retail electricity. This means that users with EVs were able to implement the strategy of purchasing electricity if the market electricity is cheaper than the nighttime electricity, and otherwise charging their EVs at night through the bidding agent's cost minimization optimization algorithm.

Figure 5 shows the execution price and the amount of electricity generated for each user and time period. It can be seen that the price did not drop significantly on the day with low power generation (26 February), and as a result, users who own EVs did not need to be contracted.

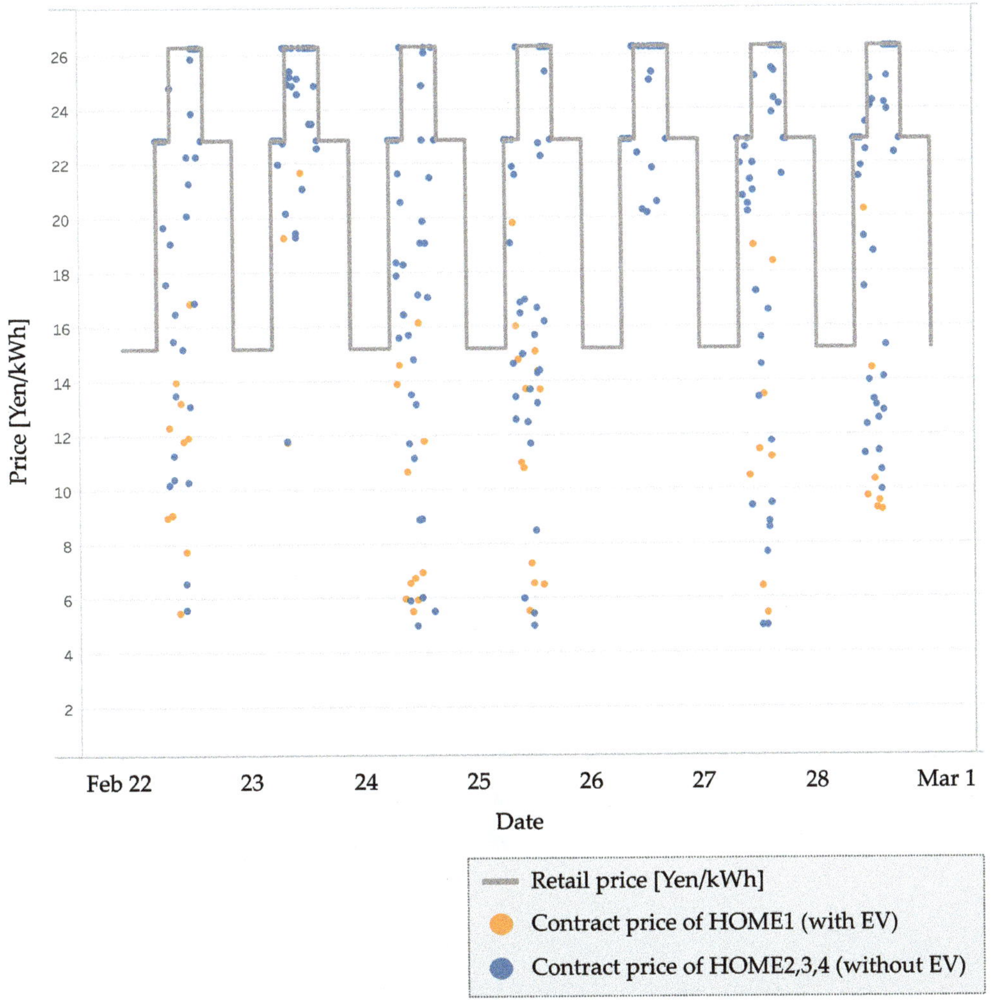

Figure 4. Contract price and retail price by user and time period.

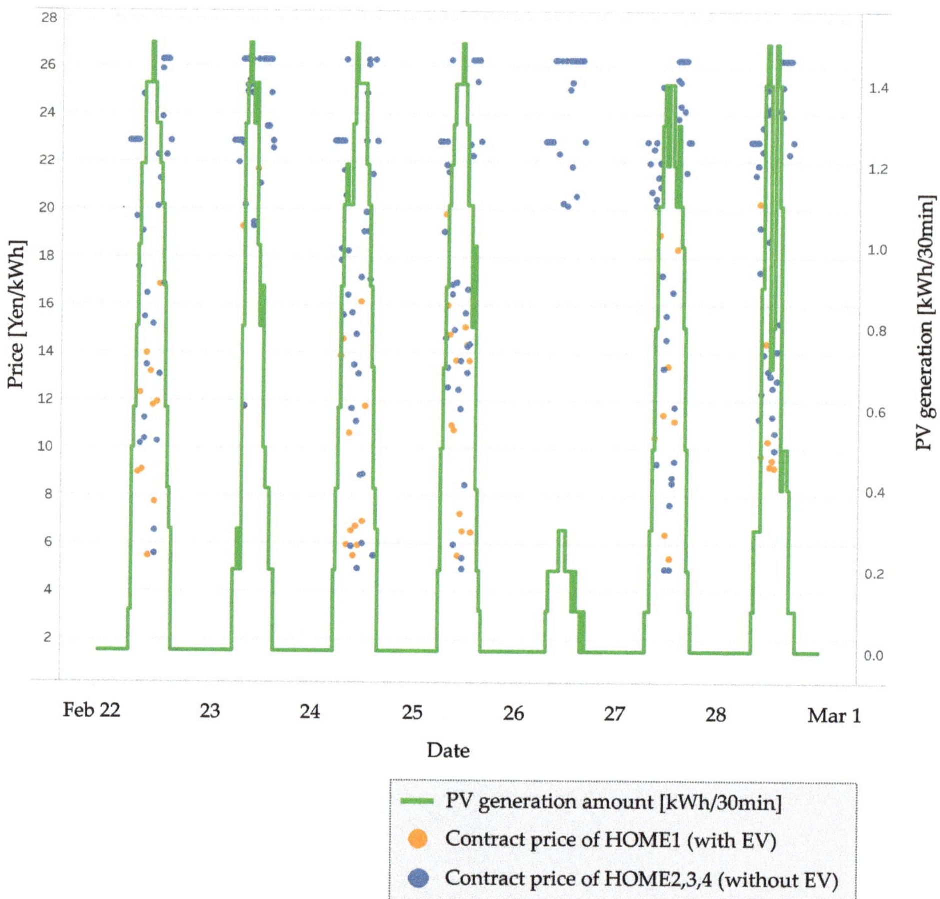

Figure 5. Contracted price by user and power generation.

Figure 6 shows the state transition of the users who own EVs. The left axis shows the amount of electricity [kWh], the right axis shows the SOC [%], the blue line shows the amount of demand, the red line shows the contracted amount in the market, the green line shows the amount of PV generation, and the orange area shows the SOC. It can be seen that the red line, the contracted amount in the market, was higher during the day, indicating that PV generation could be purchased during the day. Furthermore, during the same time period, the SOC of the orange area increased, indicating that the agent charged EVs with inexpensive PV generation during the day. In addition, if we look at the SOC, we can confirm that the EVs were being recharged not only during the daytime, but also during the late-night hours when the retail electricity price is inexpensive. Particularly, on 26 February, PV power generation was low, and thus the market price did not fall and the agent could not purchase PV during the day. In anticipation of this, we can see that the agent purchased a large amount of late-night electricity in advance, stored it in its EV, and then discharged and used it during the daytime when retail electricity prices were high. This shows that the charging/discharging optimization that takes the power generation situation into account is working well.

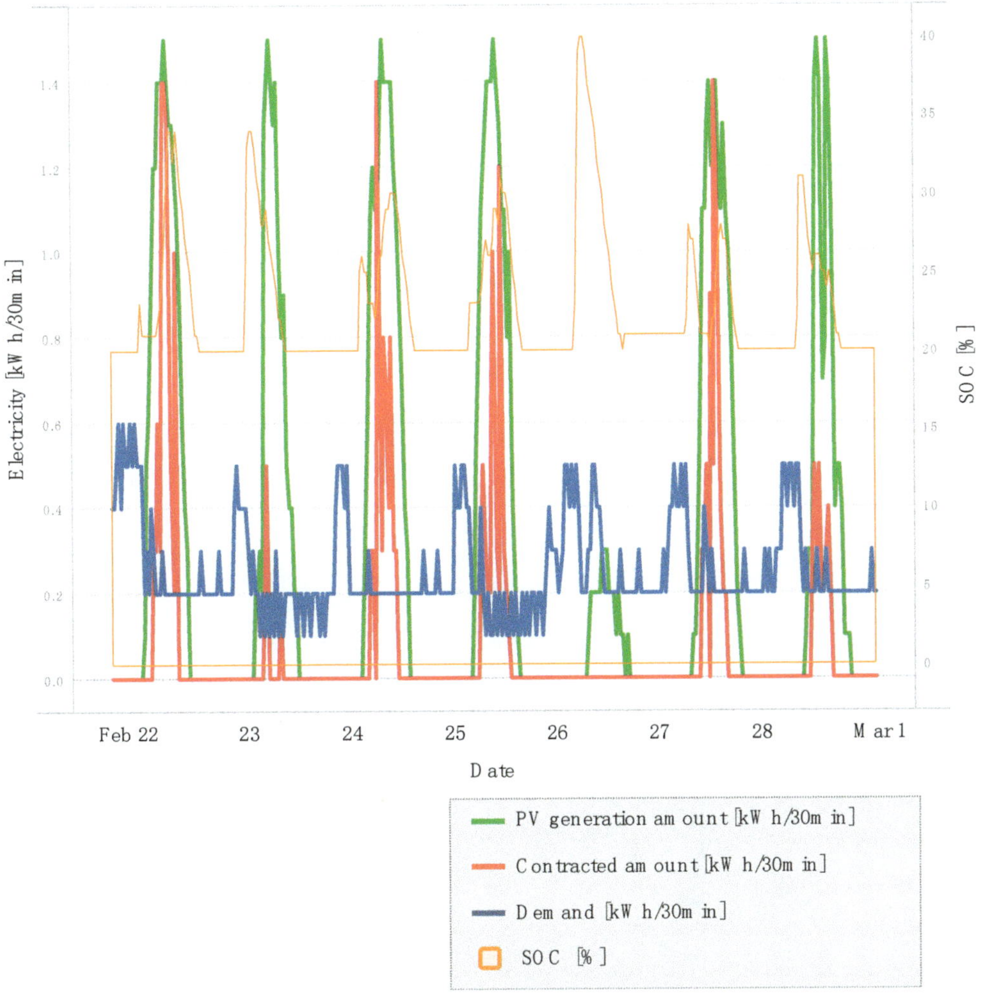

Figure 6. Changes in the status of users who own EVs in the first week.

5.2.2. Results and Discussion of the Second Week

Table 4 shows the results for the second week (green mode for EV-owning users only). Compared with the results of the first week, the average execution price for the consumer owning the EV (Home1) increased from 11.22 yen to 22.75 yen, which is slightly higher than the other users where the price was around 20–21 yen. The green mode was set to trade with the goal of achieving an RE ratio of 40% or higher, and trading has been able to exceed the target RE ratio. In addition, compared to other users who had RE ratios of around 20–30%, the green mode user who owns the EV has an RE ratio of 57.2%, indicating that the agent was able to achieve a high RE ratio at a cost of around +1 yen/kWh.

Figure 7 shows the transition of the status of the user who owns the EV in the second week. Here, it can be read that the contracted amount of electricity generated from PV in the market was large, and that this amount was being recharged into EVs during the day. It can also be seen that the electricity charged during the day was discharged and consumed during the evening and night. In addition, compared to the first week, the amount of inexpensive electricity charged at night has decreased, confirming that the green mode movement to use PV power generation as much as possible has been realized. Figure 8 also shows that week 1 had a higher SOC in the early morning than week 2.

Table 4. Summary of trading results for the second week.

User	Mode	Average Contracted Price [Yen/kWh]	Contracted Amount [kWh]	Demand [kWh]	RE Rate [%]
Home1	Green	22.75	43.9	76.7	57.2
Home2	Economy	21.0	5.0	15.8	31.6
Home3	Economy	20.75	15.7	71.7	21.9
Home4	Economy	21.6	21.4	102.3	20.9

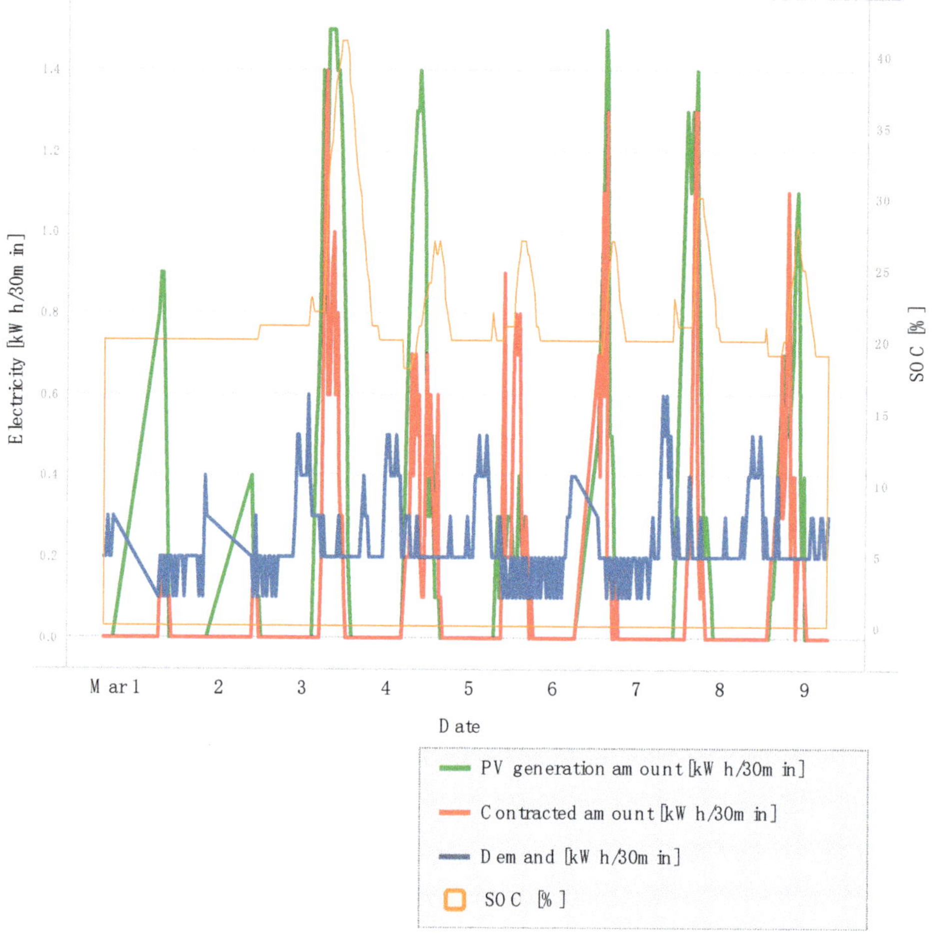

Figure 7. Changes in the status of users who own EVs in the second week.

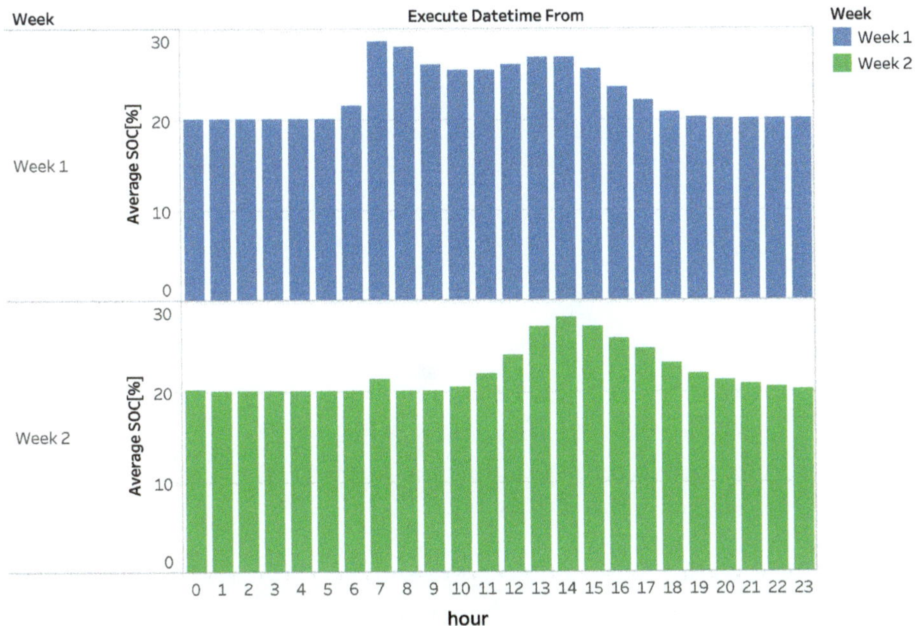

Figure 8. Comparison of average SOC of week 1 and week 2.

6. Conclusions

In this study, we developed an agent system that automatically trades electricity on behalf of users in a hypothetical power P2P trading market where consumers and generators trade with each other. The system not only makes bids based on the user's assets such as electric vehicles and solar power generators, and the user's power demand, but also takes into account the user's orientation toward renewable energy and aims to enable power trading tailored to the individual user's situation. The novelty of our research lies in two major aspects.

① The fact that individuals optimize their bidding by using energy storage facilities with the aim of maximizing profits and satisfying their own preferences regarding the ratio of renewable energy.

② By developing a bidding strategy that takes into account the individual's preference for renewable energy, we have achieved both economic efficiency and satisfaction of the individual's preference.

In order to consider users' preferences for renewable energy and costs, we developed two modes in the creation of agent bids, a green mode oriented toward achieving the desired renewable energy ratio, and an economy mode oriented toward economic efficiency, and we conducted a demonstration experiment of P2P electricity trading. In the demonstration experiment, the following results were obtained.

- The user who owns an EV and is economically oriented can trade at a lower average price than the users who do not own EVs.
- It is possible to increase the economic efficiency of storing nighttime electricity in batteries in advance by anticipating future power generation and market prices through forecasting.
- Users who own EVs and have a preference for renewable energy can achieve a high RE ratio at a cost of about +1 yen/kWh compared to other users.

- In real scale experiments, it is possible to control charging and discharging by actually issuing charging and discharging commands to electric vehicles, and to optimize the actual use of electricity according to the user's preferences.

As for future prospects, we envision devising optimization including the operation of electric water heaters as energy management that takes into account not only energy storage facilities but also demand-shifting devices, and conducting demonstration experiments. In terms of system configuration, the smart contract function of the Ethereum blockchain is currently being used to implement the electricity trading market but speeding up the processing of this function will be an issue in the future. Possible solutions include executing the market execution process in a system outside the blockchain and recording only the matching results in the blockchain.

Author Contributions: Conceptualization, D.S., K.T. and F.I.; methodology, D.S.; software, M.N., S.N. and N.A.; writing—original draft preparation, D.S.; project administration; K.T., F.I., H.S., N.T., K.S. All authors have read and agreed to the published version of the manuscript.

Funding: This research received no external funding.

Institutional Review Board Statement: Not applicable.

Informed Consent Statement: Not applicable.

Acknowledgments: The second and the third authors are supported by Grant-in-Aid for Scientific Research (A) 20H00285.

Conflicts of Interest: The authors declare no conflict of interest.

References

1. European Union. 2030 Climate & Energy Framework. Available online: https://ec.europa.eu/clima/eu-action/climate-strategies-targets/2030-climate-energy-framework_en (accessed on 29 November 2021).
2. European Union. European Climate Law. Available online: https://ec.europa.eu/clima/eu-action/european-green-deal/european-climate-law_en (accessed on 29 November 2021).
3. Abdella, J.; Shuaib, K. Peer to Peer Distributed Energy Trading in Smart Grids: A Survey. *Energies* **2018**, *11*, 1560. [CrossRef]
4. Sousa, T.; Soares, T.; Pinson, P.; Moret, F.; Baroche, T.; Sorin, E. Peer-to-peer and community-based markets: A comprehensive review. *Renew. Sustain. Energy Rev.* **2019**, *104*, 367–378. [CrossRef]
5. Morstyn, T.; Teytelboym, A.; McCulloch, M.D. Bilateral Contract Networks for Peer-to-Peer Energy Trading. *IEEE Trans. Smart Grid* **2019**, *10*, 2026–2035. [CrossRef]
6. Sorin, E.; Bobo, L.; Pinson, P. Consensus-Based Approach to Peer-to-Peer Electricity Markets with Product Differentiation. *IEEE Trans. Power Syst.* **2019**, *34*, 994–1004. [CrossRef]
7. Mengelkamp, E.; Gärttner, J.; Rock, K.; Kessler, S.; Orsini, L.; Weinhardt, C. Designing microgrid energy markets: A case study: The Brooklyn Microgrid. *Appl. Energy* **2018**, *210*, 870–880. [CrossRef]
8. Shrestha, A.; Bishwokarma, R.; Chapagain, A.; Banjara, S.; Aryal, S.; Mali, B.; Thapa, R.; Bista, D.; Hayes, B.P.; Papadakis, A.; et al. Peer-to-Peer Energy Trading in Micro/Mini-Grids for Local Energy Communities: A Review and Case Study of Nepal. *IEEE Access* **2019**, *7*, 131911–131928. [CrossRef]
9. Liu, T.; Tan, X.; Sun, B.; Wu, Y.; Guan, X.; Tsang, D.H.K. Energy management of cooperative microgrids with P2P energy sharing in distribution networks. In Proceedings of the 2015 IEEE International Conference on Smart Grid Communications (SmartGridComm), Miami, FL, USA, 2–5 November 2015; pp. 410–415.
10. Longo, A.; Markandya, A.; Petrucci, M. The internalization of externalities in the production of electricity: Willingness to pay for the attributes of a policy for renewable energy. *Ecol. Econ.* **2008**, *67*, 140–152. [CrossRef]
11. Zorić, J.; Hrovatin, N. Household willingness to pay for green electricity in Slovenia. *Energy Policy* **2012**, *47*, 180–187. [CrossRef]
12. Tabi, A.; Hille, S.L.; Wüstenhagen, R. What makes people seal the green power deal?—Customer segmentation based on choice experiment in Germany. *Ecol. Econ.* **2014**, *107*, 206–215. [CrossRef]
13. Dagher, L.; Harajli, H. Willingness to pay for green power in an unreliable electricity sector: Part 1. The case of the Lebanese residential sector. *Renew. Sustain. Energy Rev.* **2015**, *50*, 1634–1642. [CrossRef]
14. Morstyn, T.; McCulloch, M.D. Multiclass Energy Management for Peer-to-Peer Energy Trading Driven by Prosumer Preferences. *IEEE Trans. Power Syst.* **2019**, *34*, 4005–4014. [CrossRef]
15. Park, D.-H.; Park, J.-B.; Lee, K.Y.; Son, S.-Y.; Roh, J.H. A Bidding-Based Peer-to-Peer Energy Transaction Model Considering the Green Energy Preference in Virtual Energy Community. *IEEE Access* **2021**, *9*, 87410–87419. [CrossRef]
16. Park, D.-H.; Park, Y.-G.; Roh, J.-H.; Lee, K.Y.; Park, J.-B. A Hierarchical Peer-to-Peer Energy Transaction Model Considering Prosumer's Green Energy Preference. *Int. J. Control. Autom. Syst.* **2021**, *19*, 311–317. [CrossRef]

17. Kobashi, T.; Yoshida, T.; Yamagata, Y.; Naito, K.; Pfenninger, S.; Say, K.; Takeda, Y.; Ahl, A.; Yarime, M.; Hara, K. On the potential of "Photovoltaics + Electric vehicles" for deep decarbonization of Kyoto's power systems: Techno-economic-social considerations. *Appl. Energy* **2020**, *275*, 115419. [CrossRef]

18. Zepter, J.M.; Lüth, A.; del Granado, P.C.; Egging, R. Prosumer integration in wholesale electricity markets: Synergies of peer-to-peer trade and residential storage. *Energy Build.* **2019**, *184*, 163–176. [CrossRef]

19. Iria, J.; Soares, F.; Matos, M. Optimal bidding strategy for an aggregator of prosumers in energy and secondary reserve markets. *Appl. Energy* **2019**, *238*, 1361–1372. [CrossRef]

20. Iria, J.; Soares, F.; Matos, M. Optimal supply and demand bidding strategy for an aggregator of small prosumers. *Appl. Energy* **2018**, *213*, 658–669. [CrossRef]

21. Vagropoulos, S.I.; Bakirtzis, A. Optimal Bidding Strategy for Electric Vehicle Aggregators in Electricity Markets. *IEEE Trans. Power Syst.* **2013**, *28*, 4031–4041. [CrossRef]

22. Englberger, S.; Gamra, K.A.; Tepe, B.; Schreiber, M.; Jossen, A.; Hesse, H. Electric vehicle multi-use: Optimizing multiple value streams using mobile storage systems in a vehicle-to-grid context. *Appl. Energy* **2021**, *304*, 117862. [CrossRef]

23. Meteorological Engineering Center, Inc. Solar Power Output Forecasting System Apollon. Available online: https://www.meci.jp/apollon.html (accessed on 29 November 2021).

24. Breiman, L. Random forests. *Mach. Learn.* **2001**, *45*, 5–32. [CrossRef]

25. The Kansai Electric Power Co., Inc. Hapi-e-Time R. Available online: https://kepco.jp/ryokin/menu/hapie_r/ (accessed on 29 November 2021).

Article

Effectiveness and Feasibility of Market Makers for P2P Electricity Trading

Shinji Kuno [1], Kenji Tanaka [2] and Yuji Yamada [3,*]

1 Commodity Business Department, Sumitomo Corporation, Tokyo 100-8601, Japan; s.k.utokyo.l26@gmail.com
2 Department of Technology Management for Innovation, The University of Tokyo, Tokyo 113-8656, Japan; tanaka@tmi.t.u-tokyo.ac.jp
3 Faculty of Business Sciences, University of Tsukuba, Tokyo 112-0012, Japan
* Correspondence: yuji@gssm.otsuka.tsukuba.ac.jp

Abstract: Motivated by the growing demand for distributed energy resources (DERs), peer-to-peer (P2P) electricity markets have been explored worldwide. However, such P2P markets must be balanced in much smaller regions with a lot fewer participants than centralized wholesale electricity markets; hence, the market has inherent problems of low liquidity and price instability. In this study, we propose applying a market maker system to the P2P electricity market and developing an efficient market strategy to increase liquidity and mitigate extreme price fluctuations. To this end, we construct an artificial market simulator for P2P electricity trading and design a market agent and general agents (photovoltaic (PV) generators, consumers, and prosumers) to perform power bidding and contract processing. Moreover, we introduce market-maker agents in this study who follow the regulations set by a market administrator and simultaneously place both sell and buy orders in the same market. We implement two types of bidding strategies for market makers and examine their effects on liquidity improvement and price stabilization as well as profitability, using solar PV generation and consumption data observed in a past demonstration project. It is confirmed that liquidity and price stability may be improved by introducing a market maker although there is a trade-off relationship between these effects and the market maker's profitability.

Keywords: P2P electricity market; market maker; liquidity; price fluctuation; bidding strategy; artificial market simulation

---

Citation: Kuno, S.; Tanaka, K.; Yamada, Y. Effectiveness and Feasibility of Market Makers for P2P Electricity Trading. *Energies* **2022**, *15*, 4218. https://doi.org/10.3390/en15124218

Academic Editor: Peter V. Schaeffer

Received: 20 April 2022
Accepted: 4 June 2022
Published: 8 June 2022

Publisher's Note: MDPI stays neutral with regard to jurisdictional claims in published maps and institutional affiliations.

1. Introduction

Since environmental issues have been attracting worldwide attention, the Japanese government declared that it would achieve a decarbonized society by 2050. However, feed-in tariffs, which have been functioning as incentives for the introduction of renewable energy, are now being scaled down or abolished. Therefore, as a new incentive system, decentralized peer-to-peer (P2P) power trading based on microgrids is being actively explored in many countries. Various studies on P2P trading have already been conducted in the form of proof-of-concept experiments and numerical simulations [1–9] as well as investigating market mechanisms [10–14] and social implementations [15–17]. These studies have verified the effectiveness of P2P electricity trading from technical, environmental, and profitability perspectives; however, at the same time, they revealed some potential problems inherent in this market. One is extreme price fluctuation (or price volatility) caused by low market liquidity, where liquidity refers to the bidding amount in an order book. If liquidity is sufficiently high, the market price is robust to a large market order, which could significantly influence market situations because the accumulated bidding amount buffers the impact of market orders. However, liquidity tends to decrease in the P2P market because of the market's specific characteristics. First, the entire market comprises consecutive 30-min markets divided by region and time slots. Furthermore, the trading volume per participant is much smaller than that in the wholesale market because the P2P market participants are normal households or non-electric companies. For these reasons, large price fluctuations, attributed to low liquidity, can easily arise. During a proof-of-concept demonstration of P2P power trading conducted in the Urawa-Misono District, Japan (see [18] for a summary of the project and its results), excessive price fluctuations in a short time period were often detected (see Section 2). Price volatility problems are becoming a prime concern across the entire electric power industry owing to various factors, such as system shifts, abnormal climates, and soaring resource prices. Moreover, in infrastructural industries, a stable supply is of utmost importance; therefore, this volatility problem should be handled on a highest-priority basis. In this study, we introduce market makers into the P2P electricity market. Market makers are market participants who contribute to liquidity improvement and price stabilization in an exchange market (see, e.g., [19] for a market maker program introduced in the Japan Exchange Group, Inc. (Tokyo, Japan)) while securing their own profit. The objective of this study is to design bidding strategies for market makers and test them through several simulations.

Here, we introduce related studies on the application of financial functions to electricity markets. Electricity markets have stricter restrictions than other assets; for instance, electricity cannot be stored and must be generated at the time of demand. Hence, electricity market-specific solutions can be invented as follows. First, optimal bidding strategies for electricity markets have been developed in several studies [20–29]. In these references, Baltaoglu et al. [29], for example, proposed a type of arbitrage called "virtual bidding". In existing wholesale electricity markets, participants have two different markets for one specific product or a 30-min electricity delivery period: a day-ahead market and an hour-ahead market. Virtual bidding aims to make profits through buying in a day-ahead market and selling in an hour-ahead market or selling in a day-ahead market and buying back in an hour-ahead market. If the same amounts are executed for selling and buying, that is, a position is established between the two markets, then the price difference would be the profit for this strategy. Next, electricity and weather derivatives (including forwards/futures) may be considered practical applications of derivatives theory for real businesses in electricity markets [30–43]. Among them, Yamada and Matsumoto (2021) [41] and Matsumoto and Yamada (2021) [42,43] advocated weather derivatives, the payments of which depend on weather data at a predetermined place and time. Electricity utilities are constantly exposed to fluctuation risks in solar power generation and electricity demand, which are associated with solar radiation, temperature, etc. These factors greatly influence electricity prices and, in turn, their profits; therefore, implementing measures against weather forecast uncertainty is a major focus for the power industry. With a system that allows these businesses to receive insurance coverage for losses incurred by the deviation of a weather index from a predetermined range, they can hedge profit risks and stabilize their management. This type of electricity insurance has already been developed and commercialized. Finally, there are several studies on market makers in the context of enhancing liquidity and price stability for electricity markets [44–48], which are the focus of this study. For example, Bose et al. (2014) [47] explore market makers' impacts on social welfare, residual social welfare, and consumer surplus at the general Nash equilibrium in a Cournot competition model. In addition, Worthmann et al. [48] examine market makers' effect to mitigate the negative influences that come with the development of distributed electricity generation using real data from Australia.

In this study, we propose introducing market makers to solve the price volatility problem inherent in P2P markets by improving liquidity. Market makers always quote both selling and buying prices and are willing to trade at those prices at any time. Their main purpose is to make a profit; however, by repeatedly trading in large volumes, they provide liquidity to the market. This system has already been introduced in conventional markets such as stocks and commodities, e.g., [19], and its role is undertaken primarily by securities companies. To evaluate the market maker system's effectiveness against the price volatility problem, we develop an artificial P2P electricity market simulator in Python, emulating the market and participant specifications employed in the Urawa-Misono proof-of-concept project [11]. We compare and evaluate the results of three simulation case studies with and without a market maker, in which two types of market maker agents are adopted based on the market maker rules developed in [49,50] for stock trading.

This paper is organized as follows: In Section 2, we provide a detailed explanation of the motivational Urawa-Misono demonstration project and the market rules adopted in this study; in Section 3, we explain the basic configuration, information flowchart, and role of each agent in the artificial market simulation conducted in this study; in Section 4, we perform artificial market simulations based on the actual generation and demand data and compare the cases with and without market makers; Section 5 provides a comprehensive discussion based on the results of our analysis; and Section 6 provides concluding remarks and describes future research directions.

2. Motivative Experiment and Market Rules

2.1. Motivative Demonstration Project and Potential Problem

This study is motivated by the results of a P2P electricity trading demonstration project conducted in the Urawa-Misono District, Japan, from August 2019 to March 2020, which was summarized in [18]. Figure 1 shows the system used in the demonstration experiment. The market participants are photovoltaic generators (PVGs), convenience stores (consumers), residences (prosumers), and the power grid agent, which offers electricity at a price of 30 JPY/kWh or higher than that of the wholesale electricity market (JEPX). In the project, each facility with a digital grid controller (DGC) is supposed to submit orders to the market via 3G networks, where DGCs are programmable devices for reading smart meters and scheduling orders. The order schedules for the PVGs and convenience stores are determined based on the forecasted generation and demand, whereas those for the residences are decided by referring to the state-of-charge (SoC) of their battery storage. The selling and buying of orders are executed in the market according to the principle of time and price priority, and all of these activities are automatically processed and registered by the Ethereum-based blockchain ledger using smart contract programs [51,52].

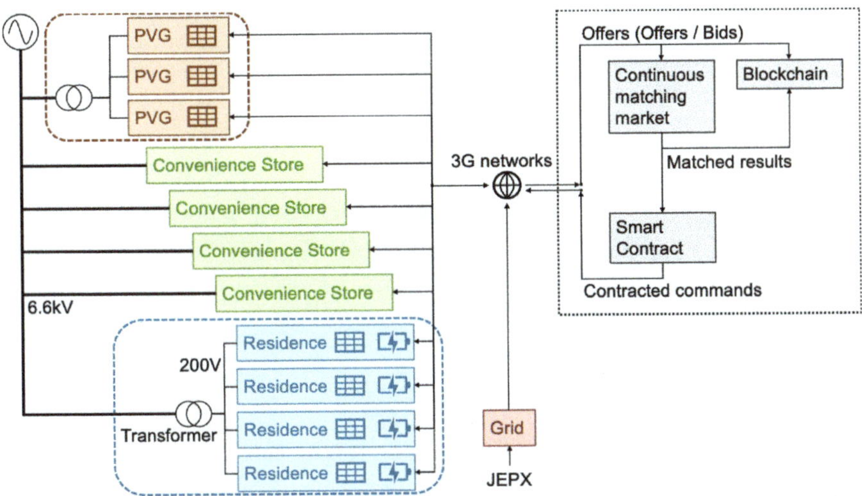

Figure 1. System for the Urawa-Misono demonstration project (reproduced with permission from Kontani et al. [18], under CC-BY license from MDPI (Basel, Switzerland), 2021).

This project was successful in the sense that it proved by quantitatively analyzing transaction data that both the selling and buying sides can benefit from trading in the P2P market. The project also showed that holding storage batteries enables owners to make advantageous contracts. However, some challenges that need to be addressed to implement the P2P electricity market in real society have also been identified. One of these potential problems is low market liquidity and the subsequent high price volatility. The proof-of-concept demonstration project in Urawa-Misono often denotes this tendency in the automatically collected data. Figure 2 shows the execution price trend for a 30-min electricity delivery period and illustrates that the electricity price sharply increases and decreases in the last two hours of the bidding period.

Figure 2. Example of large price fluctuations in the P2P electricity trading demonstration conducted in Japan. The horizontal axis denotes the time to the end of the trading period.

To observe that low liquidity can result in substantial price volatility, we consider an illustrative example of order books for electricity trading shown in Figure 3, in which the left order book has low liquidity, which means that there are few limit orders, and the right order book has high liquidity, with many orders around the current market price of 25 JPY/kWh. If someone attempts to purchase 50 kWh at once as a market order, the market electricity price soars up to 30 JPY/kWh in the low liquidity case; on the other hand, in the right high-liquidity case, all of the 50 kWh power can be procured at the same price of 25 JPY/kWh. Therefore, the difference in liquidity cost, or the cost caused by low liquidity, is 183 JPY. If liquidity costs remain high, as in the left-side case, this would be an obstacle for households and companies to enter the P2P electricity market, which, in turn, would prevent the growth of the renewable energy industry.

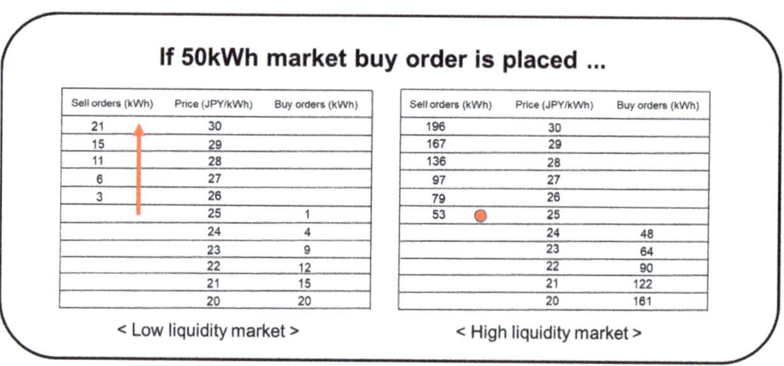

Sell orders (kWh)	Price (JPY/kWh)	Buy orders (kWh)
21	30	
15	29	
11	28	
6	27	
3	26	
	25	1
	24	4
	23	9
	22	12
	21	15
	20	20

< Low liquidity market >

Sell orders (kWh)	Price (JPY/kWh)	Buy orders (kWh)
196	30	
167	29	
136	28	
97	27	
79	26	
53	25	
	24	48
	23	64
	22	90
	21	122
	20	161

< High liquidity market >

Figure 3. An illustrative example of order books for electricity trading. The left order book has low liquidity, which means that there are few limit orders, and the right order book has high liquidity, with many orders around the current market price of 25 JPY/kWh. If someone attempts to purchase 50 kWh at once as a market order, the market electricity price soars up to 30 JPY/kWh in the low liquidity case; on the other hand, in the right high-liquidity case, all of the 50 kWh power can be procured at the same price of 25 JPY/kWh.

To solve this liquidity problem, some stock or commodity exchanges introduce liquidity-supplying market players called "market makers" by specifying market maker programs (see, e.g., [19] for the market maker program introduced by the Japan Exchange Group, Inc.). The market maker role is typically undertaken by financial institutions, and they are supposed to quote both selling and buying orders and accept deals with any other market participants while following predetermined rules, such as volumes and prices. If market makers are introduced into the left-side market in Figure 3, the market environment will become closer to the right side. In this study, we propose bringing this market maker system into the P2P electricity market to enhance liquidity and mitigate extreme price fluctuations.

2.2. Market Rules

In general, the most important rule that must be fulfilled in electricity markets is "balancing". Balancing means that supply must always be matched with demand at any time interval. In this study, we assume that supply (i.e., power generation) and demand (i.e., electricity consumption) are balanced in every 30-min window. To implement this principle, a day is divided into 48 time slots (see Figure 4), and power sales contracts are traded between a seller and buyer for each 30-min time span. Thus, 48 products can be defined per day, and the market for each product opens 24 h before the start of the 30-min period and closes 10 min before the end of the window; that is, all participants can bid for 24 h and 20 min for each time frame.

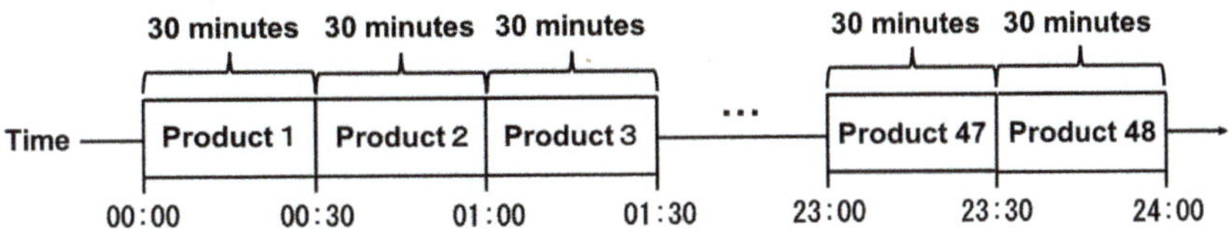

Figure 4. A total of 48 products of 30-min periods in electricity markets.

Next, we introduce the market rules adopted in our P2P market simulation, which is inspired by the demonstration project stated in the previous subsection and emulates actual electricity markets, such as the Japan Electric Power Exchange:

- The market for each product is assumed to open 24 h before the start of the 30-min electricity delivery period and close 10 min before the end of the time interval. In other words, the market accepts orders from participants for 24 h and 20 min.
- Orders in the book are executed in continuous sessions or according to the principle of price and time priority. Specifically, offers with prices lower than those of bids and bids with prices higher than those of offers are executed immediately, whereas other orders remain on the board.
- If the matched offer and bid volumes are different, the executed amount is adjusted to a smaller value.
- If multiple orders at the same price exist on the board, the earliest order is prioritized.

In P2P electricity trading markets, there exist several types of market participants: "generators," who simply sell electricity and do not consume it, "consumers," who only buy electricity and do no generate it, and "prosumers," who both generate and consume electricity themselves and offset the surplus or shortage by trading as sellers and buyers. These roles are primarily played by ordinary households and corporations. As they are not professional traders, the quantity and price of their orders cannot be determined manually. Instead, these numbers are automatically calculated based on power generation, demand data, contract history, and so on. Once bidding information is set, it is submitted to the market.

In addition, we introduce "market makers" as additional liquidity-supplying players. They themselves do not generate or consume electricity, but they continue quoting both sell and buy order prices with a specified spread while following regulations set by the market administrator (or a managing organization of an exchange). Market makers are usually introduced into a market in which investors do not need to hold real assets, such as stock markets, because they ultimately close their positions and make profits through the price spread mentioned above. However, in the P2P electricity market, participants must deliver and receive actual electricity every 30 min. Therefore, to apply this system to the P2P market, we assume that market makers own storage batteries to carry surplus electricity to the later product periods on the same or the next day.

In addition, agents can bid on up to 48 products at every bidding turn, as shown in Figure 5. Moreover, in this simulation, all orders are formally sent to the market as a limit order, but when the order matches another at the same time as bidding, it is virtually regarded as a market order.

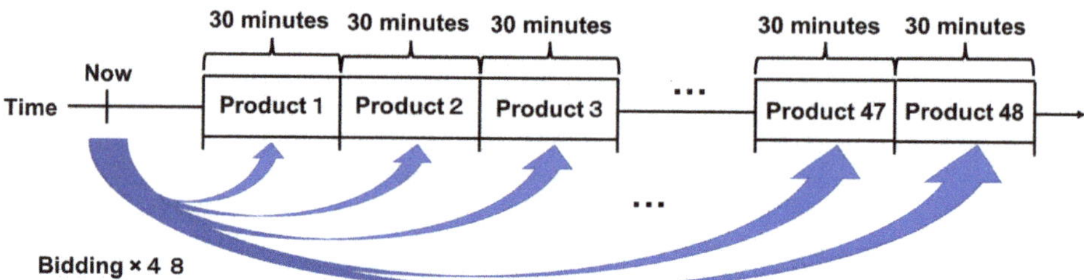

Figure 5. Simultaneous bidding on multiple markets.

3. Development of Artificial Market Simulation System

In this section, we explain the basic configuration, information flowchart, and role of each agent in our artificial market simulation.

3.1. Basic Configuration and Information Flow

Figures 6 and 7 illustrate the basic configurations and the bidding procedures with and without market makers, including the order of agents' bids in the simulators. In the case without market makers shown in Figure 6, its components can be divided into two categories. The first is the "market agent", in which orders are collected and executed. The second are the "participant agents", which automatically determine order quantities and prices and send orders to the market agent. We design both the market agent and the participant agents and construct P2P power market simulators that perform power bidding and contract processing. In the simulation, time proceeds by 10 min, and each general agent (or supply and demand agent) updates its order once every 10 min (per product) in random order, one after another.

In addition, the configuration of the simulators after the introduction of market maker agents is shown in Figure 7. In this case, general agents place their orders once every 10 min (per a product) just as the case described above, and every time one of the general agents bids, the market maker agent immediately cancels their unfilled previous orders and rebids limit orders with reference to the latest board status. This is because actual market makers update their order prices at high speed while continuously referring to limit orders in the market and conducting high-frequency trading.

In this paper, we introduce two types of market maker agents, namely the simple market maker and the flexible market maker, focusing on profitability improvement. Then we compare several simulation results with and without the market maker agents and examine their impact on the market as well as their profitability. The market maker agents are explained in detail in later subsections.

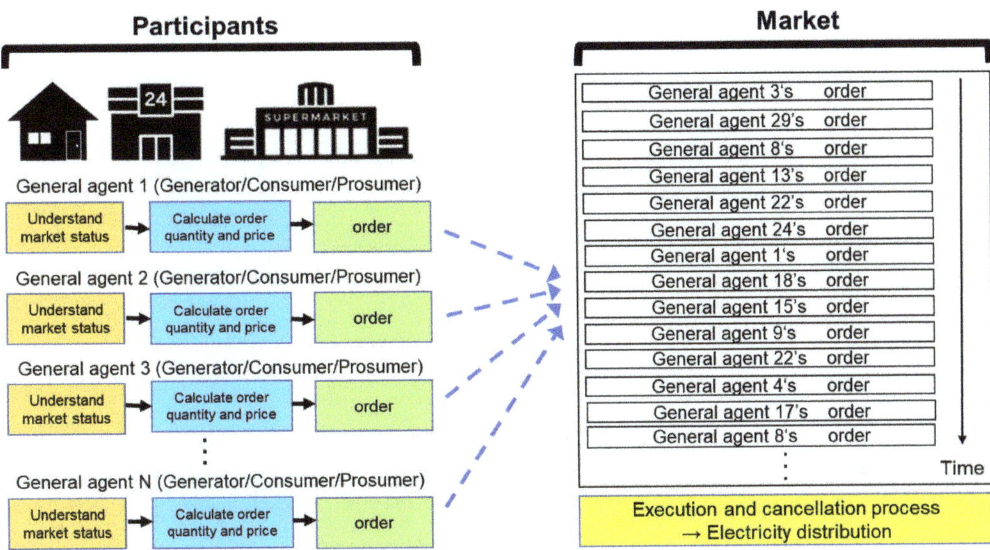

Figure 6. P2P electricity market before the introduction of market makers.

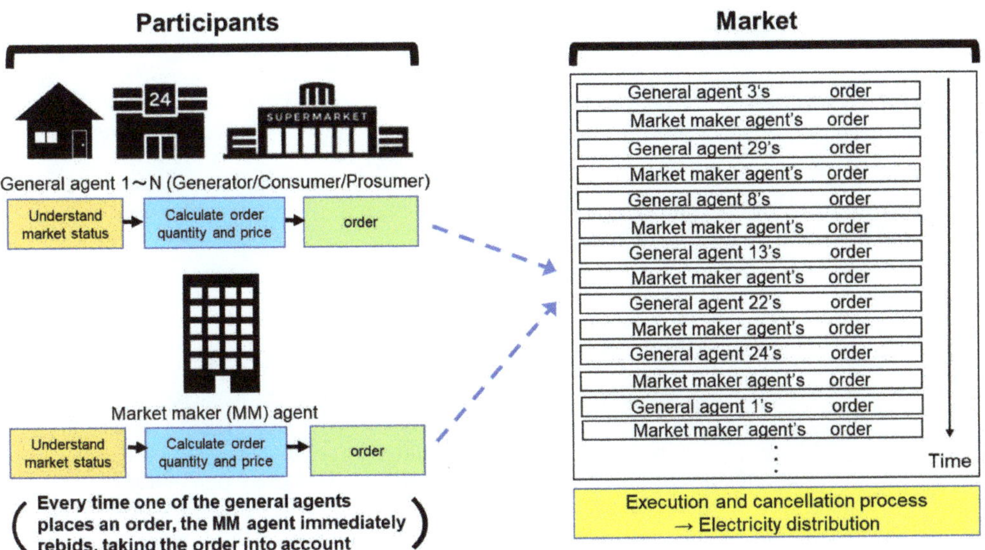

Figure 7. P2P electricity market after the introduction of market makers.

Figure 8 shows the information flowchart of this simulation. This design concept is based on the artificial market simulator constructed by Waseda and Tanaka [53], in which all agents repeat the process of determining and placing orders for all products available at each point in time. The final output of the simulation includes the final order board status, the order record, the execution record, and the bid/ask spread record during the entire simulation period.

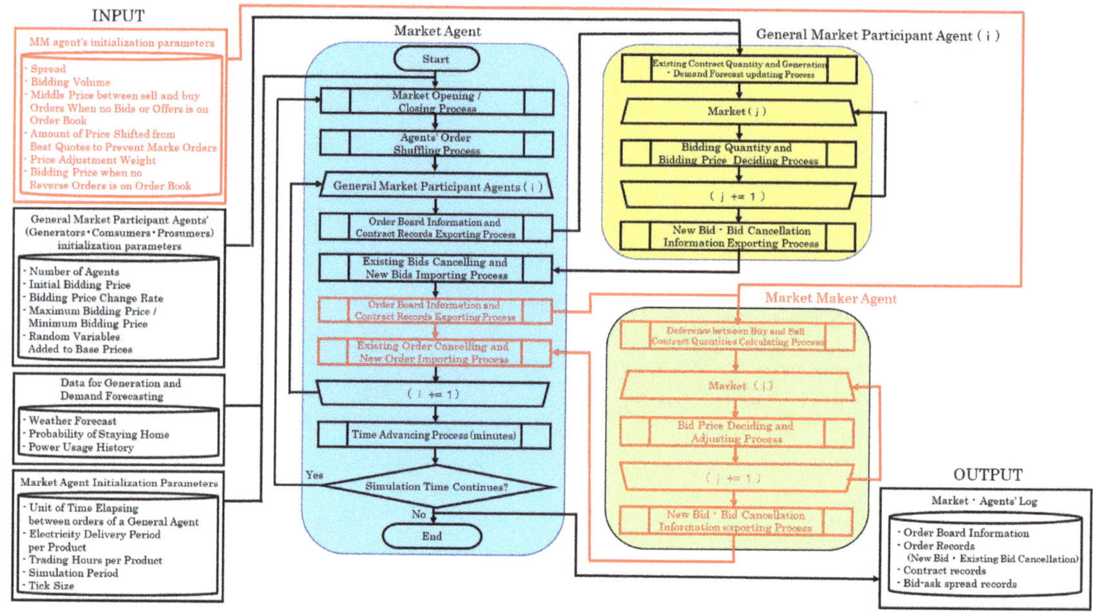

Figure 8. Information flow in the simulation system.

The blue portion, called the "market agent" in the simulation diagram below, plays a role in market management. It opens and closes markets for all products and accepts and processes orders following the market rules described in Section 2. The main functions of the market agent are as follows:

Function (1): Update the time in the simulation and open/close markets.

Function (2): Shuffle the bidding order of general agents as time proceeds to ensure fairness among them.

Function (3): Send market information, such as order data and contract records to each general agent.

Function (4): Receive orders from general agents, perform contract processing sequentially, and write the results on the order board, order record, contract record, etc.

Next, we describe the general (market participant) agents shown in yellow in Figure 8 which reflect the actual demand and supply data. These agents may be referred to as "generators", "consumers", or "prosumers. Their main functions are as follows:

Function (1): Each agent must estimate its photovoltaic (PV) power generation or electricity consumption from the weather forecast data and/or past power consumption history.

Function (2): Determine the order quantity and price for each product by considering the contract record and the elapsed time from the market opening received from the market agent.

Function (3): Update order information and send it to the market agent.

Note that information types and values used as initial parameters may be different according to agent types (i.e., generator, consumer, or prosumer) and preferences (i.e., price-oriented type, moderate type, or certainty-oriented type). This reflects the diverse needs of general market participants. In addition, regarding Function (3), each agent is programmed to cancel the unfilled previous bids and renew their orders at every turn. Appendix A provides a detailed explanation of the bidding strategies of general agents.

Finally, we explain the market maker agent shown in the green portion in Figure 8. Generally, market makers provide liquidity to markets by continuously placing both sell and buy orders while complying with various rules set by market administrators. As a result, liquidity costs are kept lower even when a large order is placed at once. Due to reduced risk, investors would be able to enter the P2P market more easily, resulting in increased trading volume and further market development. As this research focuses on how liquidity and price stability improve when market makers are introduced, we have to compare them before and after the introduction of market makers. For this reason, the strategies of general agents are kept unchanged in all cases. However, in reality, it is quite possible that the existence of market makers influences the bidding strategies of general agents. Therefore, the development of reactive bidding strategies of general agents towards the introduction of market makers could be research topics in future studies. In this paper, we first introduce the bidding strategy of the simple market maker and then describe that of the flexible market maker.

3.2. Market Maker's Bidding Strategies

3.2.1. Bidding Strategy of the Simple Market Maker

Recall that the original objective of introducing a market maker is to enhance liquidity and stabilize electricity prices. This may be achieved by adopting the simple market maker outlined below (see [49,50] for the introduction of a simple market maker into stock markets). The bidding price determination method, including its assumptions of this market maker, is illustrated in Figure 9 and is described as follows:

- The market maker derives the best quotes on the order board in each bidding turn, that is, the lowest selling quote and the highest buying quote. The selling and buying prices are then calculated using the middle price between the two best quotes. More exactly, the selling (buying) price is shifted up (down) by half of the specified spread value, θ_{sm}, from the middle price.
- If either or both sell/buy orders do not exist on the board of the market, the middle price cannot be determined. Therefore, in this case, we assume that the middle price is given a priori as an initial parameter. Specifically, we set the initial setting parameter at 25 JPY/kWh in our simulation, which is the mean value of the upper limit price of 50 JPY/kWh and the lower limit price of 0 JPY/kWh.
- In addition, since market makers are supposed to keep quoting both sell and buy prices, they are programmed to always place a limit order, not a market order, in this simulation. Therefore, in case the bidding price of our strategies may result in a market order, it is shifted up or down such that it becomes a limit order.
- That is, sell and buy prices are adjusted in the same direction by the same amount so that the spread is kept constant at the value of an initial parameter in this case as well.

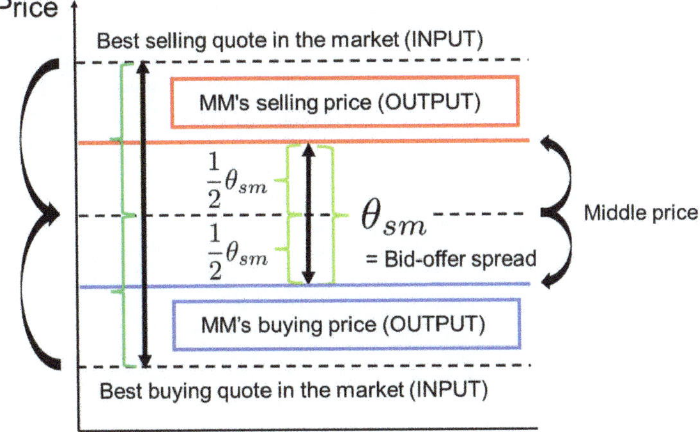

Figure 9. Price determination strategy of the simple market maker when sell and buy orders are both quoted on the order board.

Note that the simple market maker determines buying and selling order prices based on the middle price and a specified spread size θ_{sm} as "middle price $\pm\,\theta_{sm}/2$". If a bid (buy order) and an offer (sell order) are executed in equal quantity, the margin between the two prices, given by θ_{sm}, becomes a source of profits for the simple market maker.

3.2.2. Bidding Strategy of the Flexible Market Maker

As mentioned at the end of the previous subsection, a market maker can profit from the difference between selling and buying prices, whereas there is a risk of loss if the selling and buying executed amounts are unbalanced. For example, if a market maker has a larger sales position than their purchase position, they may need to buy additional electricity at a relatively high price and compensate for the shortage during the delivery period, which is called "imbalance charges" in the system. On the other hand, a larger buying position may also lead to an opportunity loss for the market maker. Hence, market makers always need to maintain their net positions close to zero and try to avoid the position imbalance. This is the reason why we introduce a new bidding price determination algorithm into the flexible market maker, based on price adjustment according to its net position of the moment.

Figure 10 illustrates the bidding algorithm when both buy and sell orders are quoted on the order board. The horizontal axis denotes the net position possessed by the flexible market maker, that is, the total of buying contracts minus that of selling contracts when bidding, and the vertical axis denotes the bidding price of the flexible market maker. The bidding price determination method, including its assumptions, is described as follows:

- The price adjustment is conducted according to the term $\left(1 - w_{fm}\left(s_{fm}^{t}\right)^{3}\right)$, where s_{fm}^{t} is the market maker's net position at time t (i.e., total executed buying volume minus total executed selling volume for all products up to time t) and w_{fm} is a weighting term. The effect based on the net position is reflected when $w_{fm} \neq 0$. For instance, if the selling contract amount is greater than the buying amount, both selling and buying prices are shifted up. When the market maker's position is net long, both selling and buying prices are shifted down according to the term $\left(1 - w_{fm}\left(s_{fm}^{t}\right)^{3}\right)$.

- If the market maker's bid price (buying order price), given by the blue line in Figure 10, were shifted beyond the best selling quote on the board, shown by the horizontal dotted line on the upper side, the order would be executed as a market order. To avoid this and make the bidding a limit order, the buying order price will be fixed just below the best selling quote by ΔP. Similarly, the selling order price will be fixed just above the best buying quote to avoid the selling order becoming a market order.

As a result of introducing the price determination strategy in Figure 10, sell orders are less likely and buy orders more likely to be executed, and the market maker's position may revert to net zero. Note that this strategy is based on the position market-maker strategy described in [49,50]. However, in this paper, we propose a P2P-electricity-market-specialized market maker's strategy using storage batteries. The P2P electricity market, whose power source is largely PV (photovoltaic) generation in Japan, usually has a supply and demand imbalance both in the daytime (when PV generation is larger than demand) and nighttime (when no PV generation occurs). Therefore, market makers have to bear a high risk of not being able to sell and buy in equal amounts. One of the solutions to this problem is installing storage batteries. They could function as a buffer and transfer electricity generated during daytime hours to nighttime periods. At the same time, market makers could avoid imbalance charges by evening up sales and purchase amounts across electricity delivery time intervals. However, storage batteries are still expensive, and the costs could be a heavy burden on market makers.

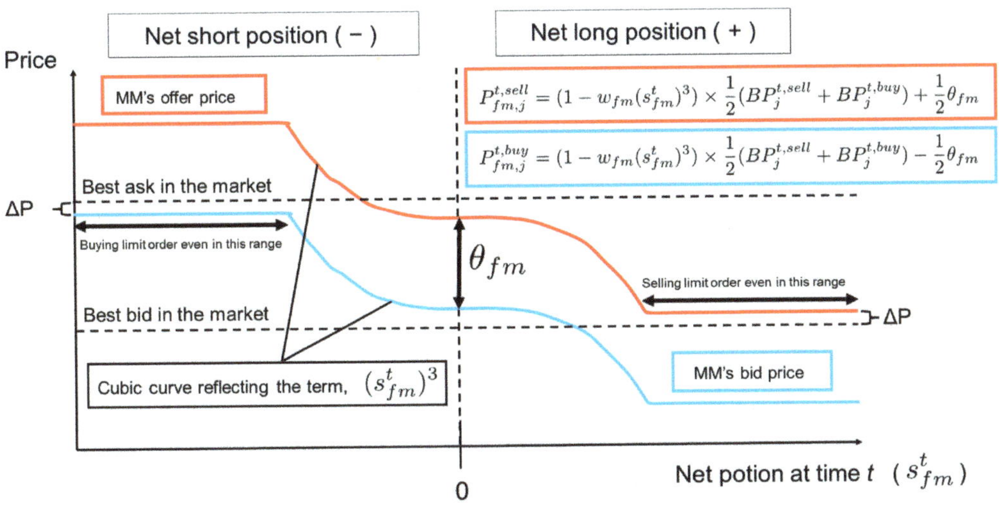

Figure 10. Price determination strategy with respect to the net position at time t, denoted by s_{fm}^{t}, of the flexible market maker when both sell and buy orders are quoted on the order board.

Considering the above discussions, this study adds the following assumptions for market makers to introduce storage batteries:

- To incentivize market makers, we assume that market makers may place an order at a favorable price when there is no other selling or buying order on the board. For example, when no other selling orders exist, which often happens at night or in the early morning with no solar power generation, the market maker can make a sell order at a relatively high price (e.g., 33 JPY/kWh in our simulation) because the market maker is the sole seller in the entire market.

- For the opposite-side order, a buy order in this case, the market maker is assumed to use the same price as the previous bidding, $P_{fm,j}^{t-1,buy}$. On the other hand, with no other buying orders, the market maker places a buy order at a relatively low price (e.g., 17 JPY/kWh in our simulation) while using the same selling price as the previous bidding, $P_{fm,j}^{t-1,sell}$. In either case, the spread between the selling and buying prices may become wider than θ_{fm}.

- When calculating s_{fm}^{t} in Figure 10, the executed volume at the above two particular prices, i.e., the selling amount at the price of 33 JPY/kWh and the buying amount at 17 JPY/kWh, is excluded to avoid the effects of the extreme imbalances during these periods.

4. Artificial Market Simulation Using Supply and Demand Data

In this section, we demonstrate the artificial market simulation using actual solar PV generation and consumption data. The entire simulation was performed in a Python environment.

4.1. Supply and Demand Data

The supply and demand data used in this study are the solar PV generation log and the power consumption log of five residential households for one day (24 h), related to the demonstration project explained in Section 2 (see Table 1 below for the description of the data). From this dataset, both or either of the two types of logs is randomly assigned to each agent (a generator, consumer, or prosumer agent). Note that these agents are originally supposed to keep predicting their generation or consumption amount (or both of them) while the market is open. However, the main focus of this study is to confirm the effectiveness of the market-maker system, and thus, we assumed that the agents' predictions are given by the same values as the actual observations, for simplicity.

Table 1. Data used in the artificial market simulation.

Items	Content
Data category	Generation and demand (kWh)
Category of participant	Residencial household
Number of households	Five
Weather	Sunny
Period	One day (24 h; 0:00–24:00)
Measurement interval	5 min
Amount of data	1440 for generation and 1440 for demand

Here, there are two important points to note. First, we normalized the total daily power generation and consumption per household to 100 kWh. In addition, because the power supply-demand ratio greatly affects the profitability of the market maker, we prepared four patterns of the ratio to test this effect (see Section 5).

The second point concerns pre-processing of the data. In the simulation, the 30-min value, which is the time interval of electricity delivery for a product, is required, although the original data consisted of 5-min values of power generation and demand. Therefore, the original data were appropriately adjusted to meet the specifications. Figure 11 shows the PV generation and consumption of five households which are converted to 30-min values and adjusted to a total of 100 kWh.

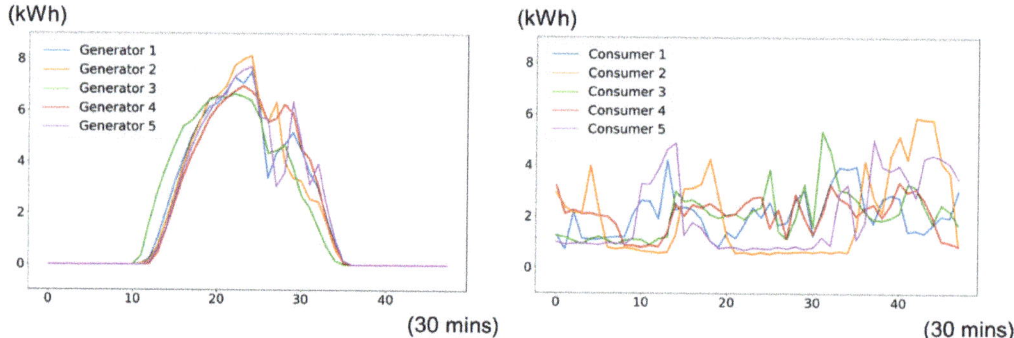

Figure 11. PV generations (**left**) and power consumptions (**right**) of five households were converted to 30-min values and adjusted to a total of 100 kWh.

In this study, P2P electricity market simulations are conducted for the following three scenarios:

Case 1. P2P market simulation without market makers.

Case 2. P2P market simulation with the simple market maker that focuses only on market liquidity and electricity price stability.

Case 3. P2P market simulation with the flexible market maker that considers its profitability, not only market liquidity and electricity price stability.

4.2. Case 1: Without Market Makers

First, we describe the initial parameters of the P2P electricity market simulation before introducing market makers. The initial parameters of the general agents (generators, consumers, and prosumers) and the market agent shown in the INPUT section of Figure 8 are summarized in Table 2 below. In addition, because this study focuses on the comparison between cases with and without market makers, these preconditions will be inherited in the simulations of Cases 2 and 3.

Table 2. Initial parameters in P2P electricity market simulation (see Appendix A for the definitions of the three types of general agents: price-oriented type, moderate type, and certainty-oriented type, and random variables added to base prices).

	Items	Values
General agents	Number of agents	18 agents in total = 6 generators + 6 consumers + 6 prosumers (6 agents for each = 2 price-oriented-type agents + 2 moderate-type agents + 2 certainty-oriented-type agents)
	Total generation per day and Total demand per day	(Generator) Generation: 100 kWh/day Demand: 0 kWh/day (Consumer) Generation: 0 kWh/day Demand: 100 kWh/day (Prosumer) Generation: 100 kWh/day Demand: 100 kWh/day
	Initial bidding price	For generators and prosumers' sell orders Price-oriented type: 35 JPY/kWh Moderate type: 31 JPY/kWh Certainty-oriented type: 27 JPY/kWh For consumers and prosumers' buy orders Price-oriented type: 15 JPY/kWh Moderate type: 19 JPY/kWh Certainty-oriented type: 23 JPY/kWh
	Bidding price change rate	For generators and prosumers' sell orders Price-oriented type: −0.0139 JPY/kWh/min Moderate type: 0.0083 JPY/kWh/min Certainty-oriented type: −0.0028 JPY/kWh/min For consumers and prosumers' buy orders Price-oriented type: 0.0139 JPY/kWh/min Moderate type: 0.0083 JPY/kWh/min Certainty-oriented type: 0.0028 JPY/kWh/min
	Maximum bidding price/ Minimum bidding price	For generators and prosumers' sell orders Price-oriented type: 15 JPY/kWh Moderate type: 19 JPY/kWh Certainty-oriented type: 23 JPY/kWh For consumers and prosumers' buy orders Price-oriented type: 35 JPY/kWh Moderate type: 31 JPY/kWh Certainty-oriented type: 27 JPY/kWh
	Random variables added to base prices	Mean: 0.0 Standard deviation: Normal distribution subject to the conditions below (In the case of "price-oriented-type agent" and "within 10 h after the 30-min delivery period starts") 6.0 (In the case of "moderate-type agent" and "within 10 h after the 30-min delivery period starts") 4.5 (Others) 3.0

Table 2. *Cont.*

	Items	Values
Market agent	Unit of time elapsing between orders of a general agent	10 min
	Electricity delivery period per product	30 min
	Trading hours per product	(Starting time) 24 h before the 30-min delivery period starts (Ending time) 10 min before the 30-min delivery period ends
	Simulation period	2 days (1 day for bidding and 1 day for delivering)
	Tick size	0.01 JPY/kWh

We first performed the artificial market simulation for Case 1, in which we demonstrate the results of trading volume and the mean, maximum, and minimum values of bid-ask spreads. A comprehensive and comparative discussion of all cases is provided in the next section.

Table 3 lists the total volume information provided in the simulation. There, the total tradable volume is 950.3 kWh, which means that if all the orders from the supply and demand agents had been executed, the total executed volume would also have been 950.3 kWh. However, only 268.7 kWh were executed in the simulation without introducing market makers; thus, the execution rate is 28.3%. This is because there are no sell orders at night (from the evening to the early morning) due to a lack of solar power generation, and on the other hand, in the daytime, solar power generation could greatly exceed demand. In other words, PV generation is limited to only daytime, and its generation in the daytime largely depends on weather and climate conditions. This can also be thought of as the reason for low market liquidity in the P2P electricity market.

Table 3. Total volumes in the simulation without market makers.

Total Tradable Volume	Total Executed Volume	Execution Rate
950.3kWh	268.7 kWh	28.3%

The mean, maximum, and minimum values of the bid/ask spread (which is the difference between the best sell and buy prices on the order board) are shown in Table 4. It should be noted that there are time periods with no bid/ask spreads on the order board from the evening to the early morning when PV generation does not occur (e.g., 00: 00–00: 30). In addition, when PV generation greatly exceeds the demand of consumers during the daytime, buy orders may disappear shortly, and the bid/ask spreads cannot be observed afterward.

Table 4. Mean, maximum, and minimum values of bid-ask spreads (without market makers).

Mean	Max	Min
3.96 JPY/kWh	16.00 JPY/kWh	0.01 JPY/kWh

4.3. Case 2: Introduction of the Simple Market Maker

We then describe the initial parameters set in the simulation with the simple market maker (see Table 5). Only the additional parameters with respect to the simple market maker are shown because the items used in Case 1 remain unchanged.

Table 5. Initial parameters in P2P electricity market simulation with the simple market maker.

	Items	Values
	Spread	3.00 JPY/kWh
Simple market maker agent	Bidding volume	(Sell volume) 10.0Wh (Buy volume) 10.0 kWh
	Reference middle price between sell and buy orders when neither bid nor offer is on the order book	25.00 JPY/kWh
	The amount of price shift from the best quotes to prevent market orders	0.01 JPY/kWh

As shown in Table 6, the total executed volume is 950.3 kWh, which is 3.54 times higher than 268.7 kWh in Case 1 and indicates that all bids by general agents have been executed; market liquidity has greatly improved from the perspective of the trading volume. This is because the market maker is assumed to own storage batteries and plays the role of balancing electricity between different points in time. In other words, the market maker becomes a seller for the time periods during which there are no other sell orders and becomes a buyer when the amount of power generation greatly exceeds the demand.

Table 6. Total volumes in the simulation with the simple market maker.

Total Tradable Volume	Total Executed Volume	Execution Rate
950.3 kWh	134.0 kWh + 1632.6 kWh/2 = **950.3 kWh** (Trading volume not involving the simple market maker) 134.0 kWh (Trading volume involving the simple market maker as a seller or buyer) 1632.6 kWh	100.0%

The bid/ask spread is shown in Table 7. We first note that the maximum value, 3.01 JPY/kWh, has become much smaller than that in Case 1, 16.00 JPY/kWh. In addition, the average value, 2.90 JPY/kWh, is also smaller than that in Case 1, 3.96 JPY/kWh, by 1.06 JPY/kWh. This shows that, because the simple market maker always holds limit orders with a spread size of 3.00 JPY/kWh, the bid/ask spread does not widen further. We see that market liquidity has improved from the perspective of bid/ask spreads, and the market environment has become more preferable for participants to trade in.

Table 7. Mean, maximum, and minimum values of bid-ask spreads (with the simple market maker).

Mean	Max	Min
2.90 JPY/kWh	3.01 JPY/kWh	0.01 JPY/kWh

4.4. Case 3: Introduction of the Flexible Market Maker

Finally, we describe the initial parameters for the simulation using the flexible market maker. Because the items in Cases 1 and 2 remain unchanged (except "Reference middle price between sell and buy orders when neither bid nor offer is on the order book" in Case 2), only the additional parameters are listed in Table 8.

Table 8. Initial parameters in P2P electricity market simulation with the flexible market maker.

	Items	Values
	Price adjustment weight	0.00005
Flexible market maker agent	Bidding price when no reverse order is on the order book	(Sell order price) 33.00 JPY/kWh (Buy order price) 17.00 JPY/kWh

As shown in Table 9, the total executed volume is 470.4 kWh when the flexible market maker is introduced. This is 1.75 times higher than the value in Case 1, 268.7 kWh, but is about half of the value in Case 2, 950.3 kWh (in which all orders are executed). This is because the flexible market maker adjusts their bidding behavior to avoid execution under unfavorable conditions to improve their profit, although market makers need to keep limit orders on the board. Nevertheless, market liquidity can be said to have improved to a certain extent in terms of trading volume compared to the case without market makers.

Table 9. Total volumes in the simulation with the flexible market maker.

Total Tradable Volume	Total Executed Volume	Execution Rate
950.3 kWh	104.2 kWh + 732.5 kWh/2 = 470.4 kWh (Trading volume not involving the flexible market maker) 104.2 kWh (Trading volume involving the flexible market maker as a seller or buyer) 732.5 kWh	49.5%

The mean, maximum, and minimum values of the bid/ask spreads are listed in Table 10. Although the maximum and minimum values remain unchanged from those before the introduction of market makers, the average value has become larger than that in Case 1. This can be explained as follows. In Case 1, bid-ask spreads are observed only for limited periods and are not calculated in the nighttime, when solar power generation does not occur, or during the daytime, when the amount of PV power generation far exceeds the demand. On the other hand, the flexible market maker continues to place limit orders based on their bidding rules even in the time periods when no other buying or selling orders exist on the order book (selling price: 33.00 JPY/kWh when there are no other sell orders; buying price: 17.00 JPY/kWh when there are no other buy orders). As a result, the mean value of the bid/ask spreads tends to be larger in Case 3, although that in Case 2 with the simple market maker is tighter than that in Case 1 without market makers.

Table 10. Mean, maximum, and minimum values of bid-ask spreads (with the flexible market maker).

Mean	Max	Min
8.03 JPY/kWh	16.00 JPY/kWh	0.01 JPY/kWh

It should be mentioned that our results above may be influenced by initial parameters. However, we set them by referring to the Japanese electricity market and the past demonstration project described in Section 2 and verified the effectiveness of the proposed methodology. It would be interesting to investigate the robustness of the results by changing some parameters with others fixed as a future study.

5. Comparative Discussions

In this section, we present comparative discussions based on the artificial market simulation results for the three case studies in Section 4 and summarize the contributions of this study.

First, we summarize the executed volumes and execution rates for the three cases, as shown in Table 11. Recall that the introduction of the flexible market maker did not tighten the bid-ask spread on average as the spread observation period was quite limited in Case 1 without market makers, as mentioned at the end of the previous section. However, we observe that the execution rate was improved by introducing market makers. In particular, the introduction of the simple market maker significantly improved the execution rate.

Table 11. Comparison of executed volumes for the three cases.

Tradable Volume if All Orders Are Executed	Case 1: without Market Makers	Case 2: with Simple Market Maker	Case 3: with Flexible Market Maker
950.3 kWh	268.7 kWh (28.3%)	950.3 kWh (100.0%)	470.4 kWh (49.5%)

A similar tendency was observed by computing and comparing the execution price change rates for the three cases, where the execution price change rates are given by the rate of change between the current and previous execution prices for the same product (i.e., electricity for the same delivery period). Figure 12 compares three histograms of the execution price change rates, where the vertical axis represents frequencies. The left-most figure indicates the results of the case without market makers, the middle with the simple market maker, and the right with the flexible market maker. Because higher change rates mean larger price fluctuations, their variance (or standard deviation) provides execution price volatility. We emphasize that high volatility in the P2P electricity market is the original motivation for introducing market makers, who are expected to mitigate price fluctuations.

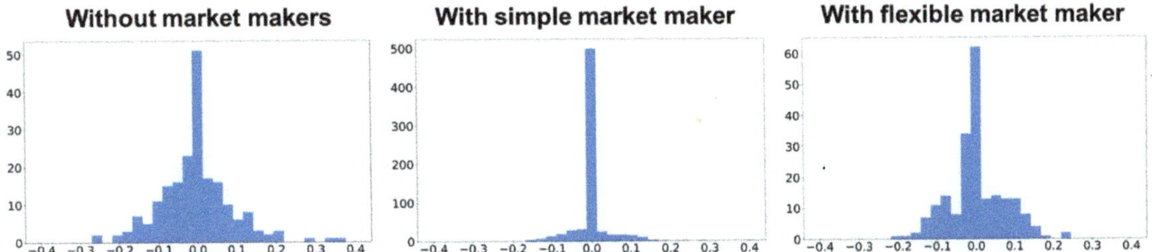

Figure 12. Histograms of execution price change rates for the three cases (horizontal axis: execution price change rate; vertical axis: frequency).

To verify that the introduction of market makers actually achieves lower volatility, we computed the mean, variance (standard deviation), maximum, and minimum values for each case, as shown in Table 12. First, we see that the price volatility given by the variance (standard deviation) is reduced in cases with market makers. Second, the improvement effect is larger in the case with the simple market maker than in that with the flexible market maker. Note that similar observation results were obtained from other statistics, such as mean, maximum, and minimum values.

Table 12. Comparison of change rates of executed prices for the three cases.

	Mean	Variance (Standard Deviation)	Maximum	Minimum
Case 1: Without Market Makers	−0.0043	0.0085 (0.0922)	0.354 (35.4% up)	−0.267 (26.7% down)
Case 2: With Simple Market Maker	0.0001	0.0018 (0.0429)	0.295 (29.5% up)	−0.164 (16.4% down)
Case 3: With Flexible Market Maker	0.0027	0.0053 (0.0729)	0.231 (23.1% up)	−0.196 (19.6% down)

Considering the comparisons above, we can conclude that introducing a simple market maker is the best among these three cases. This may be true if the profitability or risk of loss for a market maker is not examined; however, when these points are also considered, the flexible market maker is a better option. To clarify the relationship between the two types of market makers and compare their profitability, we computed the total income or loss for the simple/flexible market maker, as shown in Table 13, in which imbalance charges of 50 JPY per 1 kWh shortage are deducted. Moreover, when the market maker has a surplus position, an opportunity loss occurs due to the additional procurement cost. With regard to the simulation results, when the daily generation and demand were even (i.e., Generation/Demand $= 100/100 = 1$), the simple market maker's profit was positive. However, when the generation–demand ratio increased or decreased by 30%, the simple market maker's profit became negative, indicating that the simple market maker lost money through transactions. On the other hand, the flexible market maker was profitable even when generation and demand were unbalanced. Note that the market makers' profitability worsened when the generation–demand ratio was further decreased, but the loss of the flexible market maker was not as large as that of the simple market maker.

Table 13. Profit/loss of market makers with respect to different supply–demand ratios.

	Generation/Demand = 100/100 = 1	Generation/Demand = 130/100 = 1.3	Generation/Demand = 70/100 = 0.7	Generation/Demand = 40/100 = 0.4
Simple Market Maker	2029.80 JPY	−6281.62 JPY	−7406.00 JPY	−16773.05 JPY
Flexible Market Maker	2900.04 JPY	520.54 JPY	453.72 JPY	−3367.36 JPY

In addition, we computed the weighted average of sales/purchase prices for generators, consumers, and prosumers in all three cases and compared the results with one another, although the details are omitted here for brevity. The introduction of market maker agents affected sales/purchase prices of these general agents, but the difference between with and without market maker cases was not large on average. On the other hand, market makers provided new trading opportunities even when no PV generation was performed at night, and as a result, the executed volumes of general agents largely increased. In this sense, the market maker system could be embraced without resistance by other market participants. Therefore, we can conclude that the proposed market maker system would contribute to the development of the P2P electricity market, which could serve as a new incentive for the further spread and establishment of renewable energy power generation businesses.

6. Conclusions

In this study, we proposed the application of a market maker system to the P2P electricity market and developed an efficient market maker strategy to increase liquidity and mitigate extreme price fluctuations. To this end, we constructed an artificial market simulator for P2P electricity trading. We also designed and implemented both market and participant agents that enabled us to virtually perform power bidding and contract processes. The participant agent algorithms were built for PV generators, consumers, prosumers, and market maker agents. We prepared two bidding strategies for market makers and compared them before and after their introduction using actual solar PV generation and consumption data observed in a previous demonstration project. We confirmed that the effect of liquidity enhancement and price stability has a trade-off relationship with market makers' profitability, but all factors can be improved simultaneously without causing significant losses to other market participants. Therefore, we can conclude that the market maker system could lower the barriers to entry into the P2P electricity market and efficiently contribute to the growth of the renewable energy industry.

Finally, we describe the possible future directions of this research theme from the viewpoint of "improvement of bidding strategies for market maker agents", "improvement of bidding strategies for supply and demand agents", and "feasibility issues when introducing market makers".

First, regarding the "improvement of bidding strategies for market maker agents", we must incorporate additional factors into the current price determination algorithms. In this study, even the most sophisticated pricing method, the flexible market maker, was simply shifting half the bid offer spread up and down from the midpoint between the best quotes on the order board and adjusting them according to the net position of the moment. However, market makers also consider technical factors, such as market trends. Therefore, upgrading market-maker agents could be a topic for future research.

Second, with regard to the "improvement of bidding strategies for supply and demand agents", they should be made more flexible because, in this study, they were assumed to be fixed regardless of changes in the external environment. However, in the real world, if market makers are introduced into the P2P electricity market, supply and demand agents will react by flexibly adjusting their bidding strategies. Furthermore, installing storage batteries (such as solar storage batteries for households and/or electric vehicles) into supply and demand agents could also change their strategies and may influence the market maker's as well. These points should be considered in future studies.

Third, we would like to point out "feasibility issues when introducing market makers". In this research, we do not consider many important elements that affect the feasibility of this system, such as power loss caused by transmission, charge and discharge processes, the ideal capacity of storage batteries for market makers, and cost-effectiveness considering battery life. In our simulations, we assumed that the market maker agent was homogeneous regardless of their types and that their bidding strategy was fixed; however, it is more reasonable to expect that multiple market makers with different strategies exist in a single market.

Finally, there exist several issues related to the extension of the dataset. In this study, we have assumed that the number of agents is 18 in total and assigned both or either of the two types of logs (generation/consumption) randomly to each agent (a generator, consumer, or prosumer agent) from the original dataset. Moreover, the total daily PV generation and consumption per household were adjusted to reflect other conditions (such as weather and/or yearly trends), and we performed various simulations based on these adjusted data. A further investigation based on an enhanced dataset for a longer period and with a wider variety of participants may be interesting. Consequently, when introducing this market maker system into real-life P2P markets, discussing these issues is inevitable; thus, they could be considered potential topics for further studies.

The work in this study was primarily conducted when the first author was a graduate student in the School of Engineering at the University of Tokyo, Japan.

Author Contributions: Conceptualization, S.K., K.T. and Y.Y.; methodology, S.K.; software, S.K.; validation, S.K.; formal analysis, S.K.; investigation, S.K., K.T. and Y.Y.; resources, K.T.; data curation, K.T.; writing—original draft preparation, S.K. and Y.Y.; writing—review and editing, S.K. and Y.Y.; visualization, S.K.; supervision, K.T. and Y.Y.; project administration, K.T. and Y.Y.; funding acquisition, Y.Y. All authors have read and agreed to the published version of the manuscript.

Funding: Grant-in-Aid for Scientific Research (A) 20H00285 and Grant-in-Aid for Challenging Research (Exploratory) 19K22024 from the Japan Society for the Promotion of Science (JSPS).

Data Availability Statement: The data presented in this study are available from the corresponding author upon reasonable request. The data are not publicly available due to privacy.

Acknowledgments: We express our sincere gratitude to TATEYAMA KAGAKU CO., LTD. (Toyama, Japan), for providing the supply and demand data in Section 4. This work was supported by Grant-in-Aid for Scientific Research (A) 20H00285 and Grant-in-Aid for Challenging Research (Exploratory) 19K22024 from the Japan Society for the Promotion of Science (JSPS).

Conflicts of Interest: The authors declare no conflict of interest.

Appendix A Bidding Strategies of General Agents

In this appendix, we explain the bidding strategies of general agents, that is, the generators, consumers, and prosumers, used in our simulation.

Appendix A.1 Generators and Consumers

Generators are supposed to sell all the electricity they generate in the market without consuming any. The bidding price determination strategy is as follows. Immediately after the market opens, they place an order at a relatively high price and then gradually lower the price as time passes toward the end of the bidding period. This represents the behavioral principle that generators want to sell surplus power at the highest possible prices but do not want to waste any electricity. At the same time, the minimum sale price is predetermined, and generators will not sell below the price. This indicates that it is financially more advantageous to sell to a grid than to sell at an excessively low price in the P2P market.

Similarly, consumers do not generate or procure all the necessary amounts of electricity from the P2P market. Their bidding price determination strategy is to place an order at a relatively low price initially and then gradually increase the price over time. Consumers want to purchase the required electricity at the lowest possible price but do not want to experience a power shortage. In addition, the maximum purchase price is set, and no purchase order is placed higher than the price. This implies that consumers find it more profitable to purchase electricity from the grid than to buy it at an excessively high price in the P2P market.

In addition, a random number term is added to the prices explained above because the generation and demand conditions of each entity are not constant but, rather, are constantly subject to many uncertain factors in the real world. Thus, bidding prices should be adjusted flexibly. In the remainder of this paper, we refer to a bidding price without a random number added as a "base price".

Graphical representations of base prices for generators and consumers are shown in Figure A1. In the actual market, there are various user preferences, such as the need to sell at the highest possible prices (buy at the lowest possible prices) or the need to secure necessary electricity amounts safely. To express these differences in preference, market participants are divided into three segments: the "price-oriented type", which puts more emphasis on economic efficiency; the "certainty-oriented type", which places more importance on how fast they can ensure needed power; and the "moderate type", which falls somewhere in between these two. To differentiate all these segments, different values are used for initial bidding prices, price change rates, and limits of bidding prices.

Figure A2 shows the actual bidding price transition after random numbers are added. This allows orders to be executed even in the first 400 min when no single contract is seen in the case of bidding at base prices.

There is one more rule regarding bidding price: If the previous order was already contracted at the next bidding timing, the base price (the price without a random number added) remains the same even if some time has elapsed since the last bidding and only the random number changes. The aim is to prevent sales prices from being excessively lowered or purchase prices from excessively increasing over time, even though the market environment is such that electricity can be sold at higher prices or bought at lower prices.

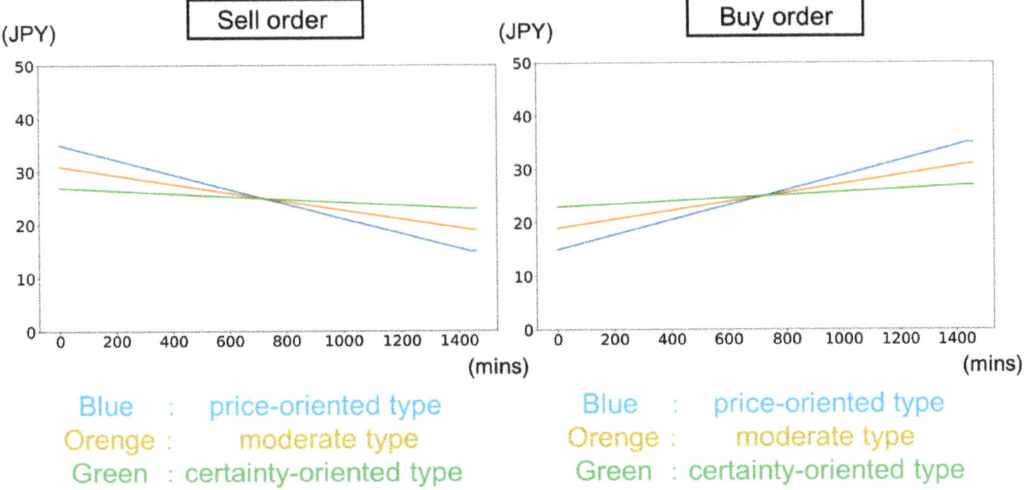

Figure A1. The base bidding price (price without a random term).

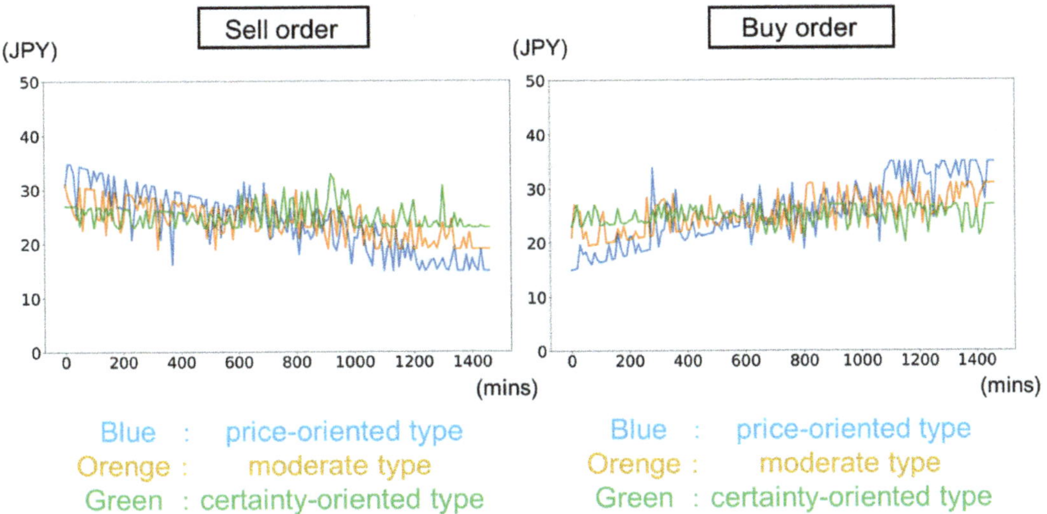

Figure A2. Actual bidding price (price with a random term).

Appendix A.2 Prosumers

Prosumers both generate and consume electricity, and they attempt to satisfy their demand with their own generation as much as possible. They sell or buy electricity in the P2P market only when there is a surplus or shortage. In other words, if the total procurement, which is the sum of predicted generation and existing buy contracts, exceeds the necessary quantity, which is the sum of the forecasted demand and existing sell contracts, the excess amount is bid as sell orders; conversely, if the procurement volume is not sufficient to fulfill the demand, a buy order is sent to the market. The bidding price determination strategy of prosumers is a combination of generators' and consumers' algorithms. Because prosumers can be both sellers and buyers, the initial prices, price change rates, and limits of bidding prices are set separately for the two sides. Furthermore, as in the case of generators and consumers, if the previous order on the same bidding side (selling or buying) has been executed by the next bidding turn, the base price remains unchanged, and only a new random number is added to it.

References

1. Mengelkamp, E.; Gärttner, J.; Rock, K.; Kessler, S.; Orsini, L.; Weinhardt, C. Designing microgrid energy markets: A case study: The Brooklyn Microgrid. *Appl. Energy* **2018**, *210*, 870–880. [CrossRef]
2. Werth, A.; Kitamura, N.; Tanaka, K. Conceptual study for open energy systems: Distributed energy network using interconnected DC nanogrids. *IEEE Trans. Smart Grid* **2015**, *6*, 1621–1630. [CrossRef]
3. Lezama, F.; Soares, J.; Hernandez-Leal, P.; Kaisers, M.; Pinto, T.; Vale, Z. Local energy markets: Paving the path toward fully transactive energy systems. *IEEE Trans. Power Syst.* **2018**, *34*, 4081–4088. [CrossRef]
4. Matsuda, Y.; Tanaka, K. EV mobility charging service based on blockchain. In Proceedings of the 2020 IEEE International Conference on Environment and Electrical Engineering and 2020 IEEE Industrial and Commercial Power Systems Europe (EEEIC/I&CPS Europe), Madrid, Spain, 9–12 June 2020.
5. Goranović, A.; Meisel, M.; Fotiadis, L.; Wilker, S.; Treytl, A.; Sauter, T. Blockchain applications in microgrids an overview of current projects and concepts. In Proceedings of the IECON 2017—43rd Annual Conference of the IEEE Industrial Electronics Society, Beijing, China, 29 October–1 November 2017; pp. 6153–6158.
6. Parka, C.; Yong, T. Comparative review and discussion on P2P electricity trading. *Energy Procedia* **2017**, *128*, 3–9. [CrossRef]
7. Sagawa, D.; Tanaka, K.; Ishida, F.; Saito, H.; Takenaga, N.; Nakamura, S.; Aoki, N.; Nameki, M.; Saegusa, K. Bidding agents for PV and electric vehicle-owning users in the electricity P2P trading market. *Energies* **2021**, *14*, 8309. [CrossRef]
8. Matsuda, Y.; Yamazaki, Y.; Oki, H.; Takeda, Y.; Sagawa, D. Tanaka, K. Demonstration of blockchain based peer to peer energy trading system with real-life used PHEV and HEMS charge control. *Energies* **2021**, *14*, 7484. [CrossRef]
9. Takeda, Y.; Nakai, Y.; Senoo, T.; Tanaka, K. Designing a user-centric P2P energy trading platform: A case study—Higashi-Fuji Demonstration. *Energies* **2021**, *14*, 7289. [CrossRef]
10. Zhang, W.; Wang, X.; Huang, Y.; Qi, S.; Zhao, Z.; Lin, F. A Peer-To-Peer Market Mechanism for Distributed Energy Resourses. In Proceedings of the 2019 IEEE Innovative Smart Grid Technologies—Asia (ISGT Asia), Chengdu, China, 21–24 May 2019; pp. 1375–1380.
11. Leong, C.H.; Gu, C.; Li, F. Auction Mechanism for P2P Local Energy Trading considering Physical Constraints. *Energy Procedia* **2019**, *158*, 6613–6618. [CrossRef]
12. Lin, J.; Pipattanasomporn, M.; Rahman, S. Comparative analysis of auction mechanisms and bidding strategies for P2P solar transactive energy markets. *Appl. Energy* **2019**, *255*, 113687. [CrossRef]
13. Azim, M.I.; Tushar, W.; Saha, T.K. Cooperative negawatt P2P energy trading for low-voltage distribution networks. *Appl. Energy* **2021**, *299*, 117300. [CrossRef]
14. Azizi, A.; Aminifar, F.; Moeini-Aghtaie, M.; Alizadeh, A. Transactive Energy Market Mechanism With Loss Implication. *IEEE Trans. Smart Grid* **2021**, *12*, 1215–1223. [CrossRef]
15. Andoni, M.; Robu, V.; Flynn, D.; Abram, S.; Geach, D.; Jenkins, D.; McCallum, P.; Peacock, A. Blockchain technology in the energy sector: A systematic review of challenges and opportunities. *Renew Sustain. Energy Rev.* **2019**, *100*, 143–174. [CrossRef]
16. Ahl, A.; Yarime, M.; Tanaka, K.; Sagawa, D. Review of blockchain-based distributed energy: Implications for institutional development. *Renew. Sustain. Energy Rev.* **2019**, *107*, 200–211. [CrossRef]
17. Nemoto, J.; Goto, M. Measurement of Dynamic Efficiency in Production: An Application of Data Envelopment Analysis to Japanese Electric Utilities. *J. Product. Anal.* **2003**, *19*, 191–210. [CrossRef]
18. Kontani, R.; Tanaka, K.; Yamada, Y. Feasibility conditions for demonstrative peer-to-peer energy market. *Energies* **2021**, *14*, 7418. [CrossRef]
19. Japan Exchange Group, Incorporated. Market Maker Program. Available online: https://www.jpx.co.jp/english/derivatives/rules/market-maker/index.html (accessed on 14 April 2022).

20. Kardakos, E.G.; Simoglou, C.K.; Bakirtzis, A.G. Optimal bidding strategy in transmission-constrained electricity markets. *Electr. Power Syst. Res.* **2014**, *109*, 141–149. [CrossRef]
21. Maceas Henao, M.; Espinosa Oviedo, J.J. Bidding Strategy for VPP and Economic Feasibility Study of the Optimal Sizing of Storage Systems to Face the Uncertainty of Solar Generation Modelled with IGDT. *Energies* **2022**, *15*, 953. [CrossRef]
22. Singh, S.; Fozdar, M.; Malik, H.; Khan, I.A.; Al Otaibi, S.; Albogamy, F.R. Impacts of Renewable Sources of Energy on Bid Modeling Strategy in an Emerging Electricity Market Using Oppositional Gravitational Search Algorithm. *Energies* **2021**, *14*, 5726. [CrossRef]
23. Qussous, R.; Harder, N.; Weidlich, A. Understanding Power Market Dynamics by Reflecting Market Interrelations and Flexibility-Oriented Bidding Strategies. *Energies* **2022**, *15*, 494. [CrossRef]
24. Khaloie, H.; Abdollahi, A.; Shafie-Khah, M.; Siano, P.; Nojavan, S.; Anvari-Moghaddam, A.; Catalao, J.P.S. Co-optimized bidding strategy of an integrated wind-thermal-photovoltaic system in deregulated electricity market under uncertainties. *J. Clean. Prod.* **2020**, *242*, 118434. [CrossRef]
25. Zheng, Y.; Yu, H.; Shao, Z.; Jian, L. Day-ahead bidding strategy for electric vehicle aggregator enabling multiple agent modes in uncertain electricity markets. *Appl. Energy* **2020**, *280*, 115977. [CrossRef]
26. Sadeghi, S.; Jahangir, H.; Vatandoust, B.; Golkar, M.A.; Ahmadian, A.; Elkamel, A. Optimal bidding strategy of a virtual power plant in day-ahead energy and frequency regulation markets: A deep learning-based approach. *Int. J. Electr. Power Energy Syst.* **2021**, *127*, 106646. [CrossRef]
27. Li, S.; Park, C.S. Wind power bidding strategy in the short-term electricity market. *Energy Econ.* **2018**, *75*, 336–344. [CrossRef]
28. Mahvi, M.; Ardehali, M.M. Optimal bidding strategy in a competitive electricity market based on agent-based approach and numerical sensitivity analysis. *Energy* **2011**, *36*, 6367–6374. [CrossRef]
29. Baltaoglu, S.; Tong, L.; Zhao, Q. Algorithmic bidding for virtual trading in electricity markets. *IEEE Trans. Power Syst.* **2019**, *34*, 535–543. [CrossRef]
30. Deng, S.J.; Oren, S.S. Electricity derivatives and risk management. *Energy* **2006**, *31*, 940–953. [CrossRef]
31. Algieri, B.; Leccadito, A.; Tunaru, D. Risk premia in electricity derivatives markets. *Energy Econ.* **2021**, *100*, 105300. [CrossRef]
32. Bevin-McCrimmon, F.; Diaz-Rainey, I.; McCarten, M.; Sise, G. Liquidity and risk premia in electricity futures. *Energy Econ.* **2018**, *75*, 503–517. [CrossRef]
33. Matsumoto, T.; Bunn, D.W.; Yamada, Y. Pricing electricity day-ahead cap futures with multifactor skew-t densities. *Quant. Financ.* **2021**, *22*, 835–860. [CrossRef]
34. Groll, A.; Cabrera, B.L.; Meyer-Brandis, T. A consistent two-factor model for pricing temperature derivatives. *Energy Econ.* **2016**, *55*, 112–126. [CrossRef]
35. Lee, Y.; Oren, S.S. An equilibrium pricing model for weather derivatives in a multi-commodity setting. *Energy Econ.* **2009**, *31*, 702–713. [CrossRef]
36. Kanamura, T.; Ohashi, K. Pricing summer day options by good-deal bounds. *Energy Econ.* **2009**, *31*, 289–297. [CrossRef]
37. Benth, F.E.; Di Persio, L.; Lavagnini, S. Stochastic Modeling of Wind Derivatives in Energy Markets. *Risks* **2018**, *6*, 56. [CrossRef]
38. Rodríguez, Y.E.; Pérez-Uribe, M.A.; Contreras, J. Wind Put Barrier Options Pricing Based on the Nordix Index. *Energies* **2021**, *14*, 1177. [CrossRef]
39. Boyle, C.F.; Haas, J.; Kern, J.D. Development of an irradiance-based weather derivative to hedge cloud risk for solar energy systems. *Renew. Energy* **2020**, *164*, 1230–1243. [CrossRef]
40. Bhattacharya, S.; Gupta, A.; Kar, K.; Owusu, A. Risk management of renewable power producers from co-dependencies in cashflows. *Eur. J. Oper. Res.* **2020**, *283*, 1081–1093. [CrossRef]
41. Yamada, Y.; Matsumoto, T. Going for derivatives or forwards? Minimizing cashflow fluctuations of electricity transactions on power markets. *Energies* **2021**, *14*, 7311. [CrossRef]
42. Matsumoto, T.; Yamada, Y. Simultaneous hedging strategy for price and volume risks in electricity businesses using energy and weather derivatives. *Energy Econ.* **2021**, *95*, 105101. [CrossRef]
43. Matsumoto, T.; Yamada, Y. Customized yet standardized temperature derivatives: A non-parametric approach with suitable basis selection for ensuring robustness. *Energies* **2021**, *14*, 3351. [CrossRef]
44. Worthmann, K.; Kellett, C.M.; Braun, P.; Grune, L.; Weller, S.R. Distributed and Decentralized Control of Residential Energy Systems Incorporating Battery Storage. *IEEE Trans. Smart Grid* **2015**, *6*, 1914–1923. [CrossRef]
45. Nguyen, V.H.; Tran, Q.T.; Besanger, Y.; Labonne, A. Complementary business models for distribution system operator in a peer-to-peer electricity market. In Proceedings of the 2021 IEEE International Conference on Environment and Electrical Engineering and 2021 IEEE Industrial and Commercial Power Systems Europe, Bari, Italy, 8–11 June 2021; pp. 1–6.
46. Huang, Q.; Xu, Y.; Courcoubetis, C. Strategic Storage Operation in Wholesale Electricity Markets: A Game Theoretic Analysis. In Proceedings of the 2019 IEEE 58th Conference on Decision and Control (CDC), Nice, France, 11–13 December 2019; pp. 2201–2207.
47. Bose, S.; Cai, D.W.H.; Low, S.; Wierman, A. The role of a market maker in networked Cournot competition. In Proceedings of the 53rd IEEE Conference on Decision and Control, Los Angeles, CA, USA, 15–17 December 2014; pp. 4479–4484.
48. Worthmann, K.; Kellett, C.M.; Grüne, L.; Weller, S.R. Distributed control of residential energy systems using a market maker. *IFAC Proc. Vol.* **2014**, *47*, 11641–11646. [CrossRef]
49. Kusada, Y.; Mizuta, T.; Hayakawa, S.; Izumi, K. Impact of Position-Based Market Makers to Shares of Markets' Volumes—An Artificial Market Approach. *Trans. Jpn. Soc. Artif. Intell.* **2015**, *30*, 675–682. (In Japanese) [CrossRef]

50. Mizuta, T.; Izumi, K. *Investigation of Frequent Batch Auctions Using Agent Based Model*; JPX Working Paper, 17; Japan Exchange Group, Inc.: Tokyo, Japan, 2016. Available online: https://www.jpx.co.jp/english/corporate/research-study/working-paper/b5 b4pj000000i468-att/E_JPX_working_paper_Vol17.pdf (accessed on 13 April 2022).
51. Abe, R.; Tanaka, K.; Triet, N.V. Energy Internet: An open energy platform to transform legacy power systems into open innovation and global economic. In *The Energy Internet: An Open Energy Platform to Transform Legacy Power Systems into Open Innovation and Global Economic Engines*; Su, W., Huang, A., Eds.; Woodhead Publishing: Cambridge, UK, 2018; pp. 241–264.
52. Tanaka, K.; Abe, R.; Nguyen-Van, T.; Yamazaki, Y.; Kamitamari, T.; Sako, K.; Koide, T. *A Proposal on an Electricity Trading Platform Using Blockchain, Volume 7: Transdisciplinary Engineering Methods for Social Innovation of Industry 4.0*; IOS Press: Amsterdam, The Netherlands, 2018; pp. 976–983.
53. Waseda, F.; Tanaka, K. Bidding agent for electric vehicles in peer-to-peer electricity trading market considering uncertainty. In Proceedings of the 2020 IEEE International Conference on Environment and Electrical Engineering and 2020 IEEE Industrial and Commercial Power Systems Europe (EEEIC/I&CPS Europe), Madrid, Spain, 9–12 June 2020.

MDPI

St. Alban-Anlage 66

4052 Basel

Switzerland

Tel. +41 61 683 77 34

Fax +41 61 302 89 18

www.mdpi.com

Energies Editorial Office

E-mail: energies@mdpi.com

www.mdpi.com/journal/energies